TEACHING AND LEARNING IN THE SCIENCE LABORATORY

Science & Technology Education Library

VOLUME 16

SCOPE

The book series *Science & Technology Education Library* provides a publication forum for scholarship in science and technology education. It aims to publish innovative books which are at the forefront of the field. Monographs as well as collections of papers will be published.

The titles published in this series are listed at the end of this volume.

Teaching and Learning in the Science Laboratory

Edited by

DIMITRIS PSILLOS

Aristotle University of Thessaloniki, Greece

and

HANS NIEDDERER

University of Bremen, Germany

KLUWER ACADEMIC PUBLISHERS

DORDRECHT / BOSTON / LONDON

A C.I.P. Ctalogue record for this book is available from the Library of Congress.

ISBN 1-4020-1018-4

Published by Kluwer Academic Publishers,
P.O. Box 17, 3300 AA Dordrecht, The Netherlands.

Sold and distributed in North, Central and South America
by Kluwer Academic Publishers,
101 Philip Drive, Norwell, MA 02061, U.S.A.

In all other countries, sold and distributed
by Kluwer Academic Publishers,
P.O. Box 322, 3300 AH Dordrecht, The Netherlands.

Printed on acid-free paper

Printed in the Netherlands.

TEACHING AND LEARNING IN THE SCIENCE LABORATORY

A book based on the European project
"Labwork in Science Education"
Co-ordinated by
Marie-Geneviève Séré

Dimitris Psillos

and

Hans Niedderer

(Eds.)

KLUWER ACADEMIC PUBLISHERS

Table of Contents

General Introduction 1

Chapter 1

APPROACHING LABWORK: FRAMES AND TOOLS

Introduction 7

Robin Millar, Andrée Tiberghien and Jean-François Le Maréchal
Varieties of Labwork: A Way of Profiling Labwork Tasks 9

Dimitris Psillos and Hans Niedderer
Issues and Questions Regarding the Effectiveness of Labwork 21

Hans Niedderer, Stefan v. Aufschnaiter, Andrée Tiberghien, Christian Buty, Kerstin Haller, Lorenz Hucke, Florian Sander, Hans Fischer
Talking Physics in Labwork Contexts - A Category Based Analysis of Videotapes 31

John Leach
Students' Understanding of the Nature of Science and its Influence on Labwork 41

Chapter 2

STANDARD LABWORK BASED ON HANDS-ON EXPERIMENTS

Introduction 49

Karine Bécu-Robinault
Modelling Activities of Students during a Traditional Labwork 51

Michel Beney and Marie-Geneviève Séré
Students' Intellectual Activities during Standard Labwork at Undergraduate Level 65

Petros Kariotoglou
A Laboratory-Based Teaching Learning Sequence on Fluids: Developing Primary Student Teachers' Conceptual and Procedural Knowledge 79

Heike Theyßen, Dieter Schumacher, Stefan v. Aufschnaiter
Development and Evaluation of a Laboratory Course in Physics for Medical Students 91

Milena Bandiera
The Biology Textbook as a Source of Ideas about Scientific Knowledge and Experimental Activity 105

Chapter 3

OPEN-ENDED LABWORK

Introduction 119

Alain Guillon and Marie-Geneviève Séré
The Role of Epistemological Information in open-ended Investigative Labwork 121

Jenny Lewis
The Effectiveness of Mini-Projects as a Preparation for open-ended Investigations 139

Jim Ryder
Data Interpretation Activities and Students' Views of the Epistemology of Science during a University Earth Sciences Field Study Course 151

Chapter 4

LABWORK AND DATA HANDLING

Introduction 163

John Leach
The Use of secondary Data in Teaching about Data Analysis in a First Year Undergraduate Biochemistry Course 165

Dimitris Evangelinos, Dimitris Psillos and Odysseas Valassiades
An Investigation of Teaching and Learning about Measurement Data and their Treatment in the Introductory Physics Laboratory 179

Chapter 5

LABWORK BASED ON INTEGRATED USE OF NEW INFORMATION TECHNOLOGY

Introduction 191

Garabet Bisdikian and Dimitris Psillos
Enhancing the Linking of Theoretical Knowledge to Physical Phenomena by Real-Time Graphing 193

Lorenz Hucke and Hans E. Fischer
The Link of Theory and Practice in Traditional and in Computer-Based University Laboratory Experiments 205

Florian Sander, Horst Schecker and Hans Niedderer
Computer Tools in the Lab –
Effects Linking Theory and Experiment 219

Christian Buty
Modelling in Geometrical Optics Using a Microcomputer 231

Alexandros Barbas and Dimitris Psillos
Evolution of Students' Reasoning about Microscopic Processes in Electrostatics under the Influence of Interactive Simulations 243

EPILOGUE

Marie-Geneviève Séré
Towards Targeted Labwork 255

Authors 260

Index 266

GENERAL INTRODUCTION

Scope of the book

There is an on-going debate regarding the role of labwork in science education, which dates back several decades and which illustrates the conviction and interest of teachers, researchers and policy-makers world-wide in the value of laboratory work for understanding science. This is evident in more recent books and studies regarding the laboratory, which mainly refer to countries with a considerable tradition in practical work in science education (Woolnough & Alsop 1985, Hodson 1993, Hegarthy-Hazel 1990, Wellington 2000). Yet in discussing research studies on labwork, several authors express their concern about its effectiveness in facilitating students' understanding of various aspects of scientific inquiry. They point out a comprehensive re-conceptualisation of the aims of labwork and, as a consequence, of investigating what the students actually learn in different contexts (Lazarowitz & Tamir 1994, Tobin & Tippins 1993, Lunetta 1998). It has also been argued that the relationship between instructional activities and student learning in labwork needs more attention than it has been given in science education research (Leach & Paulsen 1999). It appears that the case for research-based labwork emerges in several quarters in science education, particularly among researchers.

This book presents and discusses a variety of laboratory practices and their effectiveness. The studies take into account recent theoretical developments and empirical results concerning students' understanding of scientific inquiry. A whole chapter is devoted to technological advances offering new learning opportunities for the students and teaching facilities for the teacher. The authors set out to explore the potential of various ways of organising scientific experimentation, forms of data presentation and use of new information technologies for enhancing students' scientific understanding. In this respect, the book aims at making an up-to-date and substantial contribution to the discussion concerning on the one hand the differentiation of objectives and practices for labwork and on the other the effectiveness of labwork in promoting scientific understanding.

The book includes an edited collection of studies containing illustrations of teaching approaches and of students' learning acquisitions in the science teaching laboratory, presentation of new evaluation tools, theoretical frames and positions concerning laboratory work. One of the characteristic features of the book is that it focuses the discussion on the role of labwork in upper secondary and in higher education, whereas most research and discussion world-wide focuses on labwork in primary and lower secondary education. The target readers of the book are science education researchers, postgraduate students, science education teacher trainers and science education policy makers.

D. Psillos and H. Niedderer (eds.). Teaching and Learning in the Science Laboratory. 1-6.

The studies are based on research and developmental work carried out in five European countries in the context of the European Project "Labwork in Science Education" (LSE), which was funded in the context of the Targeted Socio-economic Research Programme (TSER Programme) of the European Union (Séré, Tiberghien, Paulsen, Leach, Niedderer, Psillos & Vicentini 1998; Séré 2002). The project involved 35 researchers from seven European research groups from France, Great Britain, Germany, Greece, Italy and Denmark; the project was co-ordinated by M.G. Séré (University of Paris 11) and lasted for 28 months between 1996 - 1998. The project included surveys and case studies on labwork across the countries involved, producing several working papers, communications to conferences and publications in journals. The editors of this book co-ordinated the empirical case studies on labwork, which were undertaken as a major part of this project. A number of these case studies from the LSE project form the main body of this book. This research and developmental work from five European countries provides a unique framework for gaining deeper insights into the role of experimentation in science teaching and learning, in a variety of contexts. Certainly, not all important aspects of labwork are covered in these studies that are presented in this book. But one of its distinguishing features is the fact that it presents and empirically analyses laboratory practices in quite different countries and educational settings.

For example, on the one hand there is a study about field-work for geology undergraduates in Britain, a country with a long tradition in labwork. On the other there is a study about the introduction to data treatment for Greek physics undergraduate students with negligible previous laboratory experience. And a third study discusses upper secondary students' work with computer-based models in France. These examples show that the case studies reported in the book cover a wide range of situations spread across five European countries in both upper secondary and higher education.

Although the authors of the book adopt a variety of theoretical positions and methodological approaches, they have developed certain common perspectives and identified key issues concerning the teaching and learning of science in the laboratory. Such perspectives are illustrated in the case studies and are discussed in the theoretical Chapter 1 *of this book*. One shared assumption is that setting up a laboratory situation does not necessarily imply the desired learning by the students. Accordingly, the reported investigations carefully distinguish the various teaching contexts on the one hand and research designs aimed at monitoring students' learning on the other. This means that in each study there are descriptions according to various innovative frames, e.g. illustrations of the laboratory approaches and innovations on the one hand, and empirical results concerning students' learning acquisitions on the other, followed by policy recommendations. Such a distinction between teaching in the laboratory and research into learning reveals implicit assumptions and objectives regarding science learning and teaching which influence the organisation of several types of labwork.

The first important issue, which is extensively treated in the book, is the relation of conceptual knowledge to students' practical activities (White 1996). For a long time, both in published research and in actual teaching practice science educators have often been concerned as to whether and how students use or do not use conceptual knowledge when performing experimental work. In this book, several studies make original contributions to this matter by setting up special teaching situations, focusing on students' scientific discussion (or lack of it) and actions during labwork. Several contributors search for ways to enhance the linking of the world of phenomena and procedures with the world of theory - an objective which gets high scores in science teachers' perceived priorities for labwork at both secondary and university education (Welzel, Haller, Bandiera, Hammelev, Koumaras, Niedderer, Paulsen, Robinault & von Aufschnaiter 1998).

The second important issue running through several studies in the book is the teaching and learning of scientific procedures, an issue, which is continuously debated by teachers, researchers and policy-makers. For example, measurements and data processing carried out by the students are investigated in a number of the case studies. Among the issues discussed is whether students meaningfully carry out measurements and how these are linked with the evaluation of scientific theories, which is an essential, yet not widely investigated, aspect of scientific inquiry. A number of case studies, for example, contribute substantially to the discussion on how labwork may encourage students to link theory with data.

A third issue, which emerges in the book, is the development of epistemological knowledge through labwork. In fact, this important dimension of scientific understanding has only recently started to be discussed between science educators in relation to labwork (Leach & Paulsen 1999). The authors of the various studies share the assumption that conceptual and procedural knowledge are intertwined (Séré 1999). They widen the scope of labwork, investigating and discussing its effectiveness with regard to conceptual, procedural and epistemological objectives. Often in the literature and in practice the discussion about the contribution of labwork to scientific understanding has been restricted to conceptual and procedural knowledge. How students' understanding of the nature of science influences their actions and learning during labwork is the focus of both theoretical positions and empirical investigations reported in various studies.

Too few attempts have been made so far to uncover the complex cognitive processes that take place during students' engagement in labwork: what happens and why as they carry out certain laboratory procedures. For example, one well-known observation taken up in the case studies is that students often fail to link manipulation of equipment with conceptual models or with the purpose of experimentation, often seeing labwork simply as a set of disconnected actions to be followed (Lunetta 1998). In this context, some of the case studies focus on students' cognitive constructions and models *before and after* labwork, investigating the matching between what students are intended to learn from the task and what they actually learn. Other studies focus on students' constructions *during* labwork and on the contextual factors determining what students actually do during experimentation, investigating the correlation between what students are intended to do in the task and what they actually do.

In effect, a new model of twofold effectiveness is suggested in the book, which distinguishes two main categories of labwork effectiveness leading to different sources of information in specific teaching contexts. In the first category, students' activities are related to those intended *during* labwork. In the second category, students' achievements in relation to instructional objectives are studied *after* laboratory teaching (Psillos, Niedderer & Séré 1998; Psillos, Niedderer & Vicentini 1999; Millar, Le Maréchal & Tiberghien 1998, 1999). Illustrative examples of both types of effectiveness are provided in the various studies throughout the book. In addition, two theoretical contributions in Chapter 1 focus specifically on this matter, attempting to model students' activities during labwork in relation to the intended ones.

Structure of the book

The book consists of an extended introduction, five main chapters (approaching labwork: frames and tools; standard labwork based on hands-on experiments; open-ended labwork; labwork based on secondary data; labwork based on an integrated use of new technologies), and an epilogue (towards targeted labwork).

In the introduction the two editors set out the frame and the rationale of the book and provide an overview of the various chapters.

In the first chapter of the book, four theoretical studies are included. They focus on general frames, methods and questions related to labwork and on the relation between theoretical models and experimental data from a disciplinary and learning perspective. New tools are explicitly presented for analysing laboratory tasks, for determining effectiveness, and for describing and evaluating students' activities during labwork. These tools are based on explicit hypotheses concerning the modelling of labwork and have been used in a number of empirical investigations.

The second chapter focuses on standard labwork during which students carry out hands-on experiments using standard laboratory apparatus and labguides. It may be noted that at the university level such labwork is common throughout Europe, particularly in introductory experimental courses. Some studies in this chapter investigate how students' actions and procedures as employed during experimentation are, or are not, intertwined with the theoretical knowledge which the experimental design draws upon in an attempt to shed light on the links between doing/thinking/learning during labwork. The studies focus on a variety of laboratory situations, which are set out specifically, indicating that a deep understanding of laboratory contexts/ learning interactions requires research on different levels concerning duration and task complexity. An interesting variation on standard labwork, which is presented in this chapter, comprises experimental teaching sequences or laboratory sessions based on an innovative representation and reconstruction of scientific knowledge and procedures that implies new links between the models to be taught and the corresponding experimental field. This chapter also includes a study on the presentation of experiments in ordinary textbooks, thus addressing a rather neglected, yet nonetheless important, relation between textbook and labwork, which can affect the image of scientific knowledge developed by the students, particularly in countries where labwork is not widespread.

The third chapter includes three studies focusing on open-ended labwork in which students are required to make some decisions for themselves as to how to act in various types of projects within a laboratory or in an open-field context. An important issue, which is discussed in this chapter, is what kind of scientific procedures students are required to learn in addition to conceptual knowledge and whether their epistemological understanding may be effectively improved when engaged in investigative work such as field work. Another important issue is whether any epistemological information relating to the processes and strategies of scientific investigation should be explicitly presented to the students as advance organisers of their investigative work in different contexts. How students may be helped to make the transition from set practicals to open-ended investigative work, which involves understandings of scientific procedures, is another issue investigated in this chapter.

The two investigations presented in the fourth chapter are examples of a broad conception of labwork. Both studies focus on specific phases of labwork, which do not involve planning, manipulation of apparatus and data recording. These two case studies deal with students' introductory instruction in measurement and data treatment as well as the roles and functions of measurement in science. They focus on the relation between the concrete measurement process and abstract models of that process. These issues are studied in the context of handling secondary data or data of pre-laboratory teaching.

The fifth chapter involves five case studies, which explore the new possibilities for learning science provided by the use of computers and information technology, integrated into laboratory teaching in a variety of ways. New technology is used for data collection, for analysis and graphical representation of data, for model building with appropriate software, for simulation of a physical model as well as combinations of these types of uses. The situations illustrated and studied involve, for example, the manipulation of simulated microscopic entities, the use of the computer for data capture and model building, on line and off line. These are important innovations in labwork. Their effectiveness, however, has not yet been fully explored in science education, partly because of rapid changes in the technology used. In the context of this book, new types of laboratory activities, like engagement in computer-based modelling and in real time graphing, and their learning potentials, are discussed. Whether new skills are developed or whether the improvement in linking theory to practice can be brought about are open issues, which are treated in the various studies.

It is a shared assumption of the contributors that research-based labwork may be gradually developed out of specific policy recommendations linked to research outcomes. This is why in each empirical study specific recommendations are set out. Further on a major outcome of the Labwork in Science Education Project was the advancement of the concept of "targeted labwork". Targeted labwork for upper secondary and university teaching is extensively discussed in a separate theoretical study, which is included in the epilogue of the book.

Concluding remarks

In concluding the presentation of the book, we expect that the widening of laboratory objectives, the illustrated examples of laboratory practices, the treatment of learning outcomes, the new tools, theoretical discussions and the policy suggestions will make interesting reading for a wide range of science educators. We expect the book to appeal to a wider public than researchers and postgraduate students; it could also supply valuable information to policy makers, teacher trainers and science teachers. The aim of the book is both to improve the design and organisation of innovative laboratory practices and to provide tools and exemplary results for the evaluation of their effectiveness, adequate for labwork in order to promote students' scientific understanding in a variety of countries.

References

Hegarthy-Hazel E. (Ed.) (1990) *The student laboratory and the science curriculum*. London: Routledge .

Hodson, D. (1993). Practical work in science: Time for a reappraisal. *Studies in Science Education*, 19, 175-184.

Lazarowitz, R. & Tamir, P. (1994). Research on using laboratory instruction in science. In Gabel, D.L. (Editor): *Handbook of Research on Science Teaching and Learning*. New York: Macmillan.

Leach, J. & Paulsen, A., (Eds.) (1999). *Practical work in science education*. Frederiksberg: Roskilde University Press & Dordrecht: Kluwer.

Lunetta, V.N. (1998). The School Science Laboratory: Historical Perspectives and Contexts for Contemporary Teaching. In: K. Tobin and B. Fraser (Eds.), *International Handbook of Science Education*. (Dordrecht: Kluwer), 249-264

Millar, R., Le Maréchal, J-F. & Tiberghien, A. (1998). A Map of the Variety of Labwork. Working Paper 1 from the European project Labwork in Science Education (Targeted Socio-Economic Research Programme, Project PL 95-2005), 23 pages. (http://didaktik.physik.uni-bremen.de/niedderer/projects/labwork/papers.html).

Millar, R., Le Maréchal, J-F. & Tiberghien, A. (1999). "Mapping" the domain – varieties of practical work. In J. Leach & A. Paulsen (Eds.) (1999). *Practical work in science education*. Frederiksberg: Roskilde University Press & Dordrecht: Kluwer, p. 33-59

Psillos, D. Niedderer, H. (Eds.), (1998) Case studies on labwork in 5 European countries, Working paper No 7. LSE Project, No PL 95-2005, (European Commission DG XII).

Psillos, D., Niedderer, H., Séré, M.G. (1998). Effectiveness of labwork as defined from case studies of different types of labwork. Working Paper 8 from the European project Labwork in Science Education (Targeted Socio-Economic Research Programme, Project PL 95-2005), 37 pages.

Psillos, D., Niedderer, H., Vicentini, M. (1999). Case Studies on Innovative Types of Labwork in Science Education. In: M. Bandiera, S. Caravitta, E. Torracca and M. Vicentini (eds.): *Research in Science Education in Europe*. Dordrecht: Kluwer, 201-208.

Séré, M.G., Tiberghien, A., Paulsen, A. C., Leach, J., Niedderer, H., Psillos, D., Vicentini, M. (1998). Labwork in Science Education – Executive Summary. Working paper from the European project Labwork in Science Education (Targeted Socio-Economic Research Programme, Project PL 95-2005), 16 pages, (http://edu.leeds.ac.uk/projects/lis/wp14.pdf)

Séré, M.G., (1999) Learning science in the laboratory: Issues raised by the European Project "Labwork in Science Education". In Bandiera, M. Caravita S., Torraca, E., Vicentini, M., (Eds.) *Research in Science Education in Europe*. Dordrecht: Kluwer, 165-175.

Séré, M.-G. (2002). Towards renewed research questions from the outcomes of the European project Labwork in Science Education. *Science Education* 86 (5), 624-644

Tobin, K., Tippins, D.J. and Gallard, A.J. (1993) Research on instructional Strategies for Teaching Science. In Gabel D. (Ed) *Handbook of research on science teaching and learning*. NY :Macmillan pp 45-93

Wellington, J.1998 (Ed): *Practical work in school science - Which way now*. London: Routledge.

Welzel, M., Haller, K., Bandiera, M., Hammelev, D., Koumaras, P., Niedderer, H., Paulsen, A. C., Robinault, K., von Aufschnaiter, S., (1998). Teachers' objectives for labwork. Research tool and cross country results. Working Paper 6 from the European project Labwork in Science Education (Targeted Socio-Economic Research Programme, Project PL 95-2005), 109 pages, (http://didaktik.physik.uni-bremen.de/niedderer/projects/labwork/papers.html).

Woolnough, B. & Allsop, T. (1985). *Practical Work in Science*. Cambridge: Cambridge University Press.

White, R.T. (1996). The link between laboratory and learning. *International Journal of Science Education*, 18(7), 761-774.

Chapter 1

APPROACHING LABWORK: FRAMES AND TOOLS

Introduction

The four studies in this chapter describe frameworks for constructing and analysing labwork, both theoretically and empirically. They present a frame of tools, methods and basic research results related to research and development of labwork environments. The aim is to capture and classify the varieties of laboratory work, to reflect design problems of determining its effectiveness, to offer a new method for analysing video tapes from labwork and to summarise some basic results of students' epistemological beliefs in relation to labwork.

Millar, Le Maréchal and Tiberghien give a systematic orientation regarding the central features influencing labwork in what might be called a map of labwork. The authors describe a model of the process for developing a labwork task and for evaluating its effectiveness. The description of varieties of labwork in this map can serve several different purposes related to comparing different labwork approaches, developing new labwork tasks and planning evaluation of labwork effectiveness.

Researchers and teachers around the world are concerned with the effectiveness of labwork. In the second study in this chapter, Psillos and Niedderer discuss extensively the concept of labwork effectiveness from a research perspective. Two types of effectiveness are distinguished in this study. One (effectiveness 1) examines what students actually do in the lab, while the other (effectiveness 2) is related to an analysis of the learning outcomes. This distinction is one of the results of the whole European "Labwork in Science Education" project. The same distinction is used by Millar, Le Maréchal and Tiberghien. Psillos and Niedderer present and discuss a twofold model relating effectiveness 1 and effectiveness 2, arguing that such a model is linked to the nature of labwork as a practical activity.

D. Psillos and H. Niedderer (eds.). Teaching and Learning in the Science Laboratory. 7-8.

Niedderer, Buty, Haller, Hucke, Sander, Fischer, von Aufschnaiter, Tiberghien start by giving an overview of categories used by other authors for analysing labwork. In a second part, special categories are developed for analysing effectiveness 1 of labwork in relation to the objective of "linking theory to practice". This means that special categories are developed to analyse the use of physics knowledge – or more general science knowledge – in different contexts of labwork, such as taking measurements or developing a computer model related to a special labwork task. This "category based analysis of videotapes (CBAV)" in physics labwork has been used in several detailed studies of effectiveness 1 of labwork. Four of these studies are presented in Chapters 2 and 5 of this book (Theyßen et al., Hucke et al., Sander et al., and Buty, all *in this volume*).

Leach studies how students' understanding of the nature of science influences their actions and learning during labwork. In the form of hypotheses, the author gives some basic findings about students' epistemological conceptions related to labwork. He covers students' images of data and measurement, of the nature of investigation, of the nature of theory, of the nature of explanation, and of the nature of public scientific knowledge. Altogether nine hypotheses are formulated, each with several more specific statements of details, and each based on specific research results from literature.

Varieties of Labwork: A Way of Profiling Labwork Tasks

Robin Millar, University of York, UK

Andrée Tiberghien, CNRS - Université Lyon 2, France

Jean-François Le Maréchal, CNRS - Université Lyon 2, France

Abstract

If we wish to explore the effectiveness of labwork for achieving its goals, we need to be clear about the aims of each labwork task and be able to describe its essential features in a systematic way. A model is presented of the process of developing a labwork task and evaluating its effectiveness. Two senses of 'effectiveness' are identified: the match between what students are intended to do in the task and what they actually do (effectiveness 1); and between what students are intended to learn from the task and what they actually learn (effectiveness 2). A classification scheme is then described which can be used to produce a profile of any labwork task. This provides a useful tool for exploring systematically the effectiveness of labwork tasks.

Introduction

The aim of science education is to help students develop an understanding of the natural world: what it contains, how it works, and how we can explain and predict its behaviour. So, in teaching science, we build upon students' everyday knowledge of the world around them – and augment this by providing carefully designed activities in which students observe or interact with real objects and materials. These activities are usually carried out in teaching laboratories or, in the case of some biology and earth science topics, in the field. We will use the term 'labwork' for all activities of this sort. The fundamental purpose of any labwork task is to help students to make links between two domains: the domain of real objects and observable things, and the domain of ideas (Figure 1). Through labwork, students also learn about the scientific approach to enquiry.

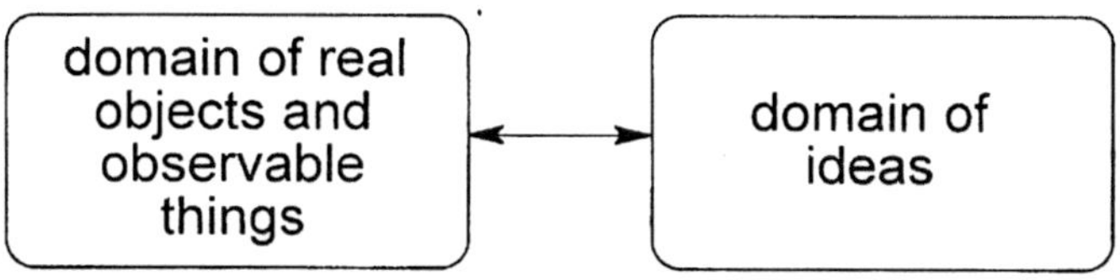

Figure 1: The fundamental purpose of labwork: to help students make links between two domains

In some countries, it is common for school students to carry out labwork tasks for themselves, usually working together in small groups. In others, there is a tradition of teacher demonstration at school level; introducing labwork for school students is often seen as a desirable reform. Where there is an established tradition of student labwork in school science, however, its effectiveness is often questioned. Students often fail to learn the things they are intended to learn. Because labwork tasks are

D. Psillos and H. Niedderer *(eds.). Teaching and Learning in the Science Laboratory. 9-20.*

carried out quickly, using rather basic equipment and often with insufficient care and precision, students frequently fail to produce the phenomena they are meant to observe. Even when they do, the features of these observations, which seem 'obvious' to the teacher can appear less so to the student. So labwork tasks can quickly become routine and purposeless to the student, rather than conveying the excitement of scientific enquiry. Hodson (1991) concludes that:

> *As practised in many schools, it [labwork] is ill conceived, confused and unproductive. For many children, what goes on in the laboratory contributes little to their learning of science or to their learning about science and its methods. Nor does it engage them in doing science in any meaningful sense. At the root of the problem is the unthinking use of laboratory work. (p. 176)*

To make labwork more effective, then, we need to think harder about its use. Labwork includes a wide variety of tasks, designed to promote quite different kinds of learning. It does not make sense, therefore, to ask about the effectiveness of labwork in general. Instead we need to ask about the effectiveness of *specific* labwork tasks for achieving *specific* learning objectives. To do this systematically, we need to be able to produce a *profile* of any labwork task. This would identify the learning objectives of the task and provide a detailed description of its key features. This chapter describes one way of producing such a profile, and discusses how it can be used to explore the effectiveness of labwork tasks.

Varieties of labwork

Science educators have suggested many different ways of classifying labwork to highlight important differences. Schwab (1962) used the idea of 'degrees of freedom' to distinguish tasks in which students follow given instructions from those where they make choices for themselves. Herron (1971) developed this to distinguish four 'levels of enquiry', from level 0 where the problem, the procedure and the conclusion are specified in advance, up to level 3 where all three are left open. Underlying these is the issue of whether a labwork task is intended to illustrate an accepted scientific idea, or to simulate some aspect of 'real' scientific enquiry. Woolnough & Allsop (1985) proposed a general classification of practical tasks into four groups: illustrations (of theory), exercises (to practice standard procedures), experiences (to give students a 'feel' for phenomena) and investigations (to allow students to experience scientific enquiry). Kirschner & Meester (1988) suggested a slightly different four-way classification of laboratory approaches: formal (to illustrate laws and concepts), experimental (open-ended), divergent (from a common start), and skills/procedures related.

More detailed classification schemes have also been proposed. Fuhrman, Novick, Lunetta & Tamir (1978) developed the Laboratory Structure and Task Analysis Inventory (LAI). Their Structure categories are: openness of the task; whether its overall approach is inductive or deductive; whether it precedes, follows, or is integrated with the related theory; the extent of student co-operation; whether the data are first- or second-hand or from a simulation. The Task Analysis categories are Planning, Performance, Analysis and Application, each subdivided into specific aspects of student performance. By answering a series of yes/no questions, the evaluator provides a characterisation of the task. Tamir & Lunetta (1978) used the

LAI to analyse laboratory activities in the Biological Sciences Curriculum Study (BSCS) teaching materials, and to compare laboratory tasks in the *PSSC Physics* course and the *Project Physics Course* (Lunetta & Tamir 1981a, b). More recently Tamir & Pilar-Garcia (1992) used the LAI along with a Laboratory Dimensions Inventory (LDI) to analyse laboratory exercises in science textbooks in Catalonia. The LDI covers eight aspects of a task: social organisation of the students, their prior knowledge, the task's relation to theory, how data are collected, instrument sophistication, form of data analysis required, time allocated and extent of concept learning involved. Another taxonomy, not directly related to LAI, has been suggested by McComas (1997), focussing on 'physical factors' (aspects of the laboratory, the curriculum) and 'personal factors' (characteristics of students and teachers).

Although the classification scheme we outline in this chapter has similarities to these earlier ones, it also differs from them in significant ways. Whereas many of these schemes were designed to explore the match between stated curriculum goals and the laboratory tasks proposed, the aim of the scheme proposed here is to provide a very general framework for exploring issues of effectiveness of labwork tasks.

The effectiveness of labwork

If we talk about the 'effectiveness' of labwork, what exactly do we mean? To answer this, it is useful to consider the process of developing and evaluating a labwork task (Figure 2). The starting point is the teacher's (or curriculum developer's) objectives for the task. These specify what the students are intended to learn from the task. Having decided the learning objectives, the teacher then designs the labwork task. Both the objectives and the task design are influenced by the teacher's views of science and of learning, and by practical and institutional factors (such as the resources available, the requirements of the curriculum, its mode of assessment, and so on).

When the labwork task is implemented we can observe what the students actually do on the task, and we can attempt to assess what they actually learn. Both of these will be influenced by the students' views of science and of learning, and by the practical and institutional setting. Many students, for instance, will concentrate on those aspects of the task, which they believe will gain them most credit in terms of course grades. The teacher's and students' views may not coincide: for example, the teacher may consider the process of practical enquiry to be very important, whereas the students are more concerned to be told the 'right answer'. As a result, the students' actions in response to the task may not be what the teacher (or curriculum developer) intended. So one measure of effectiveness ('effectiveness 1') is the extent to which the students' actions match those that the teacher intended. A second, and rather stronger, measure of effectiveness ('effectiveness 2') is the extent to which the students' learning matches the learning objectives[1].

[1] The usefulness of distinguishing these two senses of 'effectiveness' emerged during discussions in the project *Labwork in Science Education* (LSE). It is also used by Psillos et al. (1998, 1999) and by Psillos and Niedderer in chapter 1 *of this volume*.

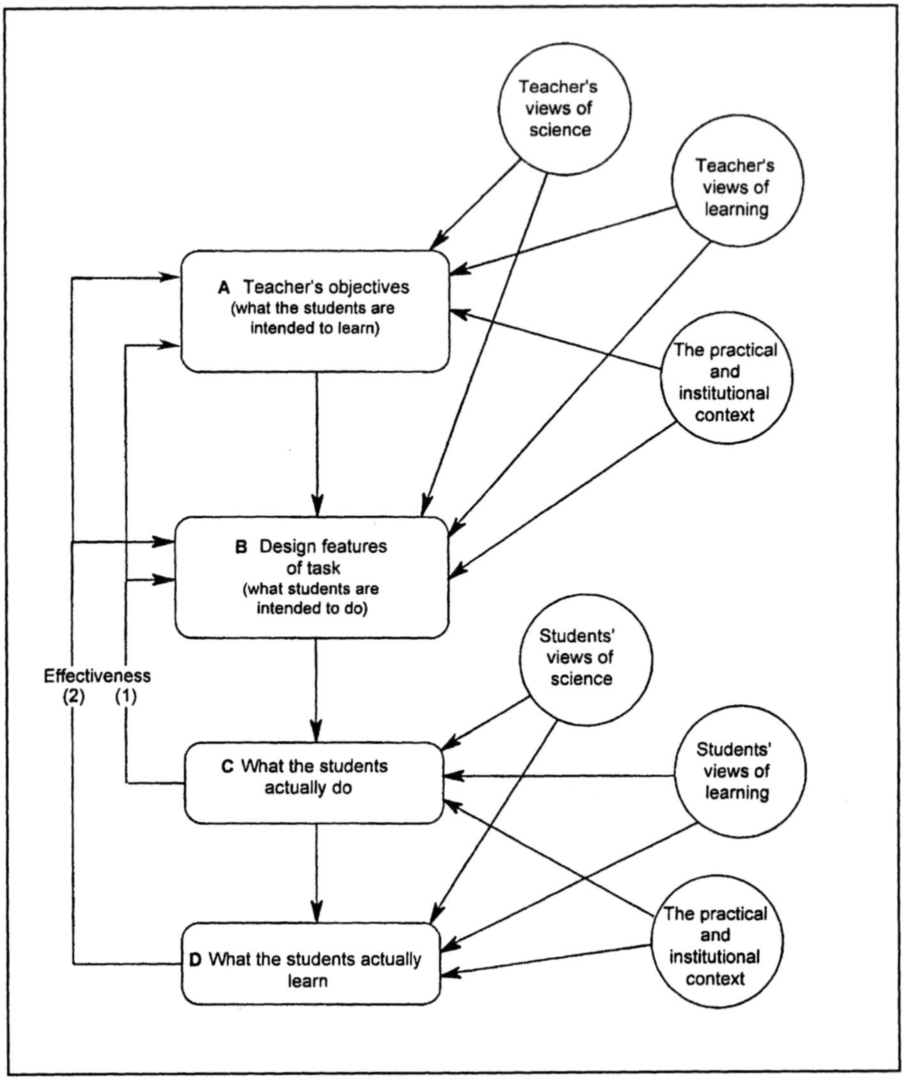

Figure 2: The process of developing and evaluating a labwork task

If a labwork task is found to be effective in one or both of these senses, it is then important to ask which aspects of the task design lead to its effectiveness. Conversely if we find that a task is less effective than we had hoped, a detailed analysis of the task design may help us see how it can be redesigned. In both cases, therefore, it is useful to have a full and systematic description of the main elements of task design.

Producing a profile of a labwork task

Overall structure of the profile

The profile, which we propose, describes the following aspects of a labwork task:

A: the intended learning outcomes (or learning objectives);

B: key elements of the task design, including:

B1: the cognitive structure of the task;

B2: the level and nature of student involvement;

B3: the practical context of the task.

Some of these can be broken down further into a set of more specific features. The profile then consists of a set of codes for each aspect (or feature). A given labwork task can then be profiled by allocating one or more codes for each feature to provide a detailed description. Each aspect will now be discussed in turn, with examples to illustrate the coding categories where this is necessary to clarify the meaning of the terms used.

Aspect A: Intended learning outcomes (or learning objectives)

A central element of the profile of a labwork task is a description of its intended learning outcomes (or learning objectives). Learning objectives divide into two main groups: content and process. The former are concerned with the learning of some aspect of scientific knowledge; the latter with learning some aspect of the process of scientific enquiry. Coding categories for Aspect A are showed in Table 1. A few terms used in this table may require brief explanation. A 'fact', for instance, means an observation statement that can be readily agreed, such as that pure water boils at (or near to) $100^{o}C$, or that common salt dissolves in water whilst chalk does not. A 'relationship' means a pattern or regularity in observations on a set of objects or materials, such as a similarity or trend.

Table 1: Coding scheme for Aspect A: Intended learning outcome

To help students to ...	(tick one or more boxes)	
Content	identify objects and phenomena and become familiar with them	
	learn a fact (or facts)	
	learn a concept	
	learn a relationship	
	learn a theory/model	
Process	learn how to use a standard laboratory instrument or piece of apparatus	
	learn how to carry out a standard procedure	
	learn how to plan an investigation to address a specific question or problem	
	learn how to process data	
	learn how to use data to support a conclusion	
	learn how to communicate the results of labwork	

Aspect B1: Task design - the cognitive structure of the task

This aspect of the labwork profile is based on the general perspective on labwork summarised in Figure 1 above. All labwork involves students in handling objects and observable things. However tasks require students to do different kinds of things, for example, to use something, or to make something, or to observe or measure something. So we can classify a task according to *what students are intended to do with objects and observable things*. The right hand box of Figure 1, however, emphasises that students are also intended to think about the real objects and materials they are handling. For example they may have to describe something, or identify a relationship, or test a prediction, or propose an explanation. So we can also classify a task according to *what students are intended to do with ideas*. Finally, we can ask whether the objects or ideas come first. That is, does the student's work with objects lead towards ideas, or does initial work on ideas lead to actions on objects. This is similar to the inductive/deductive distinction in some earlier labwork classification schemes (for example, Fuhrman, Novick, Lunetta & Tamir 1978).

Coding categories for Aspect B1 are shown in Table 2. This is the most complex part of the profile and so some of these coding categories may require a little discussion. Aspect B1.1 covers the range of things students are intended to do with objects and observables. Some labwork tasks simply require students to *use* an instrument, or a laboratory device, or a standard laboratory procedure – with the emphasis on learning how to do this correctly and well. Others ask students to *present* an object so as to *display* certain features of it clearly, for example in a dissection of a flowering plant, or the arrangement of a set of geological specimens. Some practical tasks require the student to *make* something, for example a physical object (such as an electric circuit from a given diagram) or a material (such as a chemical substance), or to make an event occur (for instance, to produce a spectrum of white light with a prism). All of these of course require the student to make observations, but the focus of the task is elsewhere: on the using, or presenting or making. A fourth, and perhaps the largest, category of labwork tasks is those where the main emphasis is on having the student *observe* something, either an object or material, or an event, or a physical variable. If an observation of a variable is quantitative, then this category means, in effect, 'make a measurement'.

To characterise a labwork task on Aspect B1.1, more than one code may be necessary. For example, measuring a physical quantity (coded as 'observe a variable') necessarily involves using a measuring instrument. But for some other tasks coded as 'observe a variable', this may not be the case. So it is the combination of codes that gives a full description of the task.

Aspect B1.2 then considers the variety of things students may be intended to do with ideas. Some labwork tasks simply require students to *report observations*, though, of course, deciding which features to observe and record is influenced by the teacher's and/or the student's ideas about the task. In other tasks the student has to *identify a pattern* in the behaviour of the objects or events observed, such as changes over time, or similarities between the case observed and one observed previously (for example, when carrying out a standard chemical test).

Table 2: Coding scheme for Aspect B1: The cognitive structure of the task

B1.1: What students are intended to do with objects and observables (tick one or more boxes)		
Use	an observation or measuring instrument	
	a laboratory device or arrangement	
	a laboratory procedure	
Present or display	an object	
Make	an object	
	a material	
	an event occur	
Observe	an object	
	a material	
	an event	
	a physical quantity (a variable)	
B1.2: What students are intended to do with ideas (tick one or more boxes)		
Report observation(s)		
Identify a pattern		
Explore relation between	Objects	
	physical quantities (variables)	
	objects and physical quantities (variables)	
Invent' (or 'discover') a new concept (a physical quantity, or an entity)		
Determine the value of a physical quantity which is not measured directly		
Test a prediction	from a guess	
	from a law	
	from a theory (or model based on a theoretical framework)	
Account for observations	in terms of a given law	
	in terms of a given theory (or model)	
	by proposing a law	
	by proposing a theory (or model)	
Choose between two (or more) given explanations		
B1.3: Objects- or ideas- driven? (tick one box)		
What the students are intended to do with ideas arises from what they are intended to do with objects;		
What the students are intended to do with objects arises from what they are intended to do with ideas		
There is no clear relationship between what the students are intended to do with objects and with ideas		

Other tasks require students to identify a *relationship between objects* (such as that the image in a pinhole camera is inverted compared to the object), or *between objects and physical quantities (variables)* (such as comparing the friction forces between different materials), or *between physical quantities (variables)* (such as the relationship between the extension of a spring and the load, or temperature and rate of a chemical reaction). Another type of practical task requires students to *'invent' or 'discover' a new concept*. This type of labwork task is rather rare, though one example is described in another case study in this book (Bécu-Robinault, Chapter 2 *of this volume*). In this task the first step is to make the students realise the need for a new parameter which will enable a model to fit the data better; the second is then to construct a meaning for this parameter so that it becomes a 'new' physical quantity, or concept. Someone who takes a realist view of theoretical terms would use the word 'discover' here, whilst a radical constructivist would prefer 'invent'. This issue is not relevant for the purposes of our classification and so we offer both.

The essential task for students in some practical tasks is to *determine the value of a quantity*, by an indirect method. This is different from direct measurement using a single measuring device. Here students have to apply a mathematical model to obtain a numerical value of the quantity. An example might be measuring the acceleration due to gravity (*g*), using a simple pendulum. Another type of task asks student to *test a prediction*. The prediction may be simply a *guess*, or it may be deduced from a more formal understanding of the situation, such as an empirical *law*, or a *theory* (or model). 'Testing' here just means looking at the match between prediction and observation. We do not want to imply that practical tasks in the teaching laboratory can provide 'severe tests' of well-established ideas. Usually the real task for the student is to 'produce the phenomenon', that is, to succeed in producing the outcome that is predicted by a well-established scientific explanation.

Finally, some practical tasks ask the student to *account for observations*, either by relating them to a given explanation or by proposing an explanation. An 'explanation' might be an empirical law, or a general theory, or a model derived from a general theory, or general principles derived from a theoretical framework. In some tasks, the explanatory ideas are given and the student has to use these to account for what is observed, perhaps extending or modifying them. A variant of this is where two (or more) possible explanations are proposed and the student has to decide which is better (or best). In other tasks, the observations come first, and the student is asked to propose an explanation using his/her existing knowledge.

Aspect B1.3 then describes the relationship between B1.1 and B1.2. Some tasks are 'objects-driven': the student has to do certain things with objects and then, it is hoped, certain ideas will emerge. Other tasks are 'ideas-driven': ideas are stated first and these direct the things students then do with objects. Of course, to some extent, the first kind are also 'ideas-driven' as all observation is guided by the ideas of the observer (or the teacher giving the instructions). But this dimension of the classification scheme can still usefully indicate the emphasis of the labwork work task (inductive or hypothetico-deductive). Here only one code is chosen as the categories are mutually exclusive.

Aspect B2: Task design - the level and nature of student involvement

One issue highlighted by previous schemes for classifying labwork is the open or closed nature of the task. Aspect B2.1 provides a full description of this (Table 3). This also recognises the possibility of an intermediate level of openness, where decisions are reached through pupil-teacher discussion. Aspect B2.2 then describes the range of levels of student involvement in carrying out the task, from observing while the teacher performs the task, to participation in a class demonstration, to actual performance of the task in groups or individually.

Table 3: Coding scheme for Aspect B2: Level and nature of student involvement

B2.1: Degree of openness/closure			(tick one box in each row)
Aspect of labwork task	specified by teacher	decided by discussion	chosen by students
Question to be addressed			
Equipment to be used			
Procedure to be followed			
Methods of handling data collected			
Interpretation of results			
B2.2: Nature of student involvement			**(tick one box)**
Demonstrated by teacher; students observe			
Demonstrated by teacher; students observe and assist as directed			
Carried out by students in small groups			
Carried out by individual students			

Aspect B3: Task design - the practical context

This final section of the labwork profile consists of a number of simple descriptive features of the task: its duration, the people with whom the student interacts, the information sources available, the type of apparatus provided, the source of the data, and the tools available for processing data. The coding categories are shown in Table 4. Most are self-explanatory. Aspect B3.5, however, may need some clarification. This has been included in order to include labwork tasks in which data is taken from a simulation or a video recording of a process or event, or even from a text record (as in a data interpretation exercise, using data previously collected in a 'real' labwork exercise). Tasks of this sort may be used where the task cannot be carried out in the laboratory for reasons of cost or safety. The intended learning outcomes are often identical to those of the corresponding 'real' labwork.

Using the labwork profile

Tables 1 - 4 together make up a Labwork Task Profile Form which can be used to produce a detailed description and characterisation of any labwork task. This could be used for several purposes. We have already used it to identify similarities and differences in the kinds of practical work used in school science courses in different European countries (Tiberghien, Veillard, Le Maréchal, Buty & Millar 2001). A detailed analysis using this labwork profile can identify types of labwork that are very common, and those that are seldom used. The balance between these can then be considered and reviewed. In a similar way, the Profile could be used to compare the types of labwork used with students of different ages or stages, or in the different science disciplines.

Table 4: Coding scheme for Aspect B3: The practical context

B3.1: Duration of task (tick one box)	
Very short (less than 20 minutes)	
Short (one science lesson, say, up to 80 minutes)	
Medium (2 - 3 science lessons)	
Long (2 weeks or more)	
B3.2: People with whom the student interacts (tick one or more boxes)	
Other students carrying out the same labwork task	
Other students who have already completed the task	
Teacher	
More advanced students (demonstrators, etc.)	
Others (technician, glassblower, etc.)	
B3.3: Information sources available to the student (tick one or more boxes)	
Guiding worksheet	
Textbook(s)	
Handbook (on apparatus), data book, etc.	
Computerised database	
Other	
B3.4: Type of apparatus involved (tick one box)	
Standard laboratory equipment	
Standard laboratory equipment + interface to computer	
Everyday equipment (kitchen scales, domestic materials...)	
B3.5: Source of data (tick one box)	
Real world: inside laboratory	
Real world: outside laboratory	
Simulation on computer or CD-ROM	
Video recording	
Text	
B3.6: Tools available for processing data (tick one or more boxes)	
Manual calculation	
Computer	

This kind of analysis can also be very useful for checking on the range of types of labwork included in a teaching sequence on a topic, and the balance between them.

If a new teaching scheme is being developed, an analysis using the Labwork Task Profile can identify varieties of labwork that are being used very frequently or are being overlooked, so that modifications can be made to redress the balance, if this is thought desirable.

The profile also enables curriculum developers to test and evaluate the impact of specific aspects of labwork task design – by helping to make these design features more explicit. So new labwork tasks can be designed, perhaps by modifying existing ones, with specific design features. These tasks can then be evaluated to see if these design features result in more effective student engagement or student learning. This more structural way of thinking about labwork tasks enables researchers to test more general hypotheses about the effectiveness of labwork, and to generalise from experience with one task (or type of task) to others with similar structural features. In this way, the labwork profile provides a better framework for science education researchers to address key questions about labwork and its effectiveness as a teaching and learning strategy.

Acknowledgements

The profile of practical work described in this article was developed by the authors as part of the work of the project *Labwork in Science Education* (LSE) (Project PL95 2005), supported by the European Commission DGXII. We are grateful for this support, and also acknowledge the contribution of our colleagues in the LSE project to our ideas.

References

Fuhrman, M., Novick, S., Lunetta, V. & Tamir, P. (1978). The Laboratory Structure and Task Analysis Inventory (LAI). (Technical Report 14). Iowa City, IA: Science Education Center, University of Iowa.

Herron, M.D. (1971). The nature of scientific enquiry. *School Review*, 79, 141-212.

Hodson, D. (1991). Practical work in science: Time for a reappraisal. *Studies in Science Education*, 19, 175-184.

Kirschner, P. & Meester, M. (1988). Laboratory approaches. *Higher Education*, 17 (1), 81-98.

Lunetta, V.N. & Tamir, P. (1981a). An analysis of laboratory activities: Project Physics and PSSC. *School Science and Mathematics*, 81, 635-642.

Lunetta, V.N. & Tamir, P. (1981b). Inquiry related tasks in science laboratory handbooks. *Science Education*, 65, 477-484.

McComas, W.F. (1997). The laboratory environment: An ecological perspective. *Science Education International*, 8 (2), 12-16.

Millar, R., Le Maréchal, J-F. & Tiberghien, A. (1998). A Map of the Variety of Labwork. Working Paper 1 from the European project Labwork in Science Education (Targeted Socio-Economic Research Programme, Project PL 95-2005), 23 pages. http://didaktik.physik.uni-bremen.de/niedderer/projects/labwork/papers.html

Psillos, D., Niedderer, H., Séré, M.G. (1998). Effectiveness of labwork as defined from case studies of different types of labwork. Working Paper 8 from the European project Labwork in Science Education (Targeted Socio-Economic Research Programme, Project PL 95-2005), 37 pages.

Psillos, D., Niedderer, H., Vicentini, M. (1999). Case Studies on Innovative Types of Labwork in Science Education. In: M. Bandiera, S. Caravita, E. Torracca and M. Vicentini (eds.): *Research in Science Education in Europe*. Dordrecht: Kluwer, 201-208.

Schwab, J.J. (1962). The teaching of science as enquiry. In J.J. Schwab & P.F. Brandwein (eds.): *The Teaching of Science*. Cambridge, MA: Harvard University Press.

Tamir, P. & Lunetta, V. (1978). An analysis of laboratory activities in the BSCS Yellow Version. *The American Biology Teacher*, 40, 353-357.

Tamir, P. & Pilar-Garcia, M. (1992). Characteristics of laboratory exercises included in science textbooks in Catalonia (Spain). *International Journal of Science Education*, 14 (4), 381-392.

Tiberghien, A., Veillard, L., Le Maréchal, J-F., Buty, C. & Millar, R. (2001). An analysis of labwork tasks used in science teaching at upper secondary school and university levels in several European countries. *Science Education*, 85 (5), 483-508.

Woolnough, B. & Allsop, T. (1985). Practical Work in Science. Cambridge: Cambridge University Press.

Issues and Questions Regarding the Effectiveness of Labwork

Dimitris Psillos, Aristotle University of Thessaloniki, Greece

Hans Niedderer, University of Bremen, Germany

Abstract

The effectiveness of labwork can be defined in two ways. A first definition is seen in comparing the actual activities of students *during* labwork to the intended activities (effectiveness 1). Determining effectiveness in this way makes sure that only the effectiveness of the labwork is determined, not of the whole teaching and learning approach. In a second definition, effectiveness is determined by comparing the actual learning outcomes *after* labwork with the aims and objectives set for a specific lab (effectiveness 2). The learning outcome of course is the ultimate goal of teaching, but in most cases it can not be attributed to the effects of labwork alone.

Introduction

As mentioned in the introduction to the present book, researchers and policy-makers worldwide are convinced of the value of labwork (Woolnough & Alsop 1985; Wellington 2000). Yet they often express their concern about its effectiveness in facilitating students' understanding of various aspects of scientific inquiry (Lazarowitz & Tamir 1995; Tobin, Tippins & Gallard 1995). It has been argued that what the students actually learn in different laboratory contexts needs more attention than it has been given in science education research (Leach & Paulsen 1999).

Scientific understanding, as promoted by labwork, involves students' learning of concepts and models of science as well as the development of their abilities to be engaged in scientific inquiry according to contextual demands. Students in labwork are involved in the world of ideas representing the world of things and are engaged in purposeful observation of and interventions into the world by using specially developed or commonly available objects and apparatus. One assumption is that conceptual and procedural knowledge are intertwined for coping with laboratory situations and both are employed if the students are to be engaged in effective experimental activities (Séré 1999).

The effectiveness of various laboratory forms has drawn the attention of researchers from several perspectives (Ganiel & Hofstein 1982; Lazarowitz & Tamir 1995; Tobin, Tippins & Gallard 1995; White 1996; Lunetta 1998). We consider that the discussion concerning the effectiveness of labwork in promoting students' scientific understanding should not focus only on what the students learn about ideas or scientific procedures. It seems equally important to focus on how the students intervene in the real world of the laboratory and handle laboratory entities. In this study we discuss such a dual approach to investigating laboratory effectiveness and describe examples related to the various laboratory forms which have been empirically explored in the present book (Psillos, Niedderer & Séré 1998; Psillos, Niedderer & Vicentini 1999; Millar, Le Maréchal & Tiberghien, 1998, 1999)

D. Psillos and H. Niedderer (eds.). Teaching and Learning in the Science Laboratory. 21-30.

Intended and actual student activities during labwork

During labwork, it is normally the intention of curriculum designers and/or teachers to engage students in several activities, which are probed by specific task features in a variety of contexts. In a traditional introductory university laboratory, for example, students learn to handle apparatus. In an other lab in secondary school, students may learn to interpret simulated models in a laboratory with new information technologies. Students are expected to be involved in doing science, their activities concern objects, ideas or data, such as calibrating an instrument, plotting a graph or predicting phenomena. Research indicates, though, that students may have their own perceptions of labwork, such as getting the right answer or getting on with the instructions, and that often there may be a mismatch between laboratory goals and actual student practice (Lunetta 1998).

It is important that this type of mismatch between intended and actual student activities be investigated in itself, since understanding of science implies student engagement in specific ways for intervening in the world and for linking their actions with the world of idea in a reliable and valid way. Evidence both of what the students do when engaged in scientific inquiry and of the structure of their actions is related to a distinctive feature of laboratory work as a practical mode of learning in science education. Evaluation of the quality of a piece of labwork on the basis of such outcomes is linked to a specific type of effectiveness, which we call **effectiveness 1** (Psillos et al. 1998, 1999; Millar, Le Maréchal & Tiberghien 1999).

Accepting effectiveness 1 as one important measure of the quality of labwork envisioned in a specific context points to a possible shift in research towards student-originated practices as worthwhile research foci and a potential constituent of developing labwork adapted to students. Previous studies have investigated student behaviour during labwork or have carried out task analysis of a given set of practicals. What we argue here is that effectiveness 1 may work as a two-way approach in revealing the complex interplay between theoretical representations and practical activities and the linking between them that takes place in a laboratory. On the one hand, effectiveness 1 involves the specification of intended actions to be developed by students, and on the other it deals with the structure and the meaning of student practices. In this way, for example, the focus on effectiveness 1 in the study by Robinault (Chapter 2 *of this volume*) has revealed implicit aims inherent in one laboratory worksheet used by the students and the relative "weight" of student activities during completion of this worksheet.

The sampling, reconstruction and assessment of the different activities taking place within a laboratory session has been either theoretically discussed from the point of continuous assessment of practicals (e.g. Fairbrother 1991) or empirically investigated by a number of researchers (e.g. Kyle, Penick & Shymansky 1979; Okebukola 1985). The former mainly concerns teachers' assessment of practicals while the latter have focused on the learning of practical skills (Niedderer et. al., Chapter 1 *of this volume*) but seldom clarify their theoretical position. We suggest that categorisation of the complex interactions taking place during labwork reflects the researchers' epistemological positions regarding the nature of scientific practices, which are exercised by the scientists in the course of experimentation. They are based upon assumptions about learning, which underlie the design of a piece of labwork by a teacher or curriculum developer and determine the specific research focus of a study. For example, in the contribution of Beney & Séré (Chapter 2 *of this volume*) the authors developed a specific categorisation of tasks and students' activities based on concepts from cognitive psychology applied to action during the hands-on phase of labwork. Student activities are shown to belong to three main sets organised in an action network, which includes actions aimed at the phenomenon, actions addressed to measurement and actions designed to accomplish conditions of feasibility. In the study by Kariotoglou (Chapter 2 *of this volume*) the meanings of students' practices when interacting with real things are investigated and discussed taking into account entities from the modelling of scientists' laboratory practice that were adapted to the student laboratory (Hacking 1992).

The research techniques appropriate for effectiveness 1 are normally based on observation of the students in action. We must note here that observation of students' actions are normally informal (Alberts 1986) and are carried out in a way that does not imply serious "interference" with teaching. In this sense, data-taking techniques differ from the specially contrived situations, are they paper-and-pencil tests or set practicals, which are used to assess the outcomes of labwork as discussed in the next section of this study. One example of a newly developed method for obtaining and analysing data has been applied to several studies (Niedderer et al., Chapter 1 *of this volume*). This method deals with what the students are doing during the labwork sessions in relation to the resources that constitute a laboratory context. The new elements in the method involve fast analysis of large amounts of videotapes from labwork according to predetermined categories, focusing on the relations of student laboratory practice with the relevant scientific theory. For example, activities involving students in making links between "theory and practice" include talking about scientific concepts or talking about relations between scientific concepts and real objects. The method provides for a step forward towards sampling and describing students' activities and, at the same time, gives interesting quantitative results about the effectiveness of different kinds of laboratory situations in regard to the intended use of knowledge during labwork. Finally, in some situations, continuous data may be triangulated by results from paper-and-pencil tasks and student assignments. For example, students may be audio-recorded or an observer may take notes of what the students are doing "naturally", according to the perceived demands of the lab guide.

Student learning outcomes after labwork

The evaluation of student learning in relation to the learning objectives is the traditional and widely used way of investigating the quality of a piece of labwork. Data are usually taken in situations, which are specifically designed by the researchers (or the teachers for that matter) for the purpose of assessing students' learning in relation to the set objectives. Data provide evidence of laboratory outcomes after students have completed a piece of labwork. Such situations may be inserted following the completion of certain phases of a piece of labwork, at the end of a laboratory session or at the end of a whole experimental sequence. Evaluation of the quality of a piece of labwork on the basis of student learning achievement after labwork is linked to another type of measuring laboratory effectiveness, which we call **effectiveness 2** in order to distinguish it from effectiveness 1.

It has been noted that in several research studies effectiveness 2 is more or less related to assessing students' conceptual acquisitions as a result of labwork (Lazarowitz & Tamir 1995). However, we must note here that effectiveness 2 is not confined to achievements regarding the concepts and models of science. The criticism levelled against this approach is that it reduces the richness of learning opportunities in labwork to learning just the conceptual part of scientific knowledge. Arguments against this position have been raised persuasively, for example by Hodson (1993), who argued in favour of evaluating understandings of scientific procedures. We emphasise that labwork may influence students' epistemological understandings, which should be one measure of effectiveness 2. This aspect of research has only recently emerged, and is producing promising results on labwork effectiveness, as shown by a number of studies in this book.

Techniques of obtaining learning outcomes involve, for example, pre/post-analysis of tests or other instruments, like concept maps, questionnaires or analysis of special reports. There is an extensive literature on the pros and cons of each technique, including lengthy discussions on the role of practical examinations. The various arguments are not going to be rehearsed here (Kind 1999, Tamir 1991). We only make a few remarks. In some studies data on students' achievements, such as interviews, are frequently triangulated with data from other sources For example, one study (Lewis, Chapter 3 *of this volume*) uses interviews with students triangulated with analysis of course documents and observation of oral presentations of mini-projects by students. Another study (Guillon & Séré, Chapter 3 *of this volume*) analyses students' initial plans and final reports concerning their open-ended projects as well as questionnaires eliciting the epistemological knowledge they employed in making the reports. Student learning pathways may also be monitored using "stroboscopic" techniques like pre-intermediate and post-interviews. For example, in the study by Bisdikian & Psillos (Chapter 5 *of this volume*) the teaching objectives for a whole teaching sequence included the acquisition of heat content knowledge and capabilities for constructing and interpreting graphs by the students. Interviews were carried out at the beginning, the middle and the end of the experimental sequence. In these interviews both students' conceptual knowledge about heat content and their abilities to construct and interpret graphs were monitored, providing data on students' development regarding both conceptual and procedural understanding.

A twofold model on labwork effectiveness

If we talk about two types of laboratory effectiveness, we must ask what the relationship between them is. To answer this, it is helpful to consider a model, which is illustrated in Figure 1. The suggested model represents both types of effectiveness, their proposed relationship and their links to the curriculum and teaching intentions inherent in a piece of labwork. The model draws upon ideas concerning the design and description of labwork activities which were developed during the "Labwork in Science Education" project (Séré, Tiberghien, Paulsen, Leach, Niedderer, Psillos & Vicentini 1998; Psillos, Niedderer & Séré 1998; Millar,

Le Maréchal & Tiberghien 1998; see also Millar et al., Chapter 1 *of this volume*). It is based on views on the types of knowledge involved in labwork (Gott & Duggan 1996; Hodson 1993; Meyer & Carlisle 1996; Hacking 1992; Kariotoglou, Tselfes, Psillos & Evagelinos 2000), and on positions regarding the assessment of students in the laboratory (Giddings, Hofstein & Lunetta 1985; Tamir 1991; Fairbrother 1991; Kind 1999). Certain aspects of the model are self-evident while others need some discussion.

In the model it is assumed that the designers, or the teacher who is involved in labwork, set out objectives and strategies that are related to specific features of the teaching context, to assumptions about learning and to the nature of science.

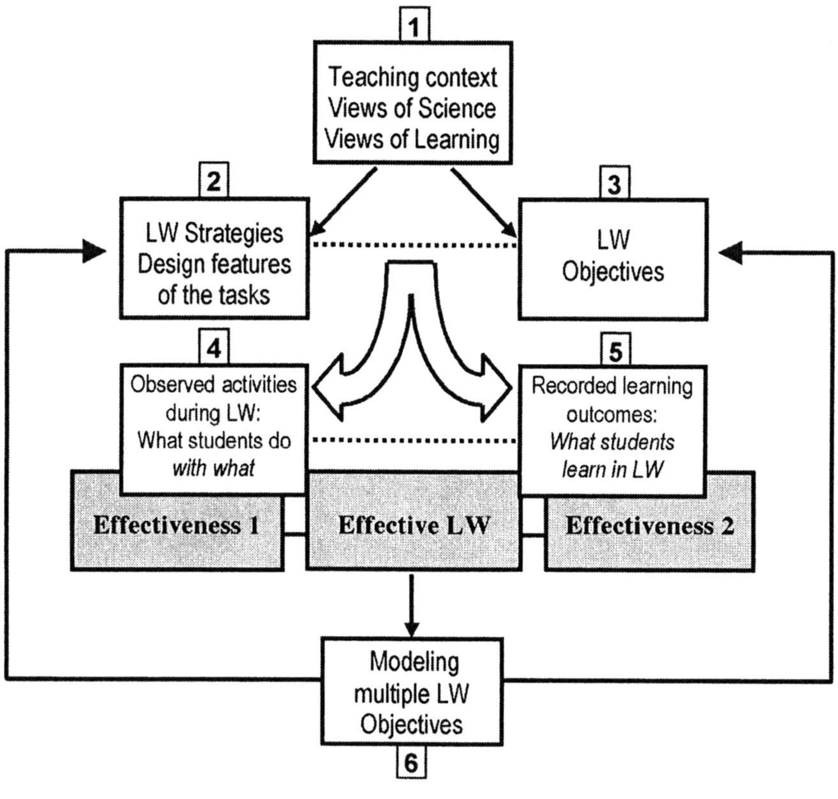

Figure 1: A twofold model of effectiveness

Inspection of the work in this book and other research indicates that the complex activities going on in laboratory situations may not actually be tuned to clearly defined objectives, which are sometimes implicit rather than explicit. Such a position is illustrated by the double arrow emerging from all three boxes rather than stemming from the objectives box, which would imply that in labwork there is a linear and explicit relationship between the perceived or suggested objectives and the design of a particular piece of labwork.

Effectiveness 1 and effectiveness 2 are represented by two different boxes related to student activities during labwork and learning outcomes with regard to intended actions and objectives respectively. We may remark that the relation between student activities in doing science and the conceptual, procedural, and epistemological outcomes after labwork is a complex one, which is currently under investigation. For example, Tamir (1991), in reviewing several studies on assessment of student practical, points out that performance in the practical mode is only weakly correlated with performance in paper-and-pencil tests. Such an undefined relation is indicated by the dotted line between the boxes representing effectiveness 1 and effectiveness 2. Acquisitions, which are signs of effectiveness 2, are in descending chronological order following student practices during labwork, which are signs of effectiveness 1. We argue, though, that descending chronological order does not necessarily imply a linear causal dependency between such outcomes. A piece of labwork may simply be effective from either or both points of view. This is why the boxes of effectiveness 1 and 2 are placed at the same level and are linked to the broader concept of laboratory effectiveness.

We suggest that both effectiveness 1 and effectiveness 2 make for effective labwork and that one type of effectiveness can not be substituted for the other. A twofold effectiveness of the type described above is a very specific feature of the practical character of labwork, which offers a balanced mix of modes of learning and provides ample opportunities for experiencing science as a mode of inquiry (Tamir 1991; Hodson 1993). It is worth mentioning that some outcomes of labwork cannot be assessed outside the laboratory context since they concern events taking place during school time. In this sense the suggested model can be called a **twofold effectiveness model.**

Research on labwork may focus either on effectiveness 1 or on effectiveness 2 or both. Depending on the study, the box on effectiveness 1 or 2 may be smaller or larger or not exist at all. While this is the general trend, some situations may be judged from both data concerning students' activities and statements on what they have learned at the end of a laboratory sequence. For example, in one study (Ryder, Chapter 3 *of this volume*) students' views on the use of geological knowledge were elicited from interviews and discussions following fieldwork as well as from selected observations on the treatment of events and measurements during field measurements. In fact, as soon as students' activities are completed in relation to a given problem statement, a labsheet or other types of instructions can be given and students can be observed, allowing effectiveness 1 to be considered. When these activities are included in a teaching sequence, effectiveness 2 can also be considered.

The suggested twofold model has a dynamic character, which is indicated by the return arrows to both the strategies and the objectives boxes. In other words, the model implies an iterative cycle of research and development by considering both types of laboratory effectiveness as a means to improving labwork in various contexts. Such an approach is in line with recent research, arguing in favour of developmental research as a means of developing research-based science teaching including labwork (Lijnse 1995; Theyssen et. al., Chapter 2 *of this volume*).

Evaluating the two kinds of effectiveness

About the relation between objectives of labwork and the two kinds of effectiveness

Since any effectiveness has to be related to aims, it is useful to look at possible aims and objectives of labwork. In general, a description was given by Millar et al. (Chapter 1 *of this volume*), and in addition the objectives of teachers were determined in the European Labwork in Science Education project (LSE) by a survey 'Teachers' objectives for labwork'. Three main domains of teachers' aims for laboratory work (and many subdomains) were found in this empirical study with nearly the same weight by Welzel, Haller, Bandiera, Hammelev, Koumaras, Niedderer, Paulsen, Robinault & von Aufschnaiter (1998, 12). These are:

(A) for the student to link theory to practice,
(B) for the student to learn experimental skills,
(C) for the student to get to know the methods of scientific thinking.

All three categories of aims could be evaluated in relation to both kinds of effectiveness. For (A) and effectiveness 1 this would mean evaluating how students link theory to practice *during* labwork activities, e.g. by making predictions or using physics concepts in the context of lab activities. For (A) and effectiveness 2, *after* the lab, students' knowledge with respect to the link between theory and practice could be evaluated with tests and interviews, e.g. asking for predictions or explanations in a certain experimental context of a given task. Similarly, for objectives of type B and effectiveness 1, either activities *during* labwork could be analysed with respect to using certain experimental skills, e.g. using an oscilloscope in an adequate way or knowledge about experimental skills could be determined for effectiveness 2 *after* the lab, perhaps in a pre-post design. Finally, epistemological aims for labwork also could be analysed on the one hand with respect to students' implicit epistemological understanding of "methods of scientific thinking" *during* labwork (effectiveness 1), and on the other as knowledge about epistemological issues evaluated *after* the lab (effectiveness 2).

Several studies reported in this book, as part of their case study, empirically determined teacher and student aims and objectives, or at least formulated specific aims for their specific piece of labwork, and then related their analysis of effectiveness to these special aims (Theyssen et al., Lewis, Ryder, Leach, all *in this volume*).

Evaluation of effectiveness 1

The idea here, which is really new, is to distinguish the two kinds of effectiveness and especially to look for effectiveness 1. This seems to represent a progress in research methodology in order to get results about effectiveness which are clearly related to the labwork itself. Whereas learning outcomes (effectiveness 2) measure the effect of the whole teaching and learning approach, with intended activities as the category of analysis we are looking directly into the lab itself. With effectiveness 1, we are looking at the effectiveness of labwork in comparison with intended activities *during* labwork.

The contributions *in this book* offer a variety of possibilities for achieving this. In several studies the intended activities are related to the objective of "linking theory to practice". This objective is by itself more related to activities *during* the lab. Of course it can be also evaluated as a learning outcome, e.g. with a given apparatus and tasks to predict or to explain. Becu-Robinault analyses transcripts from labwork with respect to the connections made by students between the world of theory and models and the world of objects and events. This analysis of activities during labwork is legitimated by an assumption about learning: "The links between the worlds of objects and events and theory-model are necessary to learn physics." Several other authors in this book have also used videotapes from labwork, but without making transcripts. They used a method called "Category Based Analysis of Videotapes from Labwork (CBAV)" (Niedderer et al., Chapter 1 *of this volume*) to analyse how often and in which contexts students talked about physics during labwork, i.e. used physics concepts related to the lab (Theyssen et al., Hucke et al., Sander et al., Buty, all *in this volume*). With this method, effectiveness 1 can be evaluated relating the amount of student verbalisations of knowledge to specific labwork contexts such as "working with the tutor" or "making measurements". In Bisdikian et al. a similar research question related to effectiveness 1 can be found: "Do students use knowledge of physics to regulate their actions during laboratory work and, if so, under what conditions?" A first general result is that different studies converge in pointing out that, during labwork, students did not to a large extent employ the intended theoretical explanations offered in their course-book or in the associated lectures, even if the experiment was considered an easy one. A second important trend is that, in quite different contexts at secondary school and university, manipulating apparatus and taking measurements are dominant activities occupying much of the intended students' time during laboratory sessions, but their contribution to allowing students to relate theory to experiments is comparatively small.

A different way of determining effectiveness 1 is related to the objective of improving "students' constructions on experimental inquiry during labwork" (Kariotoglou). He observed that students have "difficulties in using concepts to intervene in the experiment." Similarly, Ryder puts a research question related to effectiveness 1 about procedural knowledge: "In what ways do students work with data during their field course?" As one result, he finds that "when working within their peer groups, students tend to collect data at particular sites in the field study area without interpreting their data in terms of what they know of the geological history of the site as a whole." In Theyssen et al., students cognitive construction processes *during* labwork are analysed with respect to nine levels of complexity, containing levels like "objects", "events" and "principles". The implicit objective behind this is that learning should reach higher levels of complexity.

Evaluation of effectiveness 2

It was shown above that for both effectiveness 1 and effectiveness 2 different kinds of learning objectives could be relevant. Evaluation of effectiveness 2 in most cases in this book is related to a better conceptual understanding resulting from different lab approaches. Related to the specific content, specific tests, questionnaires or interviews are developed and applied. This can be done with tests like the FCI, in a pre-post design for the whole course, comparing the results in an experimental group with a new lab to results of a control group (Sander et al.). Or it can be done with pre-post testing using concept maps before and after single labs (Hucke et al.). In more qualitative studies, different levels of students' understanding can be defined to determine effectiveness 2 (Kariotoglu; Bisdikian et al., Barbas et al.). These studies, then, are able to describe conceptual learning pathways along the work of students in practicals as a special form of determining this type of effectiveness. But aims related to procedural understanding of epistemological issues of labwork can also be analysed as learning outcome (Guillon et al., Leach). The understanding of measurement errors can by itself be an objective of labwork; the related effectiveness 2 is determined in Leach and Evangelinos et al. Lewis also analyses learning outcome to a considerable extent (effectiveness 2), although her data suggest additional hints for problems relating to students' actions during labwork (effectiveness 1). It may be noted that determining effectiveness 2 by various forms may be problematical: it is often unclear which effects are due to labwork and which to other factors of the teaching and learning process.

Concluding remarks

In this study we have attempted to discuss the complex concept of laboratory effectiveness from various perspectives. We argue that evaluating effectiveness 1 in addition to the more traditional effectiveness 2 reflects the true nature of labwork as a practical activity and opens up new research possibilities as design opportunities for labwork

References

Alberts, R. V. J., Van Beuzecom, P. J. and De Roo, I. (1986). The assessment of practical work: a choice of options. *European Journal of Science Education*, 8, 361-369.

Fairbrother, B.(1991). Principles of practical assessment, in Woolnough, B. *Practical Work in Science*. Cambridge: Cambridge University Press, 153-165.

Ganiel U., and Hofstein A., (1982). Objective and continuous assessment of students' performance in the physics laboratory. *Science Education*, 66, (4), 581-591.

Giddings, G. J., Hofstein, A. and Lunetta, V. N. (1985). Assessment and evaluation in the science laboratory, in Woolnough, B. (Ed): *Practical Work in Science*. London: Open University Press.

Gott, R. & Duggan, S. (1996). Practical work: its role in the understanding of evidence in science. *International Journal of Science Education*, 18, (7), 791-806.

Hacking, I. (1992). The self-vindication of the laboratory sciences. In Pickering, A. (Ed): *Science as practice and culture*. Chicago: The University Chicago Press.

Hodson, D. (1993). Re-thinking Old Ways: Towards A More Critical Approach To Practical Work In School Science. *Studies in Science Education*, 22, 85-142.

Kariotoglou P., Tselfes V., Psillos D., and Evagelinos D., (2000) An investigation on student teachers laboratory practices during a familiarisation with phenomena phase of experimental teaching in: Behrendt, H., Dahncke H., Duit R.,. Graber W., Komorek M., Kros A., Reiska P.(Eds.) *Research in Science Education: Past, present and future*. pp 331-336. Dordrecht: Kluwer.

Kind, M., (1999). Performance assessment in science - what are we measuring? *Studies in Educational Evaluation 25*, 179-194.

Kyle, W.C., Penick, J.E. and Shymansky, J.A. (1979). Assessing and analysing the performance of students in college science laboratories. *Journal of Research in Science Teaching, 16(6), 545-551.*

Lazarowitz, R. & Tamir, P. (1995). Research on using Laboratory Instruction in Science. In Gabel D. (Ed): *Handbook of research on science teaching and learning*. NY: Macmillan, 94-128.

Leach, J. & Paulsen, A., (Eds.) (1999). *Practical work in science education*. Frederiksberg: Roskilde University Press & Dordrecht: Kluwer.

Lijnse P. L. (1995). "Developmental Research" As a Way to an Empirically Based "Didactical Structure" of Science, *Science Education* 79(2), 189-199.

Lunetta, V. N. (1998). The school science laboratory: Historical Perspectives and Contexts for Contemporary Teaching. In Tobin, K. Fraser, B. (Eds.) *International Handbook of Science Education*. Dordrecht: Kluwer, 249-264.

Meyer, K. and Carlisle, R. (1996). Children as experimenters, *International Journal of Science Education*, 18, (2), 231-248.

Millar, R., Le Maréchal, J-F. & Tiberghien, A. (1998). A Map of the Variety of Labwork. Working Paper 1 from the European project Labwork in Science Education (Targeted Socio-Economic Research Programme, Project PL 95-2005), 23 pages. (http://didaktik.physik.uni-bremen.de/niedderer/projects/labwork/papers.html).

Millar, R., Le Maréchal, J-F. & Tiberghien, A. (1999). "Mapping" the domain – varieties of practical work. In J. Leach & A. Paulsen (Eds.) (1999). *Practical work in science education*. Frederiksberg: Roskilde University Press & Dordrecht: Kluwer, p. 33-59

Okebukola, P.A., (1985). Science laboratory behaviour strategies of students relative to performance in and attitude to laboratory work. *Journal of Research in Science Education, 22,(3), 221-232.*

Psillos, D., Niedderer, H. (Eds.), (1998). Case studies on labwork in 5 European countries, Working paper No 7. LSE Project, No PL95-2005, (European Commission DG XII).

Psillos, D., Niedderer, H., Séré, M.G. (1998). Effectiveness of labwork as defined from case studies of different types of labwork. Working Paper 8 from the European project Labwork in Science Education (Targeted Socio-Economic Research Programme, Project PL 95-2005).

Psillos, D., Niedderer, H., Vicentini, M. (1999). Case Studies on Innovative Types of Labwork in Science Education. In: M. Bandiera, S. Caravitta, E. Torracca and M. Vicentini (eds.): *Research in Science Education in Europe*. Dordrecht: Kluwer, 201-208.

Séré, M.G., Tiberghien, A., Paulsen, A. C., Leach, J., Niedderer, H., Psillos, D., Vicentini, M., 1998, Labwork in Science Education – Executive Summary. Working paper from the European project Labwork in Science Education (Targeted Socio-Economic Research Programme, Project PL 95-2005), 16 pages, (http://didaktik.physik.uni-bremen.de/niedderer/projects/labwork/papers.html).

Séré, M.G., (1999). Learning science in the laboratory: Issues raised by the European Project "Labwork in Science Education". In: Bandiera, M. Caravita S., Torraca, E., and Vicentini, M., (Eds.): *Research in Science Education in Europe*. Dordrecht: Kluwer, 165-175.

Tamir, P. (1991). Evaluation of student laboratory work and its role in developing policy. In: Hegarty-Hazel, E. (Ed): *The Student Laboratory and the Science Curriculum*. London: Routledge, 242-266.

Tobin, K., Tippins, D.J. and Gallard, A.J. (1995). Research on instructional Strategies for Teaching Science. In Gabel D. (Ed): *Handbook of research on science teaching and learning*. NY: Macmillan, 45-93.

Wellington, J., (2000) (Ed): *Practical work in school science - Which way now?* London: Routledge.

Welzel, M., Haller, K., Bandiera, M., Hammelev, D., Koumaras, P., Niedderer, H., Paulsen, A. C., Robinault, K., von Aufschnaiter, S., 1998b, Teachers' objectives for labwork. Research tool and cross country results. Working Paper 6 from the European project Labwork in Science Education (Targeted Socio-Economic Research Programme, Project PL 95-2005), 109 pages, (http://didaktik.physik.uni-bremen.de/niedderer/projects/labwork/papers.html).

White, R. T. (1996). The link between the laboratory and learning. *International Journal of Science Education*, 18, (7), 761-774.

Woolnough, B. & Allsop, T. (1985). *Practical Work in Science*. Cambridge: Cambridge University Press.

Talking Physics in Labwork Contexts - A Category Based Analysis of Videotapes

Hans Niedderer, Stefan v. Aufschnaiter, University of Bremen, Germany

Andrée Tiberghien, Christian Buty, CNRS - Université Lyon 2, France

Kerstin Haller, Universum Science Center Bremen, Germany

Lorenz Hucke, Fachhochschule Gelsenkirchen, Germany

Florian Sander, Institut für den Wissenschaftlichen Film (IWF), Germany

Hans Fischer, University of Dortmund, Germany

Abstract

This study has two aims: to give some overview of methods used previously by other researchers for analysing labwork in science education and to describe a new method for analysing labwork using a category-based analysis of videotapes from labwork (CBAV). In this CBAV method, two types of categories are defined: categories for labwork contexts and categories for verbalised knowledge during work in these contexts. The method was used in five studies of labwork in France and Germany in upper secondary school and university physics classes (see contributions of Buty, Theyßen et al., Hucke et al., and Sander et al., all *in this volume*; Haller 1999). Specific results can be found there. The method among others can help to answer questions about the link between theory and practice in different labwork contexts. It can be used complementary to other methods and permits to analyse a lot of video data in a relatively short time.

Introduction

Labwork in science education is a complex situation, it involves very different kinds of activities such as making predictions, manipulating apparatus, making measurements, and talking about physics. Analysing such complex situations is very challenging. In this chapter our perspective is to present a method and its associated type of results, which aims to study the relations between the characteristics of a labwork situation and the students' activities. This method is devoted to analyse data which consist of videotapes of a group of students during laboratory activities, this is why we call it CBAV (category-based analysis of videotapes). This analysis allows us a rather fast analysis of videotapes from labwork. It should be complementary to a more detailed qualitative interpretative analysis of learning processes using transcripts from audio and videotapes (Bécu-Robinault & Tiberghien 1998; von Aufschnaiter & Welzel 1999; Fischer 1994; Petri & Niedderer 1998; Buty 2000) or to other methods (e.g. Haller 1999; Hucke 2000; Sander 2000). As any method of analysing activities, our method implies to define categories.

This study explains the CBAV method and compares it to methods previously used. It also shows some typical results. More specific results from using this method can be found in four studies in this book (see contributions of Buty, Theyßen et al., Hucke et al., and Sander et al., all *in this volume*) and in Haller (1999).

D. Psillos and H. Niedderer (eds.). Teaching and Learning in the Science Laboratory. 31-40.

Categories previously used by other researchers

Many researchers before have defined categories to analyse the complex teacher-student behaviour in observing laboratory contexts and activities. Some of them were dealing with interaction: 'It [the method] is an attempt to obtain detailed, qualitative and/or quantitative descriptions of interactions that occur during the teaching/learning process of science in the context of labwork' (Ogunniyi 1983, 195).

Other earlier investigations (Tamir & Lunetta 1981; Lumpe & Scharmann 1991; Germann, Haskins & Auls 1996) used the Laboratory Structure and Task Analysis Inventory (LAI) developed by Fuhrmann, Novick, Lunetta & Tamir (1978). This instrument was first used mainly for a content analysis of laboratory manuals, not for the observation of actual behaviour. The LAI categories are related to the four typical working phases in the laboratory: planning and design, performance, analysis and interpretation, application (Lunetta 1998, 255). This instrument, which was constructed for the analysis of labguides, particularly developed the categories dealing with manipulation of apparatus in relation with the 'performance' phase.

Abraham (1982) developed the 'Laboratory Program Variables Inventory (LPVI)'. It consists of 25 statements written on cards, to be ranked by the students. Various procedures are described like interactions between students and teachers, between students and material, and purposes and outcomes students might experience in their laboratory. Mainly, the instrument can be used for an exact description of different types of laboratories. Yet it is not a description made by an external observer, but the students themselves are asked for their opinion what the lab is like. Therefore, although this method is a step into the direction of investigating what actually happens in the laboratory, it neither helps develop categories of lab contexts nor categories of intended activities during labwork.

A third method was used by Penick, Shymansky, Filkins & Kyle (1978). They developed the 'Science Laboratory Interaction Categories - Student (SLIC)' to describe activities during laboratory work, especially interactions between learners, and between learners and the teacher. Their categories can be divided into two classes. The first class are those related the material objects and observable events: *showing, manipulating apparatus, reading, recording data, getting supplies, non lesson related behaviour.* The second class are those categories related to communication activities: *transmitting information, asking questions, listening.* Kyle, Penick & Shymansky (1979) used this instrument (SLIC) to investigate differences in students' behaviour in the laboratory in five science disciplines (Botany, Chemistry, Geology, Physics, and Zoology). Okebukola (1985) has also used the SLIC instrument in a large study with 600 eleventh grade biology students and 20 specially trained biology teachers as observers.

In a fourth approach, Stein, Friedler & Nachmias (1990) compared traditional and computer-based (MBL) science laboratory experiments by means of a cognitive task analysis. They analysed tape recordings of lab-sessions of 13 eighth grade middle school students categorising the verbal statements. They identified *off-task*, *empirical* and *conceptual* episodes as typical for laboratory work. Empirical episodes are characterised by pursuing goals related to the lower-level tasks of conducting an experiment (locating and setting up apparatus, monitoring apparatus, recording data points). During conceptual episodes students are pursuing goals directly related to the analysis of results, or drawing conclusions or inferences, thus dealing with scientific principles. Time budgets have been calculated relating the types of episodes to the typical lab phases (set-up, data-collection and analysis). The scores are represented as cumulative relative percentages for each of the three lab phases (Stein, Nachmias & Friedler 1990, 192). Stein *et al.* classify different types of episodes in each time interval.

Theoretical frameworks

In this part, first, we make explicit how this work is mainly related to one goal of laboratory activity, and how effectiveness of labwork can be defined in relation to this goal. Then we present our learning hypotheses, discuss the specificity of our method compared to others, and at last we describe the general type of our research questions.

Effectiveness of labwork in relation to objectives

Generally effectiveness of labwork can have two different aspects, one being related to intended activities *during* labwork, the other being related to learning outcomes *after* labwork. In the European project 'Labwork in Science Education', the first aspect has been named effectiveness 1, the second effectiveness 2 (Séré, Tiberghien, Paulsen, Leach, Niedderer, Psillos & Vicentini 1998, 11; Millar, Le Maréchal & Buty 1998; Psillos, Niedderer & Vicentini 1999; also Millar, Tiberghien & Le Maréchal and Psillos & Niedderer in Chapter 1 *of this volume*). In this approach, we do not take into account any data about learning outcomes after labwork, but we analyse videotapes from labwork with respect to intended behaviour during labwork. This means that results of this study are about effectiveness 1 of labwork, analysing 'what the students actually do' compared with intended activities related to important objectives of labwork. According to a European survey, teachers have rated 'to link theory to practice', 'to learn experimental skills' and 'to get to know the methods of scientific thinking' as the three most important objectives for labwork (Welzel, Haller, Bandiera, Hammelev, Koumaras, Niedderer, Paulsen, Robinault, & von Aufschnaiter 1998 a, b). Students seem to have different implicit perceptions of labwork, such as following the instructions, getting the right answer, manipulating equipment and doing measurements. Lunetta (1998, 250) therefore speaks of a 'mismatch between goals, behaviour and learning outcomes'. In view of these different possible objectives of labwork, we developed our categories for intended activities *during* labwork - 'to verbalise knowledge in different contexts of labwork' - mainly in relation to the objective 'to link theory to practice'.

Learning hypotheses

Establishing categories is based on *underlying hypotheses*. We make the hypothesis that the aim 'to link theory and practice' will be fostered if during laboratory work, the students explicitly establish such links. Therefore, we suppose that verbalisation of knowledge in the contexts of labwork is an important step towards linking theory to practice and furthermore to learning physics (Bliss 1996).

From these hypotheses two main categories are chosen, physics knowledge (KP) that is when the verbalisation involves physics concepts, and technical knowledge (KT) when the verbalisation involves apparatus or more generally material objects and perceptible events. These categories are respectively related to the 'world of theory / model' and to the 'world of objects and events ' (Tiberghien & Megalakaki 1995). A third type of category is needed to classify the relation between objects of the real world and physics concepts (KTP). They have been defined similarly to work reported by Bécu-Robinault (1997). If students during labwork talk about technical knowledge or physics knowledge as defined in these categories this indicates some cognitive processes going on. So, we assume that talking about physics means to verbalise knowledge in the different contexts of labwork, and that it is a viable indicator for cognitive processes contributing to the objective 'to link theory to practice'. Of course, there can be additional cognitive learning effects without verbalisation, which are not detected by CBAV. In spite of this, it seems relevant to analyse which contexts of labwork contribute more or less to talking physics.

Specificity of our method

The comparison to earlier category-based methods for analysing laboratory work reveals the following specific features of our CBAV method:

- All previous research in this field has been done in school laboratories and not at university. Our study relates to both school and university laboratory work.
- Previous research has often been done on analysing written materials. We focus on activities during labwork. But instead of actual classroom observations, we analyse videotapes. This offers a chance of looking at the same processes repeatedly to make statements about the reliability of the method (agreement of different raters) and to use more complex category systems, as it might be possible in an on-line rating.
- Previous research has mainly focused on one of the main goals of laboratory work: to learn practical (or inquiry) skills. Only one study (Stein, Nachmias & Friedler 1990) attempts to analyse conceptual aspects of laboratory work as expressed in the objective to link theory to practice or to learn physical concepts. Similar to this approach, our study focuses on students' verbalisation of knowledge during labwork.
- Only Stein et al. (1990) compared different types of laboratory work (traditional vs. computer-based). In our method, we compare different types of using the computer integrated into labwork and labs without computer. Additionally, with our method it is possible to focus on the relations between different contexts of labwork and students' tendency to verbalise knowledge.
- The analysis itself works more or less in real time, without transcripts and thereby allows reviewing a bigger amount of data.

Especially by this last feature, we see this method as a complementary method to analysing videotapes by using careful transcriptions and qualitative interpretative methods of analysis or to other methods like concept mapping.

General research questions

According to our learning hypotheses, the main goal is to find out how different contexts of labwork, such as manipulating apparatus or doing measurements, contribute to the amount of students' verbalisations of physics knowledge. This means to find out how effective these lab contexts are to promote talking about physics. We quantify 'talking about physics' by the 'time of talking', as 'time' is considered to be an important variable in learning processes (Berelson 1959, 509). To calculate the effectiveness, we define a new variable 'density of knowledge verbalisation in a special lab context' to have an indicator for the effectiveness of a special lab context in promoting knowledge verbalisation. It tells us which parts of time students are talking about physics (or other knowledge) while working in a special type of lab context (see below).

So, the following types of research questions can be answered by the CBAV method:

- How much time during labwork is devoted to work with the different contexts and resources?
- How much time during labwork is devoted to the verbalisation of different kinds of knowledge?
- Which of the contexts are more or less effective in the sense that they promote students to talk about physics during labwork?

For different approaches of labwork, the results along these questions can be compared.

The CBAV method

As was explained above, in developing our video analysis procedure, we defined two types of categories. First, we categorised what students do during labwork characterised by the laboratory resources they use. These context categories are defined below and they are easy to observe. Examples of these categories include 'manipulating apparatus' or 'interactions with a third person'.

The second type of categories was developed to catch the intended activities during labwork. 'Intended' means that they are related to certain objectives of labwork. As one type of intended and observable activity related especially to the objective to link theory to practice - but also to others - we see the verbalising of different kinds of knowledge, e.g. talking about physics, in the contexts of laboratory work.

The CBAV categories for contexts of labwork

Relevant categories for contexts of labwork are rather obvious. Mainly the important resources being used, such as apparatus, measuring devices, lab guides, or interaction with a tutor define them. Several other authors (see above) have defined similar categories. Table 1 shows the CBAV categories for labwork context.

The CBAV categories for types of verbalised knowledge

The CBAV method was developed primarily to examine labwork for evidence of students linking theory to practice. As was explained above, we therefore developed categories for verbalising different kinds of knowledge, such as physics knowledge (KP) or technical knowledge (KT). Table 2 gives the CBAV categories used for verbalisations of knowledge during labwork.

Table 1: CBAV categories of labwork context

Category		Description	Examples
Other	O	Activities not related to the lab.	Talking about last nights' TV
Interaction with third person	3.P	A third person can be the teacher, the tutor, other students, or similar.	Tutor helps to solve a problem and talks to the students
Labguide	LG	Using the labguide.	... to plan what to do.
Paper and pencil	PP	Using paper-and-pencil. Students are writing or reading in their own notes.	Preparing tables for measurement data, drawing a graph.
Mani-pulation of apparatus	MA	Using the apparatus and devices. Carrying out experimental set up or preparing a measurement	Building up an electrical circuit; taking a test-measurement; having a problem with the apparatus.
Measure-ment	ME	Using the apparatus to gather data and writing them down. Resources used are apparatus *and* paper/pencil	Taking the pendulum's amplitude and writing the value down
Calculation	CL	Using a (pocket) calculator or a special software like Excel for this purpose or doing a direct calculation with paper-and-pencil	Calculating a physics quantity from the measurement data
Computer measure-ment	CME	Replaces category ME in the case of computer-based measurements in labwork (MBL)	Reading the amplitude from the graph on the computer screen
Computer model building	CMB	Using a modelling software (e.g. STELLA) to create a model structure or make changes or add new relations	Building a model of an oscillating spring and incorporating a frictional force into this a model.
Computer model use	CMU	Running a simulation when a model (STELLA) is ready and only parameters in the model are changed.	To predict measurement values by the model (simulation of experiment)

Table 2: CBAV categories of verbalised knowledge

Category		Description	Examples
Physics knowledge	KP	Students use physics knowledge, e.g. using words referring to physics	Talking about how to determine the phase from an oscillation diagram
Technical knowledge	KT	Students use knowledge more related to technical apparatus. Often related to the handling of apparatus	Talking about how to operate an oscilloscope; adjusting the interface software
Technical and physics knowledge	KTP	Students use physics knowledge and technical knowledge together	Talking about how to carry out a measurement for a certain physics quantity
Mathe-matical knowledge	KM	Students use formulas in their statements or other mathematical knowledge	Describing the mathematical properties of a measured curve

Some special comments on categories and differences of their use in the five different studies can be found in the special studies (see contributions of Buty, Theyßen et al., Hucke et al., and Sander et al., all *in this volume*; and in Haller 1999).

The grid of categories

We developed a grid to facilitate recording information while watching the videotapes in real time.

Table 3: Example of a grid of categories for analysing videotapes

Time	Context										Knowledge							Comments
Min	O	3. P	L G	P P	M A	M E	C L	C M E	C M B	C M U	K T	K P	K T P	K M	K M M	K M P	K M T	
0.0			1								1							Components of a telescope
0.5			1															
1.0			1								1		1					Loading a new part of software

Generally one or more grid entries are made for every thirty seconds of playing time. Of course a specific scene can be repeated several times, if necessary. Table 3 shows a grid with some sample data.

Interrater reliability of the CBAV categories

In order to determine how consistently the categories for labwork contexts and knowledge were used, six pieces of videotapes of a length between 20 min and 45 min (corresponding to 40 and 90 coded time intervals) have been analysed by different raters. The raters were from the four different research groups, thus using all the categories in a slightly different manner and relating to different theoretical backgrounds. There are two main questions with respect to the reliability of the CBAV method:

- How consistently does one rater use the CBAV categories? To test this, each rater performed the CBAV with selected episodes of his videos once more.
- How consistently do different raters use the CBAV categories? To test this, selected episodes were exchanged between several raters and analysed again.

To calculate the interrater reliability an indicator C_{ir} was used, that specifies how many coinciding marks in percent were found between two different ratings. C_{ir} varies between 100% (two raters or one rater in two trials marked the same categories all the time) and 0% (both marked different categories all the time). C_{ir} is defined by the following formula:

$$C_{ir} = \frac{2 \cdot A}{2 \cdot A + B + C} \cdot 100$$

A: Number of cases where rater one and rater two marked the same category.
B: Number of cases where rater one marked a category while rater two did not.
C: Number of cases where rater two marked a category while rater one did not.
D: Number of cases both observers did not mark a certain category.

'Cases' relate always to the same time step.

This means, that only positive decisions of the raters (marked fields) are counted, not the empty fields (D), and that the number of all equal marks (2A) is divided by the number of all marks (2A+B+C). The results are presented in Table 4.

Table 4: Interrater reliability for categories of the CBAV method

	Labwork Context	Verbal Knowledge
C_{ir}, one rater	80 – 88%	83 – 87%
C_{ir}, two raters	70 – 83%	30 – 80%

As they are based on short pieces of videotapes the numbers can only give tendencies of interrater reliabilities. The results tell us, that the best values of interrater reliability of Cir = 80 – 88% are found with 'labwork context categories' between one rater and his or her own ratings some time later ('one rater').

Density of knowledge verbalisation

As already explained in our theoretical frameworks, we define a new variable 'density of knowledge verbalisation in a special lab context' to have an indicator for the effectiveness of a special lab context in promoting knowledge verbalisation.

To calculate the density of physics knowledge verbalisation (KP) in different contexts X, we first count all time units where students work with one context X, e.g. with the labguide (LG). Then the time units with verbalisation of physics knowledge (KP), while being in this context X, is detected and their number is counted. The ratio of the number of time units with KP divided by the total number of time units in this context X (multiplied by 100) then result in what we call the density. This results in the following formula:

$$Density(KP/X) = 100 \cdot \frac{\sum Timeunits\ KP\ in\ X}{\sum all\ Timeunits\ X}$$

Results about effectiveness of labwork related to talking physics

As can be seen in the four investigations reported in this book (see contributions of Buty, Theyßen et al., Hucke et al. and Sander et al., all *in this volume*) and in Haller (1999), data gained with this method yield some typical results:

- Manipulating apparatus and doing measurement together takes much time (between 50 – 80% of time) during labwork.
- At the same time, these lab contexts contribute rather low to the verbalisation of physics knowledge. The densities are in most cases below 10%. So, traditional laboratory work where students are to a large extent occupied by doing measurements and manipulating apparatus is clearly not the type of labwork that fosters the objective "to link theory and practice".
- On the other hand, some other labwork contexts contributes a lot better to talking physics during labwork. One of these contexts is working with the tutor. Here in some studies, we found densities of about 30 to 50%. That means that during interactions with the tutor the *students* themselves talked a lot more about physics than while doing measurements or manipulating apparatus.
- The other result is equally important: During computer model building activities integrated into labwork, we found some of the highest densities of talking physics. Densities found in this context lie between 20% and 70%. So computer model building integrated into labwork is a chance to get students to talk more about the physics background of labwork and thus can improve the link between theory and practice.

Summary

We developed a new method to analyse videotapes from labwork, called *category-based analysis of videotapes (CBAV)*. This method can be used to analyse the effectiveness of labwork by determining time budgets of different lab contexts, such as working with the labguide, working with the tutor, manipulating apparatus, and doing measurements, as well as of knowledge verbalisation, especially physics knowledge. This method permits to detect how much a specific context contributes to verbalisation of physics knowledge and thus gives a measure of the effectiveness of different labwork contexts with respect to talking about physics and therefore fostering the objective 'to link theory to practice'. The method has been useful with different theoretical backgrounds and different kinds of labwork in five different studies at four different universities in two different European countries. This shows the flexibility and value of the method to be used with different theoretical background.

References

Abraham, M.R.(1982). A descriptive instrument for use in investigating science laboratories. *Journal of Research in Science Teaching*, 19(2), 155-165.

American Association of Physics Teachers (AAPT) (1998). Goals of the introductory physics laboratory. *American Journal of Physics*, **66**(6), 483-485.

Bliss, J. (1996). Piaget und Vygotsky: Ihre Bedeutung für das Lehren und Lernen der Naturwissenschaften. *Zeitschrift für Didaktik der Naturwissenschaften*, **2**(3), 3-16

Bécu-Robinault, K., Tiberghien A. (1998). Integrating stable experiments in energy teaching. *International Journal of Science Education*, **20**(1), 99-114.

Bécu-Robinault, K. (1997). Rôle de l'expérience en classe de physique dans l'acquisition des connaissances sur les phénomènes énergétiques. Thèse, Université Claude Bernard-Lyon I.

Berelson, B. (1959). Content Analysis. In G. Lindzey (Ed.), *Handbook of Social Psychology*, Vol. I, 3rd ed. (Cambridge: Addison-Wesley), 488-522.

Buty C. (2000). Etude d'un apprentissage dans une séquence d'enseignement en optique géométrique à l'aide d'une modélisation informatique. Thèse, Université Lumière-Lyon II.

Fischer, H. (1994). *Physiklernen: Eine Herausforderung fuer die Unterrichtsforschung*. In D. Nachtigall (Ed.), Didaktik und Naturwissenschaften, Band 3. (Frankfurt: Lang)

Fuhrmann, M., Novick, S., Lunetta, V. N., Tamir, P. (1978). A laboratory organization and task analysis inventory. (Technical Report 15) Iowa City, Iowa, Science Education Center, The University of Iowa.

Germann, P.J., Haskins, S., Auls, S. (1996). Analysis of nine high school biology laboratory manuals: Promoting scientific inquiry. *Journal of Research in Science teaching*, **33**(5), 475-499.

Haller, K. (1999). *Über den Zusammenhang von Handlungen und Zielen. Eine empirische Untersuchung zu Lernprozessen im physikalischen Praktikum.* Doctoral dissertation University of Bremen, (Berlin: Logos).

Hucke, L. (2000). *Handlungsregulation und Wissenserwerb in traditionellen und computergestützten Experimenten des physikalischen Praktikums.* Doctoral dissertation, University of Dortmund, (Berlin: Logos).

Kyle, W.C., Penick, J.E., Shymansky, J.A. (1979). Assessing and analyzing the performance of students in college science laboratories. *Journal of research in Science Teaching*, **16**(6), 545-551.

Lienert, G. A., von Eye, A. (1994). *Erziehungswissenschaftliche Statistik: Eine elementare Einführung für pädagogische Berufe* (Weinheim: Beltz).

Lumpe, A.T., Scharmann, L.C. (1991). Meeting contemporary goals for lab instruction: A content analysis of two secondary biology textbooks. *School Science and Mathematics*, **91**, 231-235.

Lunetta, V.N. (1998). The School Science Laboratory: Historical Perspectives and Contexts for Contemporary Teaching. In: K. Tobin and B. Fraser (Eds.), *International Handbook of Science Education*. (Dordrecht: Kluwer), 249-264

Millar, R., Le Maréchal, J.F., Buty, C. (1998). A map of variety of labwork in Europe. Working paper 1 from the European project Labwork in Science Education (Targeted Socio-Economic Research Programme, Project PL 95-2005), 18 pages, (http://edu.leeds.ac.uk/projects/lis/wp1.pdf)

Niedderer, H., Tiberghien, A., Buty, C., Haller, K., Hucke, L., Sander, F., Fischer, H.E., Schecker, H., v. Aufschnaiter, S., Welzel, M. (1998). Category Based Analysis of Videotapes from labwork (CBAV) – the method and results from four case studies. Working Paper 9 from the European project Labwork in Science Education (Targeted Socio-Economic Research Programme, Project PL 95-2005), 41 pages, (http://didaktik.physik.uni-bremen.de/niedderer/projects/labwork/papers.html)

Ogunniyi, M.B. (1983). An analysis of laboratory activities in selected Nigerian secondary schools. *European Journal of Science Education*, **5**(2), 195-201.

Okebukola, P. A. (1984). In Search of a More Effective Interaction Pattern in Biology Laboratories. *Journal of Biological Education*, **18**(4), 305-308

Okebukola, P.A. (1985). Science laboratory behaviour strategies of students relative to performance in and attitude to laboratory work. *Journal of Research in Science Education*, **22**(3), 221-232.

Penick, J.E., Shymansky, J.A., Filkins, K.M., Kyle, W.C., jr. (1978). Science Laboratory Interaction Category (SLIC An instrument to observe and analyze the various parameters of activity-centered and vocational education classrooms. Technical Report 15. Science Education Center, University of Iowa.

Petri, J., Niedderer, H. (1998). A learning pathway in high-school level quantum atomic physics. *Int. J. Sci. Educ.*, Vol. 20, No. 9, 1075-1088

Psillos, D., Niedderer, H., Vicentini, M. (1999). Case Studies on Innovative Types of Labwork in Science Education. In: M. Bandiera, S. Caravitta, E. Torracca and M. Vicentini (Eds.): *Research in Science Education in Europe*. Dordrecht: Kluwer, 201-208.

Sander, F. (2000). *Verbindung von Theorie und Experiment im physikalischen Praktikum - Eine empirische Untersuchung zum handlungsbezogenen Vorverständnis und dem Einsatz grafikorientierter Modellbildung im Praktikum.* Doctoral dissertation, University of Bremen, Berlin: Logos

Schecker, H. (1998). Integration of Experimenting and Modelling by Advanced Educational Technology: Examples from Nuclear Physics. In: K. Tobin and B. Fraser (Eds.), *International Handbook of Science Education.* (Dordrecht: Kluwer), 383-398

Séré, M.G., Tiberghien, A., Paulsen, A. C., Leach, J., Niedderer, H., Psillos, D., Vicentini, M. (1998). Labwork in Science Education – Executive Summary. Working paper from the European project Labwork in Science Education (Targeted Socio-Economic Research Programme, Project PL 95-2005), 16 pages, (http://edu.leeds.ac.uk/projects/lis/labwork.htm)

STELLA II (1994). Hannover, NH: High Performance Systems (http://www.hps.com).

Stein, J.S. Nachmias, R., Friedler, Y. (1990). An experimental comparison of two science laboratory environments: traditional and microcomputer-based. *Journal of Educational Computing Research*, **6**(2), 183-202.

Tamir, P., Lunetta, V.N. (1981). Inquiry related tasks in high school science laboratory handbooks. *Science Education*, **65**, 477-484.

Tamir, P. (1989). Training teachers to teach effectively in the laboratory. *Science Education*, **73**(1), 59-69.

Tiberghien A. (1994). Modelling as a basis for analyzing teaching-learning situations. *Learning and Instruction*, **4**, 71-87.

Tiberghien, A., Megalakaki, O. (1995). Characterization of a modelling activity for a first qualitative approach to the concept of energy. *European Journal of Psychology of Education*, **10** (4), 369-383.

Von Aufschnaiter, S., Welzel, M. (1999). Individual learning processes: A research program with focus on the complexity of situated cognition. In: Bandiera et al. (Eds.): *Research in Science Education in Europe*. Dordrecht: Kluwer.

Welzel, M., Haller, K., Bandiera, M., Hammelev, D., Koumaras, P., Niedderer, H., Paulsen, A., Robinault, K., von Aufschnaiter, S. (1998a). Ziele, die Lehrende mit experimentellem Arbeiten in der naturwissenschaftlichen Ausbildung verbinden – Ergebnisse einer europäischen Umfrage. *Zeitschrift für Didaktik der Naturwissenschaften*, **4**(1), 29-44.

Welzel, M., Haller, K., Bandiera, M., Hammelev, D., Koumaras, P., Niedderer, H., Paulsen, A. C., Robinault, K., von Aufschnaiter, S. (1998b). Teachers' objectives for labwork. Research tool and cross country results. Working Paper 6 from the European project Labwork in Science Education (Targeted Socio-Economic Research Programme, Project PL 95-2005), 109 pages, (http://didaktik.physik.uni-bremen.de/niedderer/projects/labwork/papers.html)

Students' Understanding of the Nature of Science and its Influence on Labwork

John Leach, University of Leeds, UK

Abstract

This study considers how students' understanding of the nature of science influences their actions and learning during labwork. Hypotheses and research questions about the influence of students' understanding of the nature of science on their actions and learning during labwork are presented.

Introduction

There is now an extensive literature on how science students respond to survey and interview questions designed to investigate their ideas about the nature of science (for a recent review, see Désautels & Larochelle 1998). But what have students' ideas about the nature of science got to do with labwork? To answer this question, we need to ask ourselves about the purposes of labwork itself. Earlier in this book, a classification of the aims of labwork was presented by Millar, Le Maréchal and Tiberghien (Chapter 1 *of this volume*). The aims that were identified can be grouped into three broad areas:

- developing students' knowledge of the behaviour of the natural world, helping them to make links between the world of natural phenomena and the world of theoretical descriptions and explanations of phenomena, and thereby developing their understanding of scientific concepts;
- developing students' understanding of how scientists undertake empirical investigations to address a question or problem of interest;
- developing students' ability to use standard laboratory instruments and procedures in undertaking investigations

The first of these areas is mainly concerned with teaching scientific content (i.e. the laws, theories, concepts etc. that constitute scientific knowledge). By contrast, the main concern of the second two areas relates more to teaching about the methods used by scientists in empirical work.

Labwork with each of these aims involves students in drawing upon understandings of the nature of empirical data, the nature of scientific knowledge claims, the ways in which knowledge claims and data are related, the purposes of using techniques, procedures and instruments, and so on. Many students in teaching laboratories often work with knowledge claims already agreed as reliable within the scientific community. For example, they may be involved in work to illustrate accepted theories or to apply accepted theory in specific contexts. Their ideas about how that knowledge came to be viewed as reliable may well influence their labwork. For all these reasons, participation in labwork involves students in drawing upon epistemological understanding.

There is a good deal of evidence that the images of science drawn upon by many students during labwork constrain performance. For example, Séré, Journeaux & Larcher (1993) have illustrated how university students' understanding of the nature of data result in them taking inappropriate actions during labwork that involves measuring physical quantities. Ryder & Leach (1999) have shown how university

D. Psillos and H. Niedderer (eds.). Teaching and Learning in the Science Laboratory. 41-48.

students working on open-ended investigations sometimes draw upon understandings of the relationships between data and knowledge claims that result in inappropriate actions being taken. Furthermore, several of the case studies reported later in this book illustrate that students experience similar difficulties in open and closed labwork (e.g. Lewis; Ryder; Guillon & Séré, Chapter 3 *of this volume*), and in response to diagnostic questions (e.g. Evangelinos, Psillos & Valassiades; Leach, Chapter 4 *of this volume*).

Amongst the studies of students' ideas about the nature of science reported in the literature, few relate to students of upper secondary and university age. Furthermore, the studies do not make a link between the situations in which students' ideas about the nature of science are investigated through research, and the situations in which those ideas are used in learning. The pertinent question is: how does a student's response to a question on a researcher's survey instrument relate to how she or he will act in a specific situation such as carrying out a piece of labwork? The purpose of this chapter is to consider some of the ways in which a student's ideas about the nature of science might influence his or her activities during labwork.

Aspects of students' images of science that influence students' decisions and learning during labwork

When students engage in labwork, they have to make decisions. The kinds of decisions encountered depend critically upon the design of the task. In some labwork activities, most decisions will have been taken by the person who designed the task and students will be presented with detailed instructions about which data to collect, how to collect them, what to do with them and how to interpret them. In these cases, although students do not make many decisions for themselves it is usually intended that they appreciate the basis for the actions that they undertake. In other cases, students will have to make decisions about experimental design, data analysis and experimental interpretation for themselves. In order to make these kinds of decisions during labwork, or understand the decisions made by those who designed the labwork tasks, students have to draw upon understandings of the nature of the data and knowledge claims that they are working with, and how they relate to each other. In the following sections, hypotheses are presented linking students' ideas about the nature of science and their approaches to labwork. The hypotheses relate to 5 broad aspects of students' images of science as they relate to labwork. These are the nature of data and measurement in empirical work, the nature of investigation in science, the nature of theory in science, the nature of explanation in science and the nature of reliable public scientific knowledge.

The hypotheses were arrived at through a process of discussion within the LSE project. All the hypotheses have their origins in previous empirical work on science student's ideas about the nature of science. However, as indicated earlier much of the available literature does not make any link between the ideas about the nature of science that appear to underpin students' responses to survey or interview questions, and specific learning situations such as labwork. To this extent, the hypotheses are amenable to empirical investigation, and researchable questions are presented for each one. The hypotheses are about students' likely *performance* during labwork, rather than their *ability*. For example, previous work (e.g. Séré et al. 1993) suggests that when faced with data sets based upon repeat measurements, many students appear to work on the assumption that each measurement is 'perfect' (see Hypothesis

1, below). This does not, however, mean that students will not draw upon more appropriate views of the nature of data following appropriate teaching. Examples of teaching where student's ideas about the nature of science are developed are presented later in this book and elsewhere (e.g. Leach, Lewis, Ryder, Sére & Guillon, all *in this volume*; Roth 1995; Ryder, Leach & Driver 1998; Ryder & Leach 1999; Brickhouse, Dagher, Letts & Shipman 2000; Hind, Leach & Ryder 2001).

Some of the research questions that emerge from the hypotheses are addressed through case studies presented later in this book. In addition, a survey was designed to address some of the questions, and administered to 731 students in 5 European countries, at upper secondary school and university levels. Findings from this survey are reported in Leach, Millar, Ryder, Séré, Niedderer, Paulsen, Tselfes (1998), Leach, Millar, Ryder & Séré (2000), Ryder & Leach (2000) and Séré, Fernandez-Gonzalez, Leach, Gonzalez-Garcia, de Manuel, Gallegos & Perales (in press).

Hypotheses about students' images of data and measurement

Hypothesis 1

Many students consider that, with good enough apparatus and enough care, it is possible to make a perfect measurement of a quantity. That is, they assume that measurement can be perfectly *accurate*. Others consider that any measurement is subject to some uncertainty, and so obtaining accurate values is problematic. (Séré et al. 1993; Lubben & Millar 1996)

Why it is relevant to labwork?

During labwork, students may have to make decisions about the type of measuring device to be used, the amount of data that has to be collected and the conclusions that can be drawn from given data sets. The decisions that they make about this aspect of data collection will be influenced by their view of the nature of measurement. For example, students who see measurements as 'perfect' may join each individual data point on a graph rather than plotting lines or curves of best fit. Similarly, in deciding upon a value from a set of measurements they are likely to select the mode, reasoning that the most frequently recorded value must be the 'true' value. Others recognise that all measurements add information, and can therefore be treated as a set using statistical techniques. Amongst these students, some assume that statistical calculations will yield information about the *accuracy* of measurement, whereas others recognise that such calculations only yield information about *precision*.

Research Question

Do students see measured data as a 'perfect' copy of reality, or do they view measured data as being subject to some uncertainty? What do they see as the sources of uncertainty in measured data? How do they overcome these uncertainties and select a value? Do they recognise the difference between *accuracy* and *precision*? (This question is addressed in the case studies by Leach and Evangelinos in Chapter 4 *of this volume*).

Hypothesis 2

Some students do not recognise the kinds of empirical evidence on which scientific knowledge claims are based. In the case of measured data, they think that it is only possible to judge the quality of a measurement from a knowledge of the 'true' value, given by an authority source. That is, they do not recognise that decisions about *precision* can be made from sets of measurements. Other students think it is possible to judge the 'quality' of measured data from a set of repeated measurements. That is, they reason that data sets can be evaluated in their own terms to make decisions about accuracy and precision. (Séré et al. 1993; Lubben & Millar 1996)

Why it is relevant to labwork?
Claims about the values of measurements, whether made by students in labwork classes or authority sources in data books, are based on empirical measurements. If students do not recognise the relationships between knowledge claims and empirical evidence, they are likely to approach data collection and interpretation during labwork differently from students who believe that the quality of a measurement can be judged from a set of repeated measurements. Some students assume that mean values from sets of repeated measurements give an indication of the *accuracy* of a measurement, whereas others recognise that statistical processing of a data set only gives an indication of the *precision* of a measurement.

Research Question
Do students believe that the only way to judge the quality of a measurement is from a known 'true' result, or do they believe that the quality of a measurement can be judged from a set of repeated measurements? If so, do they distinguish the *accuracy* and *precision* of measured values? (See the case studies by Leach and Evangelinos et al. in Chapter 4 *of this volume*).

Hypothesis 3
Many students see data reduction and presentation as a process of *summarising* data and see procedures like joining data points on a graph, drawing a 'best fit' straight lines, or drawing smooth curves as *routine heuristics* - that is, they see the process as independent of theory. They believe that there are standard techniques for arriving at 'perfect' descriptions of data. Others see such procedures as a process of proposing *tentative hypotheses about a relation between variables.* That is, they believe that experimenters (and computers) make decisions during data reduction and presentation according to existing models. (Séré et al. 1993; Lubben & Millar 1996)

Why it is relevant to labwork?
The types of conclusions drawn by students during labwork will be influenced by the students' views of the nature of data handling. For example, once data points are plotted on to axes of a graph, hypotheses have to be proposed about possible relationships between the points. Students who see each data point as a 'perfect' value may well join each point. They may reject lines of best fit that do not pass through any data points. Other students who see procedures such as linear regression as routine heuristics may well apply one procedure without considering whether it is valid to do so, or whether other procedures may be more valid. In addition, they may well view scientific knowledge as akin to algorithms, leading to unique results. Similarly, data sets from laboratory work may be treated as the unique products of algorithms.

Research Question

When working with data sets, do students see procedures like joining data points with lines of 'best fit' or smooth curves as routine heuristics, or alternatively as a process of proposing tentative hypotheses? (See the case study by Leach in Chapter 4 *of this volume.)*

Hypotheses about students' images of the nature of investigation

Hypothesis 4

Some students think that the logic of proof and falsification is symmetrical: data that logically support a law 'prove' the law, in the same way that data that do not support a law logically falsify it. (Kuhn, Amsel & O'Loughlin 1988; Driver et al. 1996)

Why it is relevant to labwork?

During labwork, and particularly open-ended labwork, students' approaches to data collection and data processing may be influenced by their beliefs about the logic of proof and falsification. However, the work of Kuhn et al. (1988) and Driver et al. (1996) relates to pre-adolescent students and it is open to question whether older students in specialist science streams would think about proof and falsification in this way.

Research Question

Do students recognise the logical distinction between proof and falsification when handling empirical data?

Hypothesis 5

Some students think that most/all questions about natural phenomena are answerable by collecting observational data and looking for correlations. Explanatory theories (models) 'emerge' from this data in a logical way: there is only one possible interpretation. Other students think that prior models (theories, hypotheses) influence decisions about what data to collect and how it is interpreted, and that observation and measurement are intended to test these models. Again, the testing is based on logic: only one interpretation is possible. Others think that a data base is first collected on the basis of embryonic theories and hypotheses - more robust models are then proposed as conjecture to account for existing, and anticipated data. Then predictions derived from these may be tested by planned observations or experiments, but more than one interpretation is possible due to the conjectural nature of theory. (Driver et al. 1996; Larochelle & Désautels 1991; Aikenhead, Fleming & Ryan 1987; Niedderer, Bethge, Meyling & Schecker 1992)

Why it is relevant to labwork?

Students' approaches to data collection and data interpretation during labwork will be influenced by their views of the place and nature of theory in empirical investigation. They may not recognise the interplay between data and theory in the process of investigation, and as a result they may not accept that it is legitimate to develop the design of an experiment in the light of data already collected.

Research Question

Do students think that scientific theories 'emerge' from data, or do they think of scientific theories and data as being related in a more complex way? If so, how do they think that scientific theories and data are related? In particular, do they think that a given experiment is open to more than one interpretation? (See the case studies by Leach, Ryder and Guillon & Séré, all *in this volume.*)

Hypothesis 6

Many students see practical activities in the teaching lab as exercises to reproduce well-known results, or to illustrate important theories/models, no matter how the task is actually presented by the teacher. They do not recognise the labwork as an exercise in 'finding out'. Other students recognise that some labwork activities have an investigative component: they involve 'finding out'. Amongst these students, some assume that knowledge claims can be 'proved' or 'disproved' by a single planned intervention, whereas others assume that the process of investigation involves a sequence of interventions, which may be modified in the light of experience. (Driver et al. 1996; Lubben & Millar 1996)

Why it is relevant to labwork?

The actions of students during experimental design, data collection and data interpretation in labwork will be greatly influenced by their views of the purpose of the labwork task.

Research Question

Do students recognise the purpose of particular labwork tasks as involving 'finding out', rather than reproducing or describing phenomena?

About students' images of the nature of theory (Hypothesis 7)

Some students believe that scientific theories are really descriptions of natural phenomena: there is a one-to-one correspondence between theory and 'reality'. Such students believe that it is a straightforward empirical process to show that scientific theories are 'true'. Others believe that theories are model-like, and do not simply describe reality. However, such students still believe that it is a straightforward empirical process to show that scientific theories are 'true'. Others believe that theories are model-like, and this means that it is NOT a straightforward process to show that a scientific theory is 'true'. (Larochelle & Désautels 1991; Aikenhead et al. 1987; Driver et al. 1996; Niedderer et al. 1992)

Why it is relevant to labwork?

Students who do not recognise the conjectural nature of models in science may frame their data interpretation during labwork in terms of observational features of the phenomenon, rather than theoretical entities.

Research Question

Do students think that scientific theories are conjectural and model-like in nature, or do they think that theories are essentially descriptions of phenomena in different terms? (See the case study by Leach Chapter 4 *of this volume.*)

About students' images of the nature of explanation (Hypothesis 8)

Some students do not recognise the different levels, types and purposes of explanation that are used in science. [Examples: teleological, causal, descriptive, model-based]. (Tamir & Zohar 1991; Leach et al. 1996). However, this work refers mainly to pre-adolescent students.

Why it is related to labwork?

The types of conclusions that are drawn by students during labwork, and particularly open-ended labwork, will be influenced by the type of explanation that the student thinks is most appropriate.

Research Question

Are students able to distinguish between teleological, descriptive and model-based explanations of natural phenomena? At what ages and in which situations?

About students' images of the nature of public scientific knowledge (Hypothesis 9)

Some students think that all the knowledge claims made by science are of the same status. They do not recognise the role of the scientific community in the validation of public knowledge. Others recognise that some knowledge claims are widely accepted within the scientific community, whereas others are still the subjects of investigation and debate. (Aikenhead et al. 1987; Driver et al. 1996; Larochelle & Désautels 1991)

Why it is relevant to labwork?

If students believe that all knowledge claims in science are of equal status, this will affect their actions during labwork, and particularly open-ended labwork. If data collected during labwork are not consistent with canonical science, a number of options are open to the investigator: the option that is pursued will depend upon the investigators' beliefs about the status of the scientific knowledge in question.

Research Question

Do students recognise that different courses of action are appropriate in scientific investigations depending on the status of the scientific knowledge claim under investigation? What are the implications for teaching?

Concluding comments

At the beginning of this chapter, it was claimed that most of the work carried out to date on students' understanding of the nature of science does not give insights into the understandings drawn upon by students during learning situations. The hypotheses and associated research questions presented in this chapter are a very early attempt at setting out a research agenda on the understandings of the nature of science that are drawn upon by students when engaged in action in learning situations involving labwork. Although some of the studies in this book (Evangelinos, Psillos & Valassiades; Guillon & Séré; Leach; Lewis; Ryder; all *in this volume*) and elsewhere (e.g. Roth 1995; Roth & Bowen 2000; Ryder & Leach 1998; 1999) do begin to address these research questions, in my view we still understand relatively little about the ways in which students reason during labwork and the ways in which teaching might be developed to improve student learning.

Acknowledgements

The hypotheses presented in this study were formulated following a lengthy process of discussion involving many members of the LSE project. I particularly acknowledge helpful comments by Marie-Geneviève Séré on the text of this study.

References

Aikenhead, G., Fleming, R. and Ryan, A. (1987). High-school graduates' beliefs about science-technology-society 1: Methods and issues in monitoring student views. *Science Education*, 71, (2), 145-161.

Brickhouse, N. W., Dagher, Z. R., Letts, W. J. and Shipman, H. L. (2000). Diversity of students' views about evidence, theory and the interface between science and religion in an astronomy course. *Journal of Research in Science Teaching*, 37 (4), 340-362.

Désautels, J. and Larochelle, M. (1998). The epistemology of students: the 'thingified' nature of science. In B. Frazer and K. Tobin (Editors): *International Handbook of Science Education Research*. Dordrecht, NL: Kluwer Academic.

Driver, R., Leach, J., Millar, R. and Scott, P. (1996). *Young People's Images of Science*. Buckingham, UK: Open University Press.

Hind, A., Leach, J. and Ryder, J. (2001). Teaching high school science majors about the nature of science. Paper presented at the Annual Meeting of the American Educational Research Association, Seattle, April 2001.

Kuhn, D., Amsel, E. and O'Loughlin, M. (1988). *The development of scientific thinking skills*. London, UK: Academic Press.

Larochelle, M. and Désautels, J. (1991). 'Of course, it's just obvious': Adolescents' ideas of scientific knowledge. *International Journal of Science Education*, 13 (4), 373-389.

Leach, J., Driver, R., Scott, P. and Wood-Robinson, C. (1996). Children's ideas about ecology 3: Ideas found in children aged 5 - 16 about the interdependency of organisms. *International Journal of Science Education*, 13 (4), 373-389.

Leach, J., Millar, R., Ryder, J., Séré, M-G., Niedderer, H., Paulsen, A. C., Tselfes, V. (1998). Survey 2: Students' images of science as they relate to labwork learning. Working Paper 4, Labwork in Science Education Project. Leeds: Centre for Studies in Science and Mathematics Education. (http://edu.leeds.ac.uk/projects/lis/labwork.htm)

Leach, J., Millar, R., Ryder, J. and Séré, M-G. (2000). Epistemological understanding in science learning: the consistency of representations across contexts. *Learning and Instruction*, 10 (6), 497-527.

Lubben, F. and Millar, R. (1996). Children's ideas about the reliability of experimental data. *International Journal of Science Education*, 18 (8), 955-968.

Niedderer, H., Bethge, T., Meyling, H. and Schecker, H. (1992). Epistemological beliefs of students in high school physics. Paper presented at the annual meeting of NARST, Boston, MA: March 1992. (http://didaktik.physik.uni-bremen.de/niedderer/pubs.htm)

Roth, W-M. (1995). *Authentic school science: knowing and learning in open-inquiry science labs*. Dordrecht, NL: Kluwer.

Roth, W-M. and Bowen, G. M. (2000). Learning difficulties related to graphing: a hermeneutic phenomenological perspective. *Research in Science Education*, 30 (1), 123-140.

Ryder, J., Leach, J. and Driver, R. (1999). Undergraduate science students' images of the nature of science. *Journal of Research in Science Teaching*, 36 (2), 201-220.

Ryder, J. and Leach, J. (1998). Enseigner les pratiques effectives de la science: expériences d'étudiants en projet de recherche de licence. *Didaskalia*, 12, 39-62.

Ryder, J. and Leach, J. (1999). University science students' experiences of investigative project work and their images of science. *International Journal of Science Education*, 21(9), 945-956.

Ryder, J. and Leach, J. (2000). Interpreting experimental data: the views of upper secondary school and university science students. *International Journal of Science Education*, 22 (10), 1069-1084.

Séré, M-G., Journeaux, R. and Larcher, C. (1993). Learning the statistical analysis of measurement errors. *International Journal of Science Education*, 15 (4), 427-438.

Séré, M-G., Fernandez-Gonzalez, M., Leach, J., Gonzalez-Garcia, F., de Manuel, E., Gallegos, J. and Perales, J. (in press). Images of science linked to labwork: A survey of secondary school and university students. *Research in Science Education*.

Tamir, P. and Zohar, A. (1991). Anthropomorphism and teleology in reasoning about biological phenomena. *Science Education*, 75 (1), 57-68.

Chapter 2

STANDARD LABWORK BASED ON HANDS-ON EXPERIMENTS

Introduction

In standard labwork students normally work in small groups, carry out hands-on experiments using conventional laboratory apparatus, and are engaged in a variety of complicated yet distinctive activities. This form of organisation, which is widespread, particularly in universities, has attracted the attention of researchers as it is evidenced from numerous studies. Yet it seems that not much is known about the nature of the specific activities that constitute laboratory contexts or of how the students perceive and understand the intended learning experiences. In this context one important issue, to which the case studies in this chapter contribute, is how laboratory tasks may be described, what aspects of their structure, organisation and sequence may elucidate and clarify different models of laboratory activities. . A second issue is what knowledge, scientific or not, is used by students in carrying out laboratory experimentation at large and specific tasks in particular.

How students' actions and procedures as employed during experimentation are, or are not, intertwined with the theoretical knowledge which the experimental design draws upon relates to laboratory effectiveness and more specifically to effectiveness 1 as mentioned in various studies in this book. Several of the studies in the present chapter try to assess labwork in terms of effectiveness 1. All the studies are based on extended research programs which provide for rich data concerning students' understandings in labwork In addition, the suggested new courses are innovations that rely on extended research data.

Detailed descriptions of the various tasks at different levels of complexity and learning demands are what characterise the first two studies, which analyse specific laboratory situations in depth and provide a wealth of data. The study by Robinault and Tiberghien deals with energy teaching in secondary school. The study by Beney and Séré deals with introductory university laboratory experiments of a type that may be encountered in many universities The first study focuses on students'

D. Psillos and H. Niedderer (eds.). Teaching and Learning in the Science Laboratory. 49-50.

modelling activity in relation to teaching situations aiming at conceptual learning, while the second draws upon concepts of cognitive psychology applied to model laboratory tasks. This study investigates what happens during the action phase of labwork, examining both students and expert teachers in an attempt to shed light on the relation between doing thinking and learning.

Neither study argues in favour of eliminating standard labwork. Rather they make a case for improving the effectiveness of this type of labwork with regard to the acquisition of scientific concepts or procedures, and make a number of specific suggestions, in particular concerning the style and objectives of labsheets, which play a crucial role in guiding students' activities.

The next two studies deal with research-based innovations concerning labwork, which is addressed to university non-science majors. Prospective elementary teachers are the subjects of the first study (Kariotoglou), while medical students are involved in the study by Thyssen, Aufschnaiter and Schumacher. The related laboratory-based courses are not reduced versions of courses for science students, but involve the educational reconstruction of the scientific content, implying new links with the world of phenomena. Kariotoglou focuses on promoting student teachers' conceptual and procedural development towards a suggested scientific model and experimental method, and provides interesting insights into the effectiveness of labwork in promoting either type of knowledge as well as into their relation within specific tasks. Thyssen, von Aufschnaiter and Schumacher base their work on a specific theoretical approach to learning as fostering an increase in level of complexity. The linking of physics and medical knowledge and the development of meaning is affected by the labsheets used.

While the above mentioned studies investigate and model laboratory tasks as represented in labsheets and students' activities during labwork, the last study (Bandiera) focuses on the representation of labwork in scientific textbooks in secondary education. This study thus addresses a rather neglected, yet nonetheless important, issue: namely the relation between textbook and labwork, which can affect the image of scientific knowledge developed by the students, particularly in countries where labwork is not widespread.

Modelling Activities of Students During a Traditional Labwork

Karine Bécu-Robinault, CNRS – Université Lyon 2, France

Abstract

This case study is integrated in a research program on students' modelling activity in relation to teaching situations. This research is also a part of a research development project involving secondary school teachers and concerning energy. This case study deals with a quantitative approach of modelling energy phenomena. The data were collected by audio and video, and whole dialogues were transcribed. The transcriptions were analysed by categorising students' interventions on the basis of their modelling activities. The results obtained concern the role of the different proposed tasks in learning the targeted concepts.

Introduction

The teaching of the energy concept has been a widely debated theme within the science education community during the past twenty years (Solomon 1985; Brook & Driver 1984; Duit 1981, 1985; Lemeignan & Weil-Barais 1992). If many modifications have been proposed, few justifications have been advanced to justify the relevance of labwork to promote better understanding of this concept, or of any concept in general (Hofstein & Lunetta 1982; White 1996). In order to study the role of labwork in learning the concept of energy, we choose to examine the strategies of students when they carry out and interpret an experiment. We focus our analysis on modelling activities. This choice emerges from research carried out on modelling and cognitive activities of students (Niedderer & Schecker 1991; Martinand 1992; Wiser 1993; Meheut, Chomat & Larcher 1994; Niedderer 1996; Millar & Lubben 1996).

Our case study thus concerns two important aspects in science education, the role of experiments in learning physics and the teaching of energy. It aims at studying labwork at the second year of the French upper secondary school level (16 to17 years old). The labwork studied was carried out in regular teaching but in the framework of research development aiming at designing new teaching materials on energy. Energy is the main part of the official curriculum at this level. This study is also integrated in a research program on students' modelling activity (see the contribution of Buty in Chapter 5 *of this volume*).

Teaching approach

Framework

A similar case has already been studied where the tasks proposed to students were only qualitative (Tiberghien & Megalagaki 1995). In this case the tasks include quantitative aspects. They deal with measurements, quantitative data, concepts and experimental facts. The teaching situation is characterised by the possible sources of information given to the students: the experimental setting including measurements apparatus and the labwork sheet. We analysed the relations between these

D. Psillos and H. Niedderer (eds.). Teaching and Learning in the Science Laboratory. 51-64.

characteristics of the situation (experiments and written instructions) and students' activities. We particularly focused on the influence of written instructions given in a labwork sheet on students' activities.

This labwork session is integrated in the curriculum on energy where energy is introduced by the use of its principle of conservation. This approach of energy teaching is very different from usual approaches where energy is mainly introduced by the concept of work (Duit 1985; Bécu-Robinault & Tiberghien 1998).

As specified in the previous section, this research was imbedded in a research development, the "energy project" (Bécu-Robinault 1997; Gaidioz, Monneret, Tiberghien, Becu-Robinault, Besson, Blache, Chastan, Clavel, Colonna, Collet, Gibert, Longere, Le Marechal, Strobel & Vagnon 1998). This project involved 3 researchers and 9 teachers from different secondary schools for two years. The teaching content concerning specifically energy has been elaborated during regular meetings and under constraints of research results. All the teachers agreed to follow and use this teaching content, in which energy properties and the conservation principle are presented explicitly to students as a qualitative model of energy including an iconic representation of energy storage, transfer, transformation and conservation.

The labwork session is traditional in the sense that it is carried out in a real classroom context, but also innovative in that the structure of the worksheet was purposefully developed on research grounds. The tasks proposed all along the labwork are not those usually proposed in a classroom context.

The task is dealing with quantitative energy phenomena that students have to interpret on the basis of a previous qualitatively taught model, including the conservation principle and storage, transfer and transformation properties. The expected output is that students enrich the previous qualitative model, using the quantitative data collected during the experiment to account for quantitative aspects of energy transfer. They should thus introduce the concept of power to connect the physical quantity handled. We expect students to build the concept of power with its meaning of interaction quantity. This means that its value depends on the characteristics of the devices interacting (immersion heater and electrical supplier) and not only on one characteristic of a specific device. In our case, students should take into account both the electrical supplier and the device used to heat water (immersion heater) to explain how to modify the value of power.

Labwork description

During the labwork, students are given material (electrical supplier, immersion heater, calorimeter, water and all measurement devices). The characteristics of devices can be different from one group of students to another. In the labwork sheet, they are asked to organise and handle the experimental devices; measure energy, time, temperature, voltage and current; process data to find a relation between energy and time ($E = Pt$); assign a name associated to the meaning of the introduced quantity (P, power); represent this quantity on an iconic representation which is the energy chain; propose to modify elements to get a different value of the introduced quantity (see Table 1 for the text of the questions). All the information necessary to perform the tasks is given by the labwork sheet. During this labwork session, the role of the teacher is limited to checking the security of the electrical setting . This session has been designed in order that the students work in an autonomous way (see Table 1).

Table 1: Text of the labwork.

Task N° 1: set up these apparatus in such a way that they will correspond to the energy chain below, with all measurement apparatus

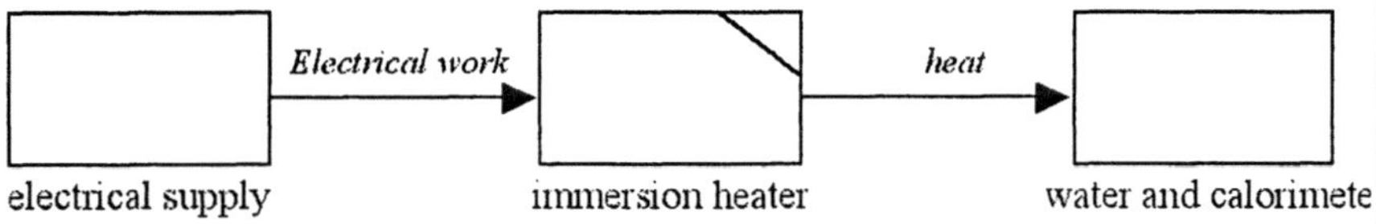

Making measurements.
Have the chronometer start when you plug in the immersion heater, note the initial temperature Θ_i and write the results in the table below.

Note U = V and I = A

We do not follow the evolution of the temperature.

Task N° 2: for this experiment, we wish to be able to predict the results without needing to make all the measurements.

	N (tours)	E (Wh)	E (J)	t (min, s)	t (s)
Θ_i =	0	0	0	0	0
	1				
	...				
Θ_f=	10				

In order to do that, look for a simple mathematical formula which would allow you to relate the quantity of transferred energy E to the duration of heating t. In this task, no other physical quantity than E and t is to be taken into account.

Write your formula E = f(t) below and briefly indicate how you got it.

Task N° 3: In this relation, you should have introduced a physical quantity, which has never been mentioned in this labwork. Find a name or an expression, which corresponds to the meaning of this physical quantity.

Task N° 4: On which part of the chain (task 1) can you insert this physical quantity? Put it on this chain!

Task N° 5: On which elements of the set up that have been represented on the chain does this physical quantity depend?

What would it be necessary to modify on this set up in order to modify the value of this physical quantity?

It might be very fruitful to analyse the main characteristics of the situation studied on the basis of some categories of the map developed by Millar et al. (Chapter 1 *of this volume*) to better grasp what students are confronted with and what they are supposed to do.

This labwork was designed to help students to learn a concept (power) and to help students to learn a relationship between this concept and its meaning in the experimental situation (concept associated to a system of connected devices). This labwork should also help students learn how to use data to support a conclusion (using collected data through measurement to find a relation).

Concerning the objects, students are intended to use observation and measurement instruments and to observe a quantity from the real world inside the laboratory. The information involved is qualitative and quantitative; the quantitative aspect is emphasised.

Concerning ideas, students are intended to report observation directly, explore the relation between physical quantities, objects and physical quantities, 'invent' a new concept, determine the value of a quantity which is not measured directly and account for observations by proposing a law. The tool used to process information is either a manual or pocket calculator.

What the students are intended to do with ideas arises from what they are intended to do with objects.

Concerning the degree of openness/closure, on the one hand, the question to be addressed, the equipment to be used and the procedure to be followed are all specified by the teacher. On the other hand, the methods of handling data collected and the interpretation of results are chosen by students.

This labwork is carried out by students in small groups. The duration is short (90 minutes). Students can interact with other students carrying out the same labwork task. The teacher is not supposed to give any information during the lab, excepted safety rules. The information sources available to students are contained in the guiding worksheet only. Finally, students use standard laboratory and everyday equipment (energy-meter).

Research questions

Basic assumptions

Epistemological positions

Modelling is intrinsic to physics functioning. It is the main activity of a physicist (Feynman 1980; Bachelard 1971; Gooding 1989; Pickering 1992; Grandy 1992). It implies that he/she establishes relations between the experimental field and the theory or model. According to the analysis of modelling in physics teaching and learning (Tiberghien 1994), we state that the students' modelling activities are situated within two worlds, the world of theories and models and the world of objects and events.

The world of theories and models is divided into four levels:

- the theory contains the explanatory system

- the models are an intermediary between theoretical and perceptual aspects (Bachelard 1989). They represent only some properties of reality.
- physics models contain all what is explicitly related to physics quantities (concepts, symbols, units, relations, properties). For instance, the sentence "it's the power, P equals E divided by t, it's in watt hour" refers explicitly to the physics model owing to the terms "power"; to the symbols P, E and t and to the units.
- numerical models contain all the mathematical tools that can be used to process data. An example of a sentence using this level is "for the figures here 12.96, we write 13". There is no explicit relation to physical concepts, units or symbols. It only deals with numerical entities.

The experimental field (world of objects and events) is divided into two levels:

- measurement;
- objects and events.

These modelling levels are reconstructed by the researcher during the analysis of students' activities. Moreover, there could be differences between the content of these levels from a student or a physics perspective.

Then, each of the levels can be expressed alone (the elements of the level are expressed for themselves), within an internal relation (different elements of a same level are linked), or within an external relation (elements of two distinct levels are linked).

For instance, when Mélanie says "wait, it's in Watt hour, it's the power", she refers to the only physics model. When Elise asks "what makes the power vary?" and then Mélanie answers "logically it's the immersion heater", they establish a relation between the physics model and the level of objects and events.

The criteria we used to analyse the modelling activities are summarised in Table 2.

Table 2: Analysis criteria

THEORY - Theory
MODEL - Physics model (ModPhy) - Numerical model (ModNum)
EXPERIMENTAL FIELD - Measurements (Meas) - Objects and events (ObEv)

Learning assumptions

- When a student performs a task, he is consistent from his own point of view. This consistency might imply factors other than knowledge involved in the task.
- The links between the worlds of objects and events and theory-model are necessary to learn physics. Thus, taking into account the different modelling levels and establishing links between these levels helps to construct the meaning of a concept.
- The use of modelling levels by students can be unconscious and/or conscious.

Specific research questions

1. Concerning the characterisation of written instructions given to students: What are the modelling levels taken into account in the questions asked to students? What are the modelling levels that the questions aim students to use?
2. Concerning activities of students: What are the modelling activities of the students? How do they take into account the modelling levels?
3. Concerning the link between characteristics of the situation and activities of students: Do the characteristics of written instructions and of the experiment lead to similar modelling activities?

Research methods

Such research questions lead us to first characterise the written instructions given in the labwork sheets: modelling levels involved and aim of instructions. This analysis is our *a priori* analysis. Then we analysed students' activities: the modelling levels used and the links elaborated between these levels. Finally, we studied the links between characteristics of written instructions and activities of students. We looked for some general results concerning the influence of information given to students through the labwork sheet on the modelling activities of students when they perform the proposed tasks.

General design

We chose to analyse traditional labwork, that is to say experiments actually carried out in classrooms. The labwork sheet has been elaborated according to research perspectives and has been adapted to teaching constraints (time, security, feasibility) by teachers. The labwork has been carried out by pairs of students in classrooms under control of the teacher.

We aimed to analyse modelling activities of students performing labwork. During two years, we collected data in nine schools chosen in Lyon's (France) suburbs. Classes with which we worked were managed by the teachers involved in the "energy project " having regular meetings with researchers.

The reported case was carried out in 4 schools, where we recorded 12 pairs of students at work for the whole duration of the labwork session (1h30). We also collected written data (one labwork report by each pair of students) from the 9 concerned schools, that is to say 116 reports the first year, and 79 the second year.

Data collection and data analysis

Data were collected by audio and video, for 12 groups of 2 students, chosen by teachers as groups usually working together, with most of the time middle or low abilities in physics. We tried as far as possible to always record the same groups for all the experiments concerned by our research. We chose to transcribe 6 whole dialogues out of the 12 recorded.

We analysed the situation proposed to students and data collected on the basis of the same analysis categories, that is to say the modelling levels. We then compared expected modelling activities to actual activities of students.

To analyse the dialogue transcriptions, we categorised the students' interventions. Each intervention including one or more modelling level constitutes what we called a students' proposition. These propositions are those that bring in one or several modelling levels. We then counted and interpreted the students' propositions in relation to our *a priori* analysis of what we expected students to do during the different tasks.

Research Results

General results concerning labwork reports

As an overview of the success of students concerning the tasks proposed, we present Table 3. It corresponds to the distribution of the student pairs in all schools according to the written answers given by the students during the first year of data collection. Notice that students could give several answers for each task.

Table 3: Distribution of answers given by the students

Task (see Table 1)	Categories of written answers	Percentage of pairs (n=116)
2 : Establishing relation	Entirely literal relation	49
	Instantiated value relation	78
3 : Giving a name	Power related to energy, energy flow rate	59
	Power related to electricity	26
	Related to energy, not compatible with flow rate	8
	Related to physics aspects other than those on energy	7
	Not related to physics	18
4 : Drawing a representation	Represented on energy transfer	77
	Represented on reservoir or transformer	8
	No representation	15
5 : Proposing modifications	Relevant	69
	non relevant	7
	both relevant and non relevant	13

It appears that students globally gave correct written answers from the point of view of the physics meaning of power. For task 3, power is mainly related to an energy flow rate or to electricity. It is represented on an energy transfer in task 4, which means that the meaning of power is related to an interaction quantity. Moreover, most of the modifications proposed by the students to change the value of the power are relevant, and many students propose to modify several characteristics of the experiment.

Analysis of modelling activities of students in transcribed dialogues

Concerning modelling activities of students that have been analysed from the dialogues, we chose to focus on the more salient activities, that is to say the plurality of activities representing, when added, more than 80% of the total amount of modelling levels taken into account by the students during each task (N). Thus, for each task, we summarised the results obtained in tables as follows:

Concerning the instructions that have been analysed in our *a priori* analysis, we indicate: on the line "in instructions", the modelling levels involved in the written instructions (in grey); on the line "aimed by the question", the minimum modelling levels students have to link to give an answer (black lines), the modelling levels the question aims students to use in their final answer.

Concerning the students' activities, on the line "used by students", we indicate the modelling levels used by students representing at least 80% of their modelling activities. The absolute number corresponds to the number of use of those levels in the dialogues for all the groups. Between brackets, we indicate the percentage of those uses, 100% being the total number of use of any of the modelling levels. On the line "linked by students", we indicate with black lines the levels linked by the majority of students (4 out of 6 groups) and above these lines, the absolute number of the corresponding links. As a matter of fact, these links represent at least 80% of the total number of links established by students. This presentation should show us the main modelling activities of students from the point of view of the levels students prefer to use as much as those they are able to link. The more there are links established by students between different levels, the more interesting is the task from the researcher's point of view. Indeed, this means that the task is rich from a modelling perspective, and, according to our learning assumptions, the task should lead to a better understanding of the underlying concepts.

Task 1: handling and measurements (N=549)

This task concerns the handling of experimental devices and measurement of time, energy, voltage, current and temperature. Students should establish relations between levels of objects and events, measurements when they collect quantitative data and they should establish relations between measurements and the physics model when they transform the measured values (Table 4).

Table 4: Results for task 1

	Modelling levels			
	ModPhy	**ModNum**	**Meas**	**ObEv**
Aimed by the question	Aimed		Aimed	
Used by the students	Used 72 (13%)	Used 92 (17%)	Used 201 (36%)	Used 180 (33%)
Linked by the students		10		5
		30		
			11	

This task is the most important concerning the time spent by students to carry out all the tasks imposed by the labwork sheet (more than 50% of the labwork time is devoted to this task). Students use the highest number of modelling levels (4 in all) and relations between levels. The measurement level is used significantly only during this task. Propositions referring to this level count for one third of the whole amount of students' propositions. Even if this level is mainly used alone, it is also related to all the other levels, excepted the theory (10 links to the numerical model, 5 to objects and events, 30 to the physics model). Relations established by students between this level and the other ones help them to convert the measured values into the conventional units, to associate qualitative and quantitative information.

Marine: It should be constant U and I

Paul: of course, because it's the mains electric current

The level of objects and events is mostly used alone, even if it is sometimes related to measurements or the physics model (respectively 5 and 11 links, to be compared to 180 uses alone).

Phil: it's that and that [shows the measurement devices], that's I, and that's V.

It appears that the numerical model is a tool constructed by the students and used as an intermediary facilitating the data processing.

Mélanie: Wait, it's written here, the conversion, one Watt is, is equal to 3600. We have to multiply by 3600 each time.

It seems that, during a handling task, it is easier for students to give meaning to the physics model in relation to the experimental situation than to give meaning to an experimental situation in relation to a physics model.

Task 2: data processing (N=158)

This task concerns looking for a relation between energy and time. Students have to read a revolution number on the energy meter, multiply this measure by a constant depending on the energy meter to obtain an energy value. The other values are directly read on the measurement devices. They then have to find a relation between this energy value and time to calculate a proportionality factor, which corresponds to the power. The kind of relation to find is not imposed by the task, but at this age, students have mostly faced proportional relations. In our a priori analysis, we thought that students would mainly use the numerical and physics models (Table 5).

Table 5: Results for task 2

	Modelling levels			
	ModPhy	**ModNum**	**Meas**	**ObEv**
Aimed by the Question	Aimed (ModPhy–ModNum)			
Used by the students	Used 59 (37%)	Used 75 (48%)		
Linked by the students	9 (ModPhy–ModNum)			

As expected, students' propositions mostly concern the levels of physics and numerical models (respectively 37% and 48%). The numerical model gives students the opportunity to deal with values without taking into account their physical meaning. For this task and the following ones, students continuously establish internal relations between elements of the physics model (not represented on the figure).

Students do not really process data coming directly from the experiment they carried out. Instead they handle their symbolic reconstruction as it appears in the data table, the real measured value corresponding to the revolution number. The huge amount of propositions related to the numerical model shows that the looking for a constant expressing a proportionality relation between physics quantities is mainly done within this level.

Charles: 12.74 divided by 36 you obtain exactly 3.53. It comes nearer and nearer to 3.6. and divided 25.7 by 7.2 and so on. You find an average of 3.6

When students finish processing data within the numerical model, they encounter a lot of difficulties in interpreting these results at a higher physical level. Very few students' propositions are related to objects and events, even if this kind of information could help them to grasp the meaning of the quantities processed. We note that this task is very poor concerning the modelling activities involved.

Task 3: assign a name (N=119)

Students are asked to find a name or a meaning for the proportionality factor they have previously introduced. This task is quite different from those usually proposed during labwork. Students are in this case explicitly requested to invent a name for the introduced quantity, which is an open-ended problem. We expected them to use information coming from the different modelling levels to give a meaning to the proportionality factor (Table 6).

Table 6: Results for task 3

	Modelling levels			
	ModPhy	**ModNum**	**Meas**	**ObEv**
Aimed by the Question	Aimed			
Used by the students	Used 70 (59%)	Used 32 (27%)		
Linked by the students	10			

About one third of the propositions concerns the numerical model. The majority of the students' groups use the numerical and physics model to express the meaning of the introduced quantity. Students thus show their preferences for using knowledge coming from physics or calculus proceedings to elaborate the physics model. This kind of knowledge seems to reinforce the validity of their productions.

Paul: constant of current flow, but constant of energy flow in function of time.

Propositions related to measurements, objects and events are not very important, whereas we *a priori* considered those levels as necessary. We thought that measurement of voltage or current could be used to verify if the word "power" corresponded to the concept previously taught during a physics lesson related to electricity. These observations show the real difficulty that students have in summoning up elements related to the material situation in order to enrich a physics model.

Task 4: Iconic representation (N=114)

Students have first to represent the energy chain corresponding to the experiment carried out, and then, they have to represent on this chain the introduced quantity, in order to show its properties of energy. To give an answer, links must be established between elements coming from the objects and events level and from the physics model. This kind of task allows students to really enrich the qualitative energy model they have been taught previously with quantitative aspects (Table 7).

Table 7: Results for task 4

	Modelling levels			
	ModPhy	**ModNum**	**Meas**	**ObEv**
Aimed by the Question	Aimed			Aimed
Used by the students	Used 62 (54%)			Used 36 (32%)
Linked by the students	28 (ModPhy–ObEv)			

As expected in the *a priori* analysis, students' propositions are related to objects and events (32%) physics model levels (54%), and frequently related (28 links). These results confirm those obtained by Tiberghien & Megalakaki (1995): the symbolic representation of qualitative aspects is favourable to the establishment of relations between the levels of objects and events and physics model.

Mélanie: No, it's related to the immersion heater. We have to put it here, because if it is powerful, it's going to decrease. If it is not powerful, the time will increase, so it's related to that [show a part of the energy chain].

Quantitative aspects (measurements and numerical data processing) that have been largely used in previous tasks are here absent. Another type of relation is frequently used: internal relations within the physics model.

Elise: because here, there is an arrow that goes down, and if it is powerful, the arrow comes down a bit more.

These relations allow students to confirm answers given for this task on the basis of other knowledge coming from the physics model.

Table 8: Results for task 5

	Modelling levels			
	ModPhy	**ModNum**	**Meas**	**ObEv**
Aimed by the Question	Aimed (linked to Meas and ObEv)			Aimed
Used by the students	Used 77 (57%)			Used 34 (34%)
Linked by the students	28 (ModPhy–ObEv)			

Task 5: Modification (N=136)

Students are asked to propose elements they could change to modify the value of the quantity previously introduced. We expected students to use elements coming from the physics model, measurement and objects and events, and to establish relations between these different levels as indicated in the Table 8.

As in the previous task, most of the propositions are related to the physics model (57%) and to objects and events (34%). The 28 links established between these two levels are generally associated with the use of the energy chain to look for what objects are likely to be modified. For three out of the six groups, we have also noticed relations between physics model and measurement. Through these relations, students recognise a strong argument to justify the changes of the experimental situation. These relations are established after the modification propositions. They thus seem to be used as a verification tool.

Annie: P=UxI thus it really depends on voltage and intensity because they are also part of the experimental setting.

It seems that it is legitimate to use measurement when the experimental situation is modified. The relation to measurement is then strongly linked to the material objects that are possible to modify.

Conclusions

The analysis criteria we have defined allowed us to analyse both labwork sheets and students' strategies on the basis of the modelling activities. We were able to draw a comparison between what students were intended to do and what they have been able to do. Our approach obtained interesting results concerning learning processes of students during a traditional labwork.

The first of our results is that the characteristics of the questions lead the students to similar modelling activities. Most of the time, students' activities are identical from one group to another. This result enables us to draw some conclusions concerning students' main modelling activities.

It appeared that students devote almost half of the time to the handling and measurement task, which is moreover, the richest task concerning modelling. Nevertheless, we pointed out that they do not establish many relations between each of the modelling levels. Handling and measuring seems then to be a relevant task in order that students immerse themselves in the diversity of modelling levels.

We also established that students use different modelling levels and elaborate relations between different modelling levels depending on what the task dealt with: they always take into account the modelling level within the written instruction, and at least the physics model. This result leads us to think that if we want students to turn their attention to a special modelling level, the first necessary condition is to introduce this level in the question addressed. In a similar way, we noticed that students do not use measurements spontaneously for verifying hypotheses and that they rarely refer to the objects and events when processing quantitative data and drawing relations between physical quantities. These strategies can be associated with low cost cognitive procedures. Indeed, they are sufficient to furnish the labwork report with correct answers. These results help us to see the labwork sheet in a new light. If we want students to refer to these levels when answering, the labwork sheet should not only be a series of questions, but also a strong guide inclining students to consider all the elements that come into play. Notice that this kind of suggestion is completely contrary to criticisms of the usual practice of laboratories (White 1996).

If students use the numerical model to process data, they do not use this level to interpret the results. It is an intermediary enabling calculations or the finding of relations between values, but it is not used to express the meaning of the relations in terms of physics. It can be interpreted as a phenomenon related to an implicit contract (Brousseau 1986) between the teacher and the students: a labwork report should necessarily contain physics terms.

At least, we noticed that all along the labwork, students do not make explicit any explanation system, which would account for the absence of theory in all the tasks solved by the students. This does not mean that students don't know the theory level exists. This level is merely not explicit in their dialogues. Arguing on the basis of an explanation system was not a pursued aim in the various tasks, but it has already been claimed that making this system explicit in the students' dialogues is very difficult (Tiberghien & Megalagaki 1995). We still have no answer concerning the tasks that could lead students to argue explicitly the theory level.

Recommendations

The obtained results are important for better understanding the role of existing labwork in learning physics. The traditional labwork sheet's information is not sufficient to lead students to establish links between the theories and models and the experimental field. At the same time, students are able to establish that kind of link when a task imposes this activity. If we expect students to learn physics in that way, we have to think about tasks guiding (more or less strongly) students to establish relations between theoretical knowledge and experimental activities. Improving the texts accompanying existing labwork could enable students to use varied modelling activities and thus improve learning of concepts underlying the labwork.

These improvements might be reinforced by the teachers' guidance, which can focus students' attention on the elements that they do not usually take into account. It is thus very important to insist on the difficulties students meet with when interpreting experiments during teachers' initial or continuing training.

References

Bachelard, G. (1971). *Epistémologie, textes choisis*. Ed. PUF.

Bécu-Robinault, K. (1997). Rôle de l'expérience en classe de physique dans l'acquisition des connaissances sur les phénomènes énergétiques, Thèse présentée le 12 mars 1997 devant l'Université Claude Bernard - Lyon I pour l'obtention du diplôme de Doctorat.

Becu-Robinault, K. & Tiberghien, A. (1998). Integrating experiments into the teaching of energy. *International Journal of Science Education*, Vol 20, n° 1, pp. 99-114.

Brook, A. & Driver, R. (1984). Aspects of secondary students' understanding of energy: full report, *Children's Learning in Science Project*, November 1984.

Brousseau, G. (1986). Fondements et méthodes de la didactique des mathématiques. In *Recherches en Didactique des Mathématiques*, Vol. 7, n° 2, pp. 33-116.

Duit, R. (1981). Understanding energy as a conserved quantity. In *European Journal of Science Education*, Vol. 3, No. 3, 291-301.

Duit, R. (1985). In search of an energy concept, paper prepared for "conference on teaching about energy within the secondary school science curriculum" Leeds, March 1985.

Feynman R. (1980). *La nature de la physique*. Ed. du Seuil.

Gaidioz, P., Monneret, A., Tiberghien, A., Becu-Robinault, K., Besson, G., Blache, M.-J., Chastan, J.-M., Clavel, C., Colonna, A.-M., Collet, G., Gibert, D., Longere, G., Le Marechal, J.-F., Strobel, M.-P., Vagnon, H. (1998). *Introduction à l'énergie*, Collection Appliquer le programme, Lyon : Centre Régional de documentation pédagogique de Lyon.

Gooding, D. (1989). Introduction : some uses of experiment. In Gooding, G., Pinch, T., Schaffer, S. (Eds.), *The uses of experiment - Studies in the natural sciences*. Cambridge University Press.

Grandy, R. (1992). Information, observation, and measurement from the viewpoint of a cognitive philosophy of science. In N. Giere (Ed.), *Cognitive models of science*. Minneapolis : University of Minnesota Press.

Hofstein, A. & Lunetta, V. N. (1982). The role of the laboratory in science teaching : neglected aspects of research, *Review of Educational Research*, 54, 201-217.

Lemeignan, G. & Weil-Baris, A. (1992). L'apprentissage de la modélisation dans l'enseignement de l'énergie. In *Enseignement et apprentissage de la modélisation en sciences*, équipe INRP/LIREST, INRP.

Martinand, J.L. (1992). *Enseignement et apprentissage de la modélisation en sciences*. INRP, Paris.

Meheut, M., Chomat, A. & Larcher, C. (1994). Construction d'un modèle cinétique de gaz par des élèves de collège: jeux de questionnement et de simulation. In M. Caillot (Ed.), *Actes du Quatrième Séminaire National de la Recherche en Didactique des Sciences Physiques*. Amiens, IUFM de Picardie, pp. 53-71.

Millar, R. & Lubben, F. (1996). Knowledge and action : student's understanding of the procedures of scientific enquiry. In *Research in Science Education in Europe, Current issues and themes*. Welford G, Osborne J., Scott P. (Eds.), London, The Falmer Press.

Niedderer, H. (1996). Teaching quantum atomic physics in college and research results about a learning pathway. In proceedings of the *International Conference on Undergraduate Physics Education*, University of Maryland, July 31 - August 3.

Niedderer, H. and Schecker, H. (1991). Towards an explicit description of cognitive systems for research in physics learning. In *Research in Physics Learning: Theoretical Issues and Empirical Studies*, proceedings of an international workshop, University of Bremen, March 4 - 8 1991, pp. 74-98.

Pickering, A. (1992). Living in the material world. In Gooding G., Pinch T., Schaffer S. (Eds), *The uses of experiment - Studies in the natural sciences*. Cambridge University Press, pp. 275-297.

Solomon, J. (1985). Learning and evaluation: a study of school children's views on the social uses of energy. In *Social Studies of Science*, vol. 15, 343-371.

Tiberghien, A. (1994). Modeling as a basis for analyzing teaching - learning situations. *Learning and instructions*. Vol. 4, pp. 71-87.

Tiberghien, A. & Megalakaki, O. (1995). Characterization of a modelling activity for a first qualitative approach to the concept of energy. *European Journal of Psychology of Education*, Vol. X, n°4, 369-383.

White, R.T. (1996). The link between the laboratory and learning. *International Journal of Science Education*, Vol 18, n°7, 761-774.

Wiser, M. (1993). Naive theory, conceptual models, and mental models in thermal physics. Communication at *Biennial Meeting of the Society for research in Child Development*, New-Orleans, L.A., March 25.

Students' Intellectual Activities During Standard Labwork at Undergraduate Level

Michel Beney, University of Bretagne Occidentale Brest, France

Marie-Geneviève Séré, University Paris - Sud XI, France

Abstract

The purpose of this case study is to shed light on the links between doing-thinking-learning during standard labwork. In this context, the only phases, which have been observed, are the phases of action. Observations have been carried out for novices (students) in real class situations. They have been carried out for experts (teachers) as individual interviews. Be they novices or experts, experimenters' cognitive activities are described with concepts borrowed from cognitive psychology. The description highlights the difference between what is expected by teachers and what students actually do. It also puts forth that knowledge of procedures is lacking in students. Recommendations underline that labwork is irreplaceable in learning procedural knowledge. Thus recommendations point not to the suppression of this type of labwork, but to its improvement by a renewed definition and selection of objectives.

Introduction

This study concerns standard labwork. This type of labwork is spread throughout Europe and is frequent in every country (Psillos, Niedderer & Séré 1998). On one hand, teachers seem to expect considerable learning results from it, as has been shown by surveys (Hedewig 1990). On the other hand, teachers and researchers criticise this type of labwork (Claugh & Clark 1994, Gangoli & Gurumurthy 1995) as being impossible to improve labwork sessions in which students are guided and carry out experiments with ready-to-use devices. In our opinion, the first step to improving this sort of low cost labwork is to know more about it and to know more about the phases which are specific to it, namely the phases of practical action, because they differ from the phases of any other academic activity.

In this study we will analyse careful observations of students in the labs themselves during the action phases (excluding data processing, calculations and report writing). The aim of the analysis is to give a detailed description of what students do, both with their hands and with ideas (Millar et al., Chapter 1 *of this volume*). To this purpose we have developed a specific tool to understand the links between doing and thinking as well as doing and learning. A survey at European level showed that teachers' expectations are in terms of both conceptual and procedural learning (Welzel, Haller, Bandiera, Hammelev, Koumaras, Niedderer, Paulsen, Bécu-Robinault. & von Aufschnaiter 1998). We will question the effectiveness of such kind of labwork.

Teaching approach

This study has been carried out in the University of Bretagne Occidentale, where, at undergraduate level, physics laboratory sessions are always standard labwork with

D. Psillos and H. Niedderer (eds.). Teaching and Learning in the Science Laboratory. 65-78.

the following characteristics: guidance through a labsheet, students working in pairs for three hours, apparatus available from the beginning of the session. One tutor (for 20 students) or two (for 35) are present to help students complete the tasks. They are just supposed to help, not to teach anything specifically, though they may seize the opportunity to teach, on their own initiative. It is in the labsheets that students find the necessary items of knowledge to understand the experiment, namely descriptions of the apparatus in order to handle it, the relevant theory, the development of the required tasks, the type of required achievements such as data processing, on which students spend a lot of time.

Conceptual knowledge is carefully selected to help understanding and is supposed to be learned through the laboratory activity. As to the procedural knowledge, it is part of the description of the experiment. It is not valued for itself but rather is included in the particular case of the experiment. It appears as exemplified, not explicitly exposed. In another part of this study that we will not report here, we checked that students did not clearly identify these procedures for themselves and retained a poor consciousness of having used them after the labwork sessions. However, the role of procedural knowledge is to monitor the active phases of labwork. In a study concerning doing and intellectual activities, namely thinking and learning, it was necessary to identify and make clear the procedures used. We established a list of the procedures intervening in the phases of action of the sessions we observed. The list can be found in Appendix 1. Some of them are common to different sessions

Research questions and design of the study

As said in the introduction, the main data consist of observations of students during the active phases of labwork sessions. We limited ourselves to three of the ten sessions of a semester, as being representative:

- the measurement of the speed of light in different media (optics with large use of electronics);
- transfer of heat;
- electromagnetic induction.

Nothing was changed in the usual lab sessions for the sake of the research. The observer being a tutor, the others behaved exactly as usual during the observed sessions. The study focuses on the specific elements of labwork: setting-up the apparatus, adjusting and taking measurements by applying written instructions (excluding data processing). The same observations cannot be collected for all the students during the whole time. We recorded observations from various students who would be possible to analyse in terms of cognitive activities, difficulties in carry out the experiment, use of contextual conceptual and procedural learning (Beney 1998; Séré & Beney 1997). 40 student pairs were observed.

Making the hypothesis that labwork sheets do not reflect the logic of what students do during labwork, the link between doing – thinking can be split into three research questions :

1. What is the logic of the practical tasks during a given labwork session (RQ1)

2. How is it possible to categorise the intellectual activities performed by students with concepts taken from cognitive psychology, applied to action, sometimes used in ergonomy (RQ2) (Piaget 1974; George 1983; Hoc 1990; Richard 1990)
3. What are the respective roles of procedural and conceptual knowledge in these activities (RQ3).

The above hypothesis being confirmed, we wondered if this was a characteristic deriving from students or from the very nature of any academic experimental task. Considering students as 'novices', we wondered how 'experts' manage with such tasks. This is the reason why we also carried out observations of teachers, considered experts, completing the same practical tasks in situations as similar as possible to that of students.

For experts, the research questions are similar. The logic of the practical tasks being the same, they concern the analysis of the intellectual activities during the phases of action of the same sessions (RQ2'), as well as the role played by procedural and conceptual knowledge in these activities (RQ3').

In another part of this study, we explored the link between doing and learning, using interviews and questionnaires. This will be not presented at length here. It will only be referred to in this chapter.

Research Methods

Observation of students during labwork

During each session, the verbal exchanges between students and teachers were audio recorded[1]. The student-student discussions were recorded only if the teacher was present. The observer strove to speak as much as possible with students when they were asking for help or explanation, as well as through simple routine questions:

'What are you doing now ?' 'Are you managing ?' 'What happens now ?' 'What do you expect ?'

40 student pairs were involved. We obtained by this method a sort of 'collection' of intelligible events occurring during labwork, since it was impossible to get a continuous series of events given the long intervals during which students seem to say and do nothing.

The analysis of observations made of actions, gestures, fast thinking, must not be only in terms of conceptual thinking. Erroneous or not, it is not sufficient to give account of the decisions and actions of students. Our study is a contribution to a different understanding of processes of thought directed to action. For this sort of study, the cognitivist paradigm (students do have conceptions in mind which they retrieve in order to interpret experiments) is not sufficient. As will be shown below, in the context of action during labwork, students' conceptual knowledge did intervene, but somewhat seldom. Our analysis will mainly point at other ways of interpreting how students decide to carry out specific actions.

[1] Compared to audio recording, video recording did not provide significantly different information. Pictures were only helpful in transcripting and analysing students' and teachers' information.

Observation of 'experts in labwork' (teachers), doing the same experiments as the observed students

We found eight volunteer teachers, considered experts, who agreed to carry out alone, for the first time, exactly the same experiments, in conditions as similar as possible to those for students, the same labsheet being at their disposal. Each of them agreed to 'think aloud' and to be recorded during the active phases. This means that, conversely to students, we obtained a consistent series of events for each of the eight teachers. We described and analysed the intellectual activities of experts with the same research tools used for students. This permitted comparison of the role of conceptual and procedural knowledge during the phases of activity for students and teachers. It also allowed comparison of conditions of initiative as well.

Research results

Analysis of the practical tasks during a given labwork session (RQ1)

Students carry out actions in order to set up the apparatus, adjust and take measurements. We call an action a gesture during the experiment. For example *'to move a lens', 'to turn off a switch', 'to push a button to start heating'.* Each of these gestures provoke events: *To move a lens modifies the intensity of the image. To turn off a switch breaks a circuit and makes the ballistic galvanometer needle move. To push a button makes the thermometer move and the time counter start.* The information on the evolution of each of these events is provided by a 'signal'. When acting, experimenters (experts or novices) have constantly to watch for a signal, which informs them of the evolution of the experiment. More than one gesture correspond to a given signal as required in the labsheet. A choice is necessary. This is a source of difficulty.

The actions having an effect on the signal can be divided into three sets , as shown by Figure 1, which gives the generic 'action network' of a physics experiment.

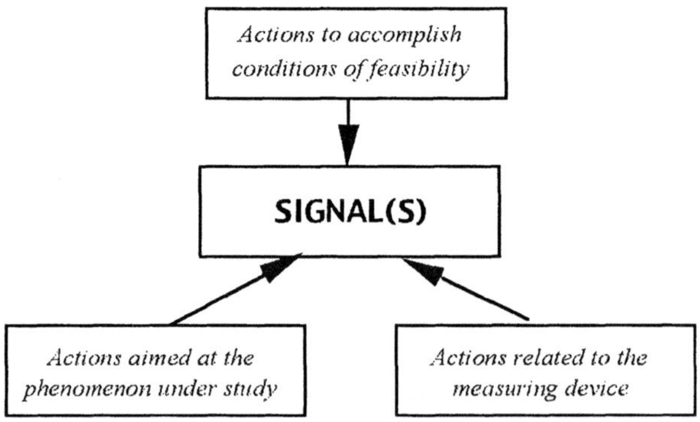

Figure 1: The generic shape of the 'action network' for an experiment.
The three types of possible actions on the signal(s) are grouped into three poles: The Phenomenon Pole (Phen-Pole) - The Measurement Pole (Mea-Pole) - The Feasibility Pole (Fea-Pole)

Those actions are produced on objects which have been displayed for different purposes: some vary the parameters of the phenomenon, others allow measurement. The role of the third set is to match the phenomenon and the measurement instruments (feasibility). This schema makes clear that quite often more than one action have the same effect on the signal. The links between the different actions and the signal are more complicated in a real experiment, as shown in Appendix 2.[2] In particular it may happen that two signals have to be checked;

Students' hands-on and cognitive activities (RQ2)

The development of an action network for each observed session, allowed the classification of each student gesture into one of the poles. The following step of the analysis is to understand why each student, watching the signal, chose one pole or another to act on the signal. It was made possible thanks to the concepts of *goals, rules* and *criteria of control.*

The choice of a pole (hands-on activity) depends on the *goals* (cognitive activity) students elaborate. They sometimes take a goal directly from the labsheet. Frequently, however, they develop goals by themselves. They also recall and use *rules* to proceed. They develop *criteria* to control what they have just done and decide if an action has been successful or if it is necessary to do something else. (Hoc 1990; Richard 1990)

From the different recorded protocols, we extracted goals, rules and control criteria that students developed. They can easily be attributed to one of the poles in the action network.

Definition of goals

We observed that students developed goals not required by the labsheet, arising from the actual events occurring during the sessions. If 'touching' a button on an apparatus produces a change, students may develop a sort of causality centred on one object only that they directly apply to act on the signal. By doing this, students think that they are in one pole, acting on the *Phen-pole* for instance, even though they are in the *Mea-pole.* The confusion of poles is a source of mistakes. Some examples follow:

- During the session *'Measurement of the speed of light'* (abbreviated to 'Speed of light') several student pairs happened to 'play', as they said, with a button on the oscilloscope to change the relative positions of the curves on the screen (*Mea-pole*). They thought they had obtained a different path for the light (*Phen-pole*), erroneously.
- During the session *'Transfer of heat'*, when acting on the thermostat, students could see the time counter (signal) start (*Fea-pole*). They thought that the heating process (*Phen-pole*) to be studied had been triggered and they adopted this way of starting a measurement, neglecting the necessity of obtaining equilibrium. In a way, they tried to replace the process of regulation by the

[2] We attempted to produce such an analysis of the actions to be carried out during an experiment, in different domains. The monitoring of actions by one or more signal appears to be specific to quantitative physics experiments. This analysis is relevant as long as both a measuring instrument and the apparatus demonstrating a phenomenon are present.

thermostat, manually. In fact, students were not conscious of the procedure of regulation.

In order to choose the correct pole, when defining a goal during the course of action, the best solution for the experimenters is often to be aware of the *procedure* in play. For instance, in *'Speed of light'*, the labsheet requires the experimenter to adjust the positions of two mirrors to produce equal phases. The underlying reason is that the measurement is done by difference. If this procedure *'measurement by difference'* (see Appendix 1) is not identified, the corresponding measurement becomes a list of unintelligible actions. Procedures provide a structure to the tasks, which is a correct representation [3] of them.

Our observations plead in favour of the necessity of procedural knowledge to define correct goals and to choose the correct 'action pole'. Of course, *conceptual knowledge* intervened as well. It happened that erroneous conceptions were the cause of difficulties. For instance, during the session *'Transfer of heat'*, several students stated that convection was occurring only during the period of cooling and not during the period of heating. Identifying convection with 'cooling', they had difficulty in applying a procedure: *to study one phenomenon (convection) when another (radiation) occurs at the same time.* In these cases, students define sub-goals according to their own personal pre-existing conception.

Concerning the respective uses of procedural/conceptual knowledge (see p. 73) during the course of action, it can be said that the use of conceptual knowledge was not frequent to define a goal and to reach it, through the action in a pole. Students generally tended to avoid to use conceptual knowledge. An example of students avoiding retrieval of what they had already learned was provided during the session *'Speed of light'*. One of the goals as defined in the labsheet was to make a parallel beam. Some students associated that with an image at infinite distance and moved the mirrors as far apart as possible. Some were puzzled and wondered what to do, because in their minds the use of a lens is associated:

- with obtaining an image;
- with magnification;
- not with a specific shape of the beam.

These are rudimentary mental associations and not retrieval or use of conceptual knowledge. It is worth pointing out that it was mainly during the adjustment phase, implying one pole only (*Fea-pole*), that the need for conceptual knowledge was most felt by students.

In most of the preceding examples, students 'translated' some of the goals of the labsheet into personal goals and sub-goals. The more complex the action network is, the more inadequate this translation is. In other words, the difficulty stems from a lack of a correct representation of the experiment. At a less global level, the more unknown the procedures in play are, the more inadequate this translation is. In other words, the difficulty stems from a lack of representation of the successive tasks. Students carry out successive actions which are poorly structured.

---

[3] We use the word REPRESENTATION in the expression 'REPRESENTATION OF THE TASK', which is rather different from its meaning in 'CONCEPTUAL REPRESENTATION', that we replace by the word CONCEPTION.

Consistently, when the action network was simple, close to the generic action network (see Figure 1) students could manage easily with little procedural knowledge, but also little conceptual knowledge. An example occurred during the session *'Electromagnetic induction'* , the network of which is very simple (one signal, three types of clearly distinguishable actions. See Appendix 2) . Students could manage without using much conceptual knowledge, the drawback being that they learned nearly nothing during this session (according to a post-questionnaire).

Developing and recalling rules

The goal being defined, students need rules to go ahead. It seems that most of the rules they use, which are misleading, come from previous experiments. For instance, when variation of the electric current is produced by induction, physics does not explain it only by Ohm's law *('Electromagnetic induction')*. Students had in mind this law, transforming it into a rule, and consequently they tried to change the current in the small coil (*Phen-pole*) by using a resistance different from the critical resistance of the galvanometer circuit (*Fea-pole*).

Some rules were attached to instruments already used: for instance a zero button for time counters, in the session *'Transfer of heat'*, made some students think that this button is only a switch. They pushed it twice, the first time to display zero on the counter, and the second time to start measurement. They were disappointed that the indication remained zero (all the time the thermostat kept heating switched off). An arbitrary rule arouses from a lack of representation of the role of the thermostat.

The use of the oscilloscope was also the source of numerous rules in the session *'Speed of light'*. For instance students strove to get an a priori aspect of the curves on the screen.

Another tendency was to adopt certain tasks, as defined in the labsheet, as rules:

> *'You have to be very fast with this switch. You just have to be, that's all'*
> *['Electromagnetic induction'].*

This rule of rapidity produced a confusion between two kinds of time: the time during which they switch the interrupter and the time during which the galvanometer spot is moving.

This makes it clear how much students prefer to utilise ready-to-use rules rather than to rely on conceptual knowledge. We observed a very limited place, not to say no place at all, for conceptual knowledge in rules. Students do recall and develop rules in a very automatic and fast way.

Development of control criteria

The experimenter constantly has to make decisions in order to decide whether what has been done is correct or not, or whether it is relevant to continue. Theoretical reasons to continue or to do it again would work. Nevertheless, students create control criteria, which do not come from physics.

Most of the controls are done on the values obtained and are simplistic opinions:

> *'It is too small '*
> *'It does not change enough from one measurement to another'*
> *'It should be symmetrical'* [Any session].
> *'The curves must fill the* screen' ['Speed of light']

The criteria for two signals to be in phase were varied: the figure was said to be a straight line, or an ellipse, or a circle, or two curves intersecting at the point (0 , 0) in mode X-Y.

Criteria sometimes simply come from what can be called the implicit didactical contract, which governs any session where students and a teacher are working together. For instance:

'Other students do not have the same result'.

'The labsheet states that 9 measurements must be made. We are wrong as we only achieved 8 when moving the mirror ' ['Speed of light'].

It appears that students' control criteria, at least those we have been able to observe, did not employ much conceptual knowledge.

Experts' cognitive activity when carrying out the experiments (RQ2')

We made the same classification of the teachers' actions according to the action network, as for the students. A first result is that the action phases are not so easy. Teachers, considered as experts, are likely to succeed easily, without hesitation, in data processing and interpretation phases. Adjustments, organising data collection, managing the three poles and their respective links is not a simple task.

Then, as for the students, we were able to use the concepts of goals, rules and control to account for the cognitive activities underlying the hands-on activities.

We will only give results of the comparison of novices (students) versus experts (teachers):

The similar cognitive activities: rules

Teachers did have similar ways of recalling rules as students and sometimes had the same difficulties as them, namely acting in one pole instead of another to obtain a given change of the signal. For instance, in the session *'Transfer of heat'*, a teacher confused the thermostat button with a simple switch, exactly like students (see paragraph 5.2.1), confusing the respective *Mea-pole* and *Phen-pole*. In carrying out the experiment *'Speed of light'*, another teacher also confused the same two poles. Contrary to the instructions, he adjusted the front part of the block of plastic through which light was travelling, right to the beginning of the scale because

... measurement always begins at the zero position .

Another example concerns the experiment *'Electromagnetic induction'* where the resistance of the induced circuit, a small coil connected with a galvanometer, has a predominant influence on the signal. The total value of the resistance must be equal to the critical resistance of the galvanometer. A teacher made a mistake when adjusting the value of the resistance

...because a galvanometer is always used with such a value .

When doing so, he used a rule, directly, with no representation of the task in the particular case.

The cognitive activities which differ: goals and controls

There were discrepancies in the experts actions with respect to the labsheet, because most of them were defining rather different goals to achieve the experiment. During *'Transfer of heat'*, one of them decided to heat the plate first, erroneously, because

...we have to measure a loss and so we need first to get heat .

His personal goal led him not to follow instructions. During the same experiment, a different experimenter stated that

We have to initialise the apparatus, in order to make it acquire data by itself

This sort of autonomy cannot be explained by the use of more sophisticated conceptual knowledge. Surprisingly, teachers used and evoked very little physics. But they had, available to them in their minds, numerous experimental procedures which were of help. For example, one of them did not follow instructions ('*Speed of light*') because

...it is not necessary to adjust the scale of time (of the oscilloscope) because we will have to compare values (procedure 'measurement by difference') and it is not necessary to get exact measurements

They were bold enough to try a process because they knew that it corresponded to a procedure to measure by difference, by comparison, etc. Even if, from time to time, the chosen procedure was not adapted to the experiment in hand, it seems that this awareness of procedures gave them considerable autonomy. A difference with students was that they did not exhibit erroneous conceptions influencing the definition of goals. Wrong goals came occasionally from a fuzzy representation of a task (use of the thermostat, for example).

Globally, teachers were successful with personal goals and more or less erroneous rules, or in a trial and error process, that some of them carried out brilliantly, independently from the labsheet, because they had good effective control criteria. An example is this teacher (*'Electromagnetic induction'*) who moved the receptor to check the evolution of the signal and said

It is OK, in physics a lot of values behave like that.

In doing this, she confirmed that the way she conducted her actions was correct. Generally, the control criteria were not those of students (see 5.2.3). They were more sophisticated and helped them to handle problems created by their autonomy. They were mainly judgements on approximations, which showed a good knowledge of measurement procedures. For instance :

'Transfer of heat' - *We do not have to take into account every fluctuation of the values of the signal, we will take an average.*

'Speed of light' - *By putting mirrors far from the source of light we will not have to take into account every parameter because this factor (distance) is very high compared to the others.*

In a nutshell, the experiment required efforts from the experts. Their cognitive activity was characterised by autonomy and awareness of procedures useful to develop goals and control criteria, driving to success in spite of approximate rules, retrieved speedily.

The use of conceptual and procedural knowledge (RQ3 and RQ3'): synthesis from the observations of novices (students) and experts (teachers)

A first conclusion of our observations is that, in spite of apparently strict guidance from the labsheet, experimenters do have a certain autonomy. Experts use it more widely, thanks to the knowledge of procedures, whereas students manage with what is implicit in labsheets by making decisions. We described them with the concepts of goals, rules and control criteria.

At least for the students we observed, it can be said that in this narrow margin of decision, students avoid using conceptual knowledge, or use it in a clumsy, approximate way, exhibiting from time to time erroneous conceptions. In fact we saw that rules are not criticised, that students use various elements in the elaboration of goals and criteria, which have to do with rationality but little with physics. Experts do not use much physics to manage. They exhibit no erroneous conception.

In students' discourses, there is no place for procedures. For teachers, procedures are omnipresent, in goal definition and also control criteria. They are guided by them, in spite of functioning very similarly to students, the similarity being mainly for the rules, which are applied as recalled from memory, with little criticism. But again a good comprehension of the procedures in play, and the availability of others, contribute to good controls. Teachers are helped more by procedural than by conceptual knowledge.

Conclusion

In order to obtain a better knowledge and improvement of very ordinary standard labwork, we observed what happens during the *action phases* only, putting aside any other phase of labwork (experimental design, data processing, interpretation, etc.). In other words we decided to study the links which are supposed to exist between DOING, THINKING AND LEARNING.

DOING:

We produced an analysis (RQ1) of the tasks necessary for each labwork session according to the labsheet, using the idea of the action network. We listed the procedures monitoring the actions, though not implicitly presented in the labsheets. This double analysis concerns the experiment itself, independently of the experimenters. However, it has a direct influence on the difficulties met by the experimenters, expert or novice.

DOING – THINKING:

Observations of novices (students) and experts (teachers) allowed to develop categories of cognitive activities. Students 'think in order to do', (RQ2, RQ3)

- by developing goals and sub-goals, which, for some of them, are personal, coming from the representation they have of the task. This is the place for initiative.
- by recalling rules coming from fuzzy knowledge, from other experiments, from other uses of the same apparatus in a different context.
- by using control criteria some of which are simple common sense. Others, arise from previous experience. Others are likely to come from the 'didactical contract' (demand for quick achievement and success of the experiment, types of explanations, types of assessment, status of mistakes, etc.).

Observing experts (RQ1' and RQ2') doing the same experiments for the first time and with the same labsheets, we demonstrated the importance of procedures to manage this sort of work. Be they teachers or students, experimenters use, as little as possible, their own theoretical knowledge and the theoretical knowledge offered in the guidebook. Having already used and being aware of a large range of experimental procedures seems to be more effective. Experts, in case of difficulty, are able to try procedures, even if not adapted, and carry out effective controls.

Both types of observations make clear the differences between novices and experts: the course of action and the poor use of conceptual knowledge in physics are surprisingly similar; it is the important use of procedures by experts that makes the main difference as compared with novices.

Obviously, to develop realistic objectives for labwork, it is useful to get realistic comparisons of expert and novice practice. Conceptual learning through labwork, more precisely through this type of labwork, must not be the main expectation. It has been confirmed through interviews and questionnaires post teaching: effectiveness with respect to conceptual learning is not nul but it is poor. To foster and support conceptual teaching, other types of tasks must be promoted during labwork, with teacher-student interaction rather than guidance by labsheets eliciting what has to be done and achieved.

Concerning procedural knowledge, the effectiveness is even worse, since students are seldom conscious of having put into operation such and such procedure. However, these procedures are the experts' key reason of success. If the objectives for procedural knowledge are not taken seriously, standard labwork will be limited to objectives like familiarity with apparatus, experimental skills and motivation. They represent minimal benefits.

Recommendations

Following this work, we do not condemn standard labwork. Obviously it has at least one advantage: once the apparatus is bought, the cost is minimum in terms of teachers' time. Conversely to what is generally said, they are not merely 'cookbook activity' (Claugh & Clark 1994). They allow a modest initiative to students, who have to develop goals, retrieve rules and elaborate criteria of control. But we have demonstrated their poor effectiveness in term of conceptual and procedural learning. Other pedagogical methods are more effective in achieving conceptual learning: tutorials (McDermott 1996), problem solving, didactical software, etc. But for procedural knowledge, they are irreplaceable. No 'theoretical' teaching of procedures is effective, to our knowledge. So our recommendation is improvement rather than suppression. It can be obtained in two directions:

- Students should be given a better *global representation* of the experiment and the corresponding tasks. This can be easily done through the identification of the different poles of the action network. It could be a remediation to the lack of meaning which comes from the fact that the experiment is imposed, not designed by students (Bandiera, Torracca & Rossi 1998). It would avoid random or trial and error action.
- A stronger emphasis should be put on procedures used in each experiment. From one session to another, it would be useful to point at the common and different experimental procedures to be studied and used. For instance, the procedures 'choosing a time basis' or 'measurement by comparison' have to be explained and studied in different domains of physics

All this can be obtained only from a radical change in the style of labsheets and teachers' explanations at the beginning of the sessions. This should be the subject of further research.

References

Bandiera M., Torracca, E, Rossi A. (1998) *Textbooks as a Source of Experiments.* In : Psillos D. Niedderer H. & Séré M.-G (1998) The mains results of case studies: about the effectiveness of different types of labwork. Working paper 8 of the project 'Labwork in Science Education'. DGXII SOE2 CT 95 2001.

Beney M. (1998) Contribution à l'analyse des phases manipulatoires des Travaux Pratiques de physique en 1er cycle universitaire . Les apprentissages possibles à travers l'action. PhD dissertation presented to University Paris Sud XI.

Claugh M.-P. & Clark R. (1994). Cookbooks and constructivism – a better approach to laboratory activities. *The science Teacher* , 61 , 2, pp 34-37.

Gangoli S.-G. & Gurumurthy C. (1995). A study of the effectiveness of a guided open-ended approach to students experiments. *International Journal of Science Education*, vol. 17, n°2, pp. 233-241.

George, C. (1983). *L'apprentissage par l'action.* Paris, PUF.

Hedewig, R. (1990) Bericht der Arbeitsgruppe "Experimentieren im Biologieunterricht" In: Killerman (Ed.), Methoden des Biologieunterrichts. Aulis Verlag.

Hoc J.-M. (1990). Les démarches ascendantes et descendantes dans la planification. In J.-F. Richard, C. Bonnet & R. Ghiglione (Eds), *Traité de psychologie cognitive tome2.* Paris, Dunod, pp. 228-229.

McDermott, L.C. (1996). *Physics by inquiry.* New York: Wiley and sons

Millar, R., Le Maréchal, J.-F., & Buty, C. (1998) A map for characterising the variety of labwork in Europe. Working paper 1 of the project 'Labwork in Science Education'. DGXII SOE2 CT 95 2001.

Psillos D. Niedderer H. & Séré m.-G (1998). The mains results of case studies: about the effectiveness of different types of labwork. Working paper 8 of the project 'Labwork in Science Education'. DGXII SOE2 CT 95 2001.

Piaget, J., (1974). *Réussir et comprendre* . Paris, PUF.

Richard, J.-F. (1990). *Les activités mentales.* Paris, Armand Colin.

Séré, M.-G., & Beney, M. (1997). Le fonctionnement intellectuel d'étudiants réalisant des expériences. Didaskalia. 11, pp. 75-102.

Welzel, M., Haller, K., Bandiera, M., Hammelev, D., Koumaras, P. Niedderer, H., Paulsen, A.C. Bécu-Robinault, K. & von Aufschnaiter, S. Teachers'objectives for labwork. Research tool and cross country results. Working paper 6 of the project 'Labwork in Science Education'. DGXII SOE2 CT 95 2001.

APPENDIX 1: The three experiments and the procedures they put in play

Electromagnetic induction

The apparatus comprises an inducing circuit (solenoid) and an induced circuit (a small coil). The aim of the experiment is to measure the induction field generated in the solenoid by creating a variation of its flux through the induced circuit. The measuring device is a ballistic galvanometer, which gives the value of the total charge **q** circulating in the induced circuit during a time Δt. This time must be short for correct use of the galvanometer. Students have to act: on a switch K_1 (breaking the circuit as quickly as possible), on a variable resistor, and to look at the spot on the galvanometer. Two different times are to be distinguished: Δt and the time of deflection of the galvanometer spot.

Transfer of heat

The apparatus comprises a black painted plate, which is maintained at a given temperature by a thermostat. The energy supplied is electrical. When in normal operation a heating cycle is established: the electric current is switched off and on, alternately by the thermostat. The experimenters measure the duration of the total cycle as well as the time during which electrical energy is provided. The ratio of these two times, as well as the electric power provided, allows the transferred heat to be determined.

The interpretation relies on the fact that both convection and radiation has to be taken into account. The theory of radiation is taken for granted and data are collected for an interpretation in terms of convection.

The measurement of the speed of light in different media.

The aim is to measure the speed of light by measuring the time light takes to go through a given path. An optical signal is transformed several times (modulated first, then transformed into an electronic signal, modulated again, mixed with a signal of a different frequency). Finally an adjustment must be made on both the optical and the electronic signals, the latter depending on the former.

The apparatus available is: a source of light, two lenses (L1 and L2) and two mirrors to guide the beam from the source to the receptor system, different systems of electronic modulation, different media such as water and plastic in addition to air.

The procedures put into play in the three experiments.

- Measurement of a quantity independent of time through a phenomenon depending on time. (Two methods.)
- Choosing parameters to balance the drawbacks and advantages of amplification.
- Taking into account several phenomena, which contribute, to the same signal in order to exhibit and measure one of them.
- Control by regulation.
- Measurement by compensation.
- Choice of a scale.
- Measurement by comparison:
  when the zero of the quantity to be measured cannot be obtained;
  with a known quantity when the calibration of the measuring device is not possible.
- Calibration of a measuring device by comparison with a reference.
- Obtaining the best quality of a measurement by choosing a special value for a parameter.
- Adaptation of the signal (e.g. its frequency) to the performances of the measuring device (e.g. the range of frequencies).

Appendix 2: The action networks of the observed labwork sessions.

The measurement of the speed of light in different media

This action network is rather complicated, because there are two signals to monitor the actions. *Each arrow means: "... is able to change ....".*

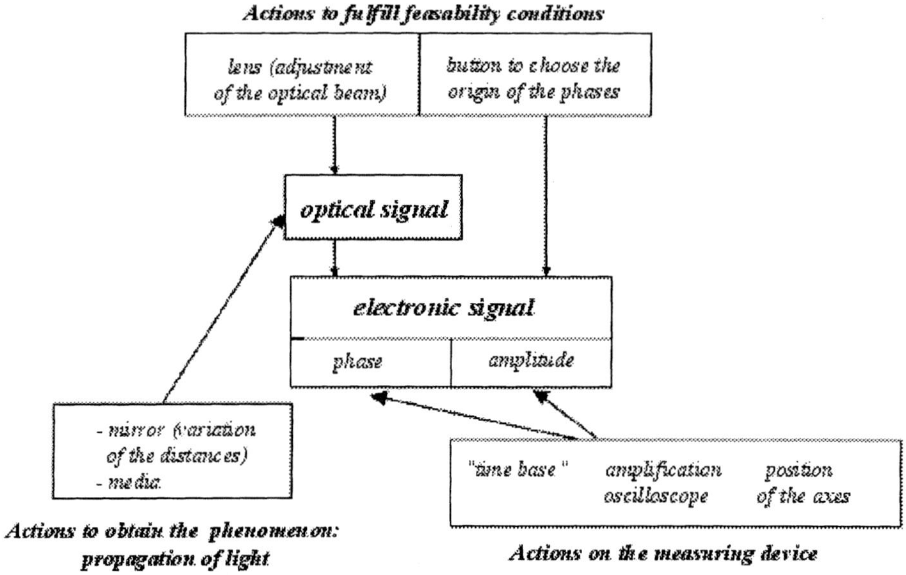

Electromagnetic induction

The action network is closed to the prototypical action network (Figure 1)
Signal (linked to three poles): deflection D of the spot of the galvanometer
Actions to obtain the phenomenon: production of an e.m.f. : on switch K1 (breaking of the solenoid circuit)
Actions on the measuring device: action on the galvanometer and choice of the ballistic function
Actions to accomplish feasibility: choice of the value of the resistor (R2 = Rcritical) and action on switch K1 to use the galvanometer on a ballistic way.

Transfer of heat

The action network is complex because of a control signal
Signals (linked to three poles): heating time, overall time of measuring.
control signal (linked to one pole only) : plate temperature.
Actions to obtain the phenomenon: choice of the type of plate, action on the thermostat (to choose the temperature).
Actions on the measuring device: choice of the time scale, action on a switch to start measuring.
Actions to accomplish feasibility: action on a transformer (to adapt time of heating to the values of the signals).

A Laboratory-Based Teaching Learning Sequence on Fluids: Developing Primary Student Teachers' Conceptual and Procedural Knowledge

Petros Kariotoglou, Aristotle University of Thessaloniki, Greece

Abstract

This work is part of a long-term research programme concerning the design, development, application and evaluation of a laboratory-based teaching learning sequence (TLS) dealing with fluids. We have developed and discuss here an approach to the teaching of the conceptually demanding topic of fluids, which focuses on promoting student teachers' conceptual and procedural development towards a suggested scientific model and experimental method. In this study we focus on the design aspects of the TLS as well as on selected results concerning students' conceptual and, mainly, procedural knowledge.

Introduction

During the last two decades a huge amount of work has been published concerning students' understanding and conceptual difficulties regarding scientific concepts and phenomena (Gabel 1994). Less research has been devoted to pupils'/ students' views on the nature of science and their scientific practices (Driver, Leach, Millar & Scott 1996). Recently, there has been growing research interest in the study of several different aspects of students' procedural knowledge and their learning during labwork (Wellington 1998; Leach & Paulsen 1999). However, if we look at instruction, teaching approaches often treat such aspects of scientific knowledge in a rather isolated manner. Recently, more holistic teaching /learning approaches concerning the various aspects of scientific knowledge have been attempted through the development of didactical structures (Lijnse 1995) or teaching learning sequences (Psillos & Kariotoglou 1999). Teaching learning sequences (TLS) are considered as medium-scale curriculum development, as well as a product of developmental research (Lijnse 1995). Important factors affecting the design of TLS are pupils' /students' views, content transformation (Duit 1999), educational constraints and the choice of experimental field. The last factor refers to the choice of phenomena / experiments which are appropriate to facilitate learning procedures.

In this context we have been involved in a long-term research and development programme in the area of fluids concerning the teaching and understanding of aspects of conceptual knowledge and scientific inquiry. The programme involves the development and study of medium-term laboratory-based teaching sequences at different levels of education. In the present study, we present design features of a 12-hour TLS (Kariotoglou 1998), as well as results concerning aspects of conceptual learning and, mainly, aspects of learning of scientific inquiry by prospective primary teachers.

Designing a laboratory-based teaching sequence

Research results on students' domain-specific conceptions and reasoning, on the one hand, and content analysis on the other are brought into play in order to develop the

D. Psillos and H. Niedderer (eds.). Teaching and Learning in the Science Laboratory. 79-90.

fluids model to be taught and define the intellectual demands on students. We hold that the models accepted by the scientific community need to be transformed in order to be intelligible to students. As a result we reconstructed the scientific model to be appropriate for teaching, a research product influenced heavily by investigations on students' domain specific knowledge. Details on the content transformation are reported in previous work (Kariotoglou, Koumaras, & Psillos 1993), while basic elements of the content to be taught are included in the left column of Table I.

From previous studies (Engel-Clough & Driver 1985; Kariotoglou & Psillos 1993) and our own data, it emerges that on the one hand some elements of students' conceptions do succumb to learning opportunities. On the other hand, other conceptions may act as deep-level conceptual obstacles with regard to understanding scientific models about the macroscopic treatment of fluids. With regard to the states of matter, for some pupils fluids are considered to be a fourth state of matter between solids and liquids while a few of them identify fluids with liquids (Kariotoglou & Koumaras 1997). For most students, important intuitive conceptions containing the germ of scientific concept development are that pressure is related to the depth and the nature of the liquid, although many of them consider pressure to be influenced by the amount of the liquid in the container. This emerges from our own and others' research and indicates the non-differentiation between the two concepts, thus lending the characteristics of force to the concept of pressure (Kariotoglou & Psillos 1993; Engel-Clough & Driver 1985). So, it seems that the most important conceptual obstacle towards scientific modelling derives from the meanings that pupils attribute to the concept of pressure, which is usually confused with the pressing force. It is worth noting that most pupils do not recognise the transmission of pressure through liquids, while those who understand the transmission of actions are not sure if the pressure or the force or even the volume of liquid is transmitted. At the level of tertiary education, results from a questionnaire addressed to 80 of our students in the context of the present study showed that they do not hold significantly different views from pupils. However, the students' replies were richer in the terminology used and the range of justifications provided. Based on the above thoughts, the interplay between the scientific model to be taught and students' views led us to the construction of a conceptual change strategy dealing with the conceptual differentiation of force and pressure (Smith, Carey & Wiser 1985; Kariotoglou & Psillos 1993).

The basic objective of our approach is to interlace content and methodology in a unified corpus of knowledge overcoming any gap that may exist between theory and practice (Howe & Smith 1998). For this reason, we developed qualitative and quantitative experiments concerning empirical laws in order to engage the students in data-processing activities (Howe & Smith 1998). With the qualitative experiments, we aim mainly at the students' familiarisation with the phenomena, concepts and experiments under study. With the quantitative ones, we aim mainly at promoting students' learning of scientific inquiry. We focus on four aspects of scientific inquiry: performance of experiments, distinction and control of variables, taking and processing measurements and making inferences (Millar 1998). We expect that our students, as prospective primary teachers, recognise the value of testing their conceptions. So their engagement in such work would reveal the role of experimental work during science instruction. On the other hand, the negotiation of experimentation, e.g., data processing or making inferences, involves students in using concepts scientifically, while leading also to the development of principles and laws such as Pascal's principle (Germann & Aram 1996). So the two levels of negotiation of knowledge, i.e., the conceptual and procedural, are not separated but interlaced, so that students are led to the knowledge of one level by making use of the knowledge of the other, and vice versa (Millar 1998).

The familiarisation and practice of our students in the process of scientific inquiry is based on a "usually used approach" to scientific methods. In this context, we consider a cyclical procedure functioning on two levels: a level of representation and the level of the material world. The questions to be answered, the hypothesis formed from both prediction and interpretation, and the design of the experiment are included in the first level of the procedure. The performance of the experiment, the gathering and processing of data and the results of the experiments are included in the second level of the procedure. Our TLS exploits aspects of scientific inquiry, attempting on the one hand to verify/falsify the students' conceptions as learners (distinction and control of variables) and, on the other hand, to familiarise them, as prospective teachers, with aspects of experimental methodology (performance of experiments, taking and processing of measurements, making inferences).

Teaching approach

The TLS consists of four units, each of them lasting approximately three hours. The successive steps of our teaching approach with regard to conceptual and procedural development may be seen in Table I. The conceptual goals and content (left column) are interlaced with the designed procedural aims and content to promote students' conceptual and methodological understanding (right column). The objective of the 1st unit is the students' familiarisation with the phenomena / concepts and experiments under study. It is also the unification by the students of the experimental fields of liquids and gases into the unified category of fluids (Kariotoglou, Koumaras & Psillos 1995, Psillos & Kariotoglou 1999). With the term experimental field we mean the choice of experiments/phenomena, which are appropriate to facilitate students' learning. At the start of this unit, the students had to hypothesise the outcomes of three experiments and interpret them. The 1st experiment involved the flow of water out of a pierced bottle containing water. The 2nd experiment involved the piercing of an inflated balloon and the 3rd experiment concerns the compression/extension of air or water in a syringe. After the realisation of these experiments they had to discuss the results and compare them with their predictions.

The familiarisation phase is very important for two reasons. First, because it contributes to the creation of a broad conceptual and methodological framework of thinking and intervening for our students who are not familiar with such scientific procedures. Second, because it helps the students to familiarise themselves with the first steps of experimental/scientific inquiry (performance of experiments, distinction and control of variables, taking and processing measurements and making inferences) and their discussion at group level. This phase helps students to feel more prepared for the next step of the TLS application, which involves the negotiation of the quantitative experiments.

The objective of the second unit are both the enhancement of the students' intuitive views about pressure, e.g., the relation of pressure to depth and their introduction to quantitative experimentation focusing on the distinction and control of variables, taking and processing of measurements and making inferences. These objectives are pursued by engaging the students in experimental work, e.g., to control the variables affecting hydrostatic pressure: i.e., depth, type of liquid, vessel shape, amount of the liquid.

Table 1: The steps of the teaching approach and units of the TLS

	Conceptual Aim & Content	Procedural Aim & Content
UNIT 1 1st STEP	*Familiarisation with phenomena and concepts* Prediction and interpretation of three experiments Unification of experimental fields	*Familiarisation with qualitative experiments and inquiry* Making hypothesis on the three experiments Performance of the three experiments
UNIT 2 2nd STEP	*Enhancement of intuitive conceptions* Basic law of hydrostatics Hydrostatic pressure as an intensive quantity	*Introduction to quantitative experimentation* Distinction and control of variables Experimental verification of the empirical law for hydrostatic pressure
UNIT 3 3rd STEP	Creation of a cognitive conflict Distinction between pressure and force Introduction of the relation between pressure and force	*Comparison application in order to verify/ falsify hypothesis* Pressure's comparison in a wide / narrow vessel; Force's comparison to detach two suckers
UNIT 4 4th STEP	*Application of the knew knowledge to investigate principles and laws* Discovery of Pascal's principle and Boyle's / Mariotte's law	*Application of quantitative experiments to investigate relations between variables* Measurement of pressure in different points of a bottle, after increasing the pressure by air-pump

The objective of the 3rd unit is the distinction by the students between pressure and pressing force. We introduce the new knowledge about pressure, force and their relationship, taking advantage of the students' dissatisfaction with their initial, undifferentiated conception of pressure – force. In this unit there is no significant experimental work from the point of view of procedural knowledge. The two experiments carried out in this unit are quite simple: a) measurement of the pressure at the same depth in a large and a narrow vessel; b) comparison of the forces required to detach two suckers of different size. The role of these experiments was to facilitate cognitive conflict rather than promote procedural knowledge.

The 4th unit is an application of the results of the 3rd one and it concerns the investigation by the students of Pascal's principle and Boyle's/Mariotte's law. As regards procedural knowledge, the 4th unit aims at the application of the experimental methodology introduced in the 2nd unit, targeting the development of the students' skills with respect to the performance of an experiment and making inferences. We used a specially developed experimental apparatus, which consisted of three digital manometers connected at three different points in a closed vessel containing water. In the top of the vessel there is an air pump in order to create an increase in pressure, which is transmitted to every point of the water. The students were asked to take measurements with the manometers after the creation of a new pressure difference. Then the students were asked to process these data, i.e., to make abstractions of the values of the pressure at each level before and after the creation of the pressure change. We wanted our students to recognise that the differences are equal and then to infer that pressure is transmitted invariantly to any point of the liquid.

During all units the students predict, interpret and carry out experiments and discuss in groups and in front of the class the results of the experiments. In this way, the students feel at ease to express their views and to try experimental approaches at the peer-to-peer level. Later on, having acquired experience and feeling more confident, they participate in class discussions in order to improve their understanding and verify their skills. During group experimentation, the teacher offers only technical help to the students, while during whole class work, the teacher has a double role. On the one hand (s)he demonstrates some basic experiments, beyond those carried out by the students. On the other hand (s)he co-ordinates the discussion, helping the students to clarify the concepts of the scientific model and the steps of experimental inquiry.

Research questions and research methods

The basic research questions of this work are derived from a need to assess our students' development and our basic teaching-learning assumptions underlined above. These are the following:

- What are the learning pathways of the students' attending TLS on the topic of fluids, following laboratory-based teaching strategies?
- What are the students' constructions on experimental inquiry during labwork?
- What is the influence of the students' existing theoretical knowledge on both procedural and conceptual development?

In the first phase, data were collected during autumn 1996, from a pilot application with a small group of 10 students. The majority of our students were non-science majors and were selected after a written test and an individual interview in order to secure a group of mixed ability with a variety of alternative conceptions concerning fluids and pressure. The final application took place during autumn 1997, after some minor changes in the content and the articulation of the sequence, derived from the pilot application. The sample of the final application was similar to that of the pilot one.

Video and audio recordings of whole class teaching were used for monitoring aspects of classroom interactions providing evidence of "on-task" student constructions during experimental activities or demonstrations. The conceptual development and the improvement in experimentation of individual students is captured in a stroboscopic manner on selected tasks/experiments, during the whole sequence. With these methods we managed to describe the reactions of each student on the crucial points and experiments, such as, the control of variables affecting hydrostatic pressure (2nd unit), the recognition of cognitive conflict (3rd unit), or the way of understanding the transmission of pressure (4th unit). From the sample of 10 students we selected three (3) students (see conceptual domain, next page) who were representative in recognising the cognitive conflict and consequently in understanding the differentiation between pressure and force. We thus traced the successive steps of 3 students towards the expected learning.

Individual semi-structured interviews were carried out with all students at the beginning and with the three selected at the end of the teaching. Results from these interviews were complemented by a written questionnaire.

From the above descriptions it can be seen that this research is qualitative in that it aims at the description of conceptual and procedural learning rather than at the measurement of success in a quantitative way. We consider that our research and its results are valid and reliable because we use multiple sources of data, we exploit a panel of experts to test both our research materials and our conclusion from the data. We also follow participatory modes of research in the frame of the whole group project and because this research is part of a long-term programme of our group (Kariotoglou 1998).

Research Results and Discussion

The design, development, application and evaluation of this TLS is a long-term research programme (Kariotoglou et al. 1993; Kariotoglou et al. 1995; Psillos & Kariotoglou 1999). In the present study we make reference to the results of our previous work on students' conceptual development and will focus mainly on the description and analysis of students' constructions concerning the aspects of procedural knowledge mentioned previously, during the application of the TLS. We also discuss the relation between the students' prior theoretical knowledge and their achievements.

Conceptual domain

During the familiarisation phase (1st unit) our students display in their explanations evidence of a transitional phase from the phenomenological level to the desired model, although this transition is not always accompanied by scientifically sufficient explanations. With respect to the unification of the experimental field, the majority of our students seem to unify the concepts of liquids and gases as fluids at the conceptual level, in a roughly speaking homogeneous way. Three out of the ten students interpret phenomena at a level complying sufficiently with the desired model. The rest of the students are less successful in using the concept of pressure, but they eventually use it as a concept to describe liquid and gas phenomena in a consistent way. As an example, one student employs pressure to interpret both liquid and gas experiments, besides using the concept of pressure incorrectly ("... *the exerted pressure...*") to classify the phenomena.

With respect to the distinction between pressure and pressing force, the complete discussion of the recognition of cognitive conflict is treated in detail elsewhere (Kariotoglou 1998; Psillos & Kariotoglou 1999). Here we present the main results with regard to the learning pathways (Niedderer 1997) described in our previous studies. Our students approach the distinction between pressure and force at three levels. An example of the first level was when a student with very good initial knowledge distinguished between the two quantities using her knowledge of the relevant theory and formulas. However, in applying her knowledge to the experimental situation, she attributed to pressure the properties of force, i.e., she considered it additive. This suggests that pressure and force may be undifferentiated even by those students who provide correct interpretations to relevant tasks in the initial questionnaire. On the other hand, perhaps, the student's very good knowledge made her seek stronger arguments, such as proving experimentally that pressure is not an additive variable. This led us to the conclusion that the student perceived the proposed cognitive conflict at a different level than the one designed. She required more evidence in order for her cognitive conflict to be resolved.

The 2nd level approach we classify as that of the three students who understood the cognitive conflict as predicted by the design of the TLS. At this point we should note that the conflict revealed to the students by the teacher is disunity between two representational systems, the first being the undifferentiated notion of pressure/force and the second the results of the two experiments referred to in the 3rd unit (see above).

Finally, the 3rd level we classify as that of the three students who realised the difference between pressure and force at a different level than that initially aimed at and, what is more, in contrast to our initial design. According to the latter, the difference that could be shown experimentally and be comprehensible, as indicated by the analysis of content, was the dependence of force, as opposed to pressure, on the quantity of the matter. Despite this, these students appreciated a difference originating in the TLS ("there is pressure", "force is exerted"), that we didn't expect to lead to differentiation.

As regards the remaining three students, two retained their initial knowledge and one student (totally three out of ten) answered using a different way of thinking for each task, so he is difficult to classify.

Concluding the results on the conceptual domain of the TLS, we may claim that it achieved its target conceptual aims (seven out of the ten students), although the teaching design did not predict two of the above three pathways to the conceptual target. Concerning the relation between the students' achievements and their prior knowledge, we can remark that all the students (3 individuals) with initial rich knowledge improved their knowledge to an acceptable level or even higher, whether or not they followed the pathway designed. Some of the students with poorer initial knowledge improved their knowledge, whether or not through the pathway designed (4 individuals). Finally, the others with poor initial knowledge did not improve it (3 individuals).

Procedural domain

We shall describe the students' constructions concerning the four aspects of procedural knowledge: a) performance of experiments b) distinction and control of variable, c) taking and processing of measurements, d) making inferences. For these aspects we study our students' (in)efficiency in understanding or intervening in experimentation.

Performance of Experiments

Our students confronted several difficulties during the performance of experiments due to their lack of practical experience. We provide an example regarding the use of a digital manometer, which consists of a long thin pipe connected to the main measurement unit. When the pipe is inserted into water, pressure is transmitted by the air contained in the pipe, thus resulting in an indication on the manometer's display. In most cases the students' difficulties using the instrument lead them ask for the teacher's help

*S (*student*) 1: what is this? (*pointing at the thin pipe*). Where do we put it?*

Or, even for the simplest instruments, like a ruler for measuring the depth at the point of measurement:

S2: ... this is a ruler ... not an instrument ... err ... oh yes. It is the instrument for the measurement of depth ...

Another example concerns the performance of the experiment to investigate Pascal's principle. In this case, the lab guide provides detailed instructions for the realisation of the experiments. It appeared from the observations that the students did not have significant difficulties in performing this experiment. The problem in this case arose in the representational part of the experimental methodology proposed. For example, initially our students did not find it plausible that by using the air pump they increased the pressure above the water in the closed vessel. Instead they understood it as increasing the air in the vessel:

T (teacher): What did you achieve by using the air pump?
S1: We put (more) air in the vessel.
T: Why? What did we gain by this?
S1: Err ... to increase the air, hum ... the atmospheric pressure. ...
S2: I think (that) normally it is possible to increase the air (he means the pressure).

Putting the above extract in the overall framework of this research we can state that, although our students have understood the concept of pressure at the conceptual level (see the two previous pages), they face difficulties in using this concept, as an entity, to intervene in the experiment.

Concluding this paragraph we can remark that our students face difficulties in realising the experiments, mainly in understanding the role of the instruments or the reason for an intervention.

Distinction and control of variables

It seems that our students did not face particular difficulties, when they had to decide how they should proceed to distinguish and control the variables affecting hydrostatic pressure:

*S1: (*reading the lab guide*) ... what are the factors affecting hydrostatic pressure? ...*
S2: ... the depth ... eeeer ... the kind of liquid....
S3: ... the shape of the vessel.... hum... or the quantity of liquid? ...
S1: ... let's try to check (test) them...

Or for the control of variables

S1: ... how should we check the variation of pressure with depth? ...
S2: ... err ... we measure pressure at different depths, hmm ... with the manometer....
S3: ...Yes, but we also need to know the depth...err.... Have we got a ruler?

Or for the next variable:

S1: ... now we should check the other variable (that means) ... the cross section of the vessel...
S2: ... err ... we shall take two vessels, one large and one narrow and we shall pour water into them, ... eeeer ... up to the same level ...
S3: ... the same level ... err ... or the same quantity?

S2: ... if we do that we will not know what caused the effect.... I mean the shape of the vessel or the level of the water?

S1: .... You are right ... err ... when you check a variable ... all the others must be constant ... so let's use the same level ...

We consider that the distinction and control of variables consist of two parts: a representational, where students choose the variable and decide how they will proceed the test and an interventional one when they realise the experimentation. Taking this remark into account, we can note here that the distinction and control of variables are easy tasks for our students, concerning the representational part of these procedures. But as pointed out in the previous section, the interventional part of the procedure is not so easy, given the difficulty in choosing and using the instruments.

Taking and processing measurements

The greatest difficulty is created when our students are about to process the measurement data coming from the measurements of pressure at various depths. The three groups of students found small differences in the pressure values at the same level, e.g., at 10 cm, the 1st group found 30 (arbitrary) pressure units, the 2nd 32 and the 3rd 31 units. The students discussed these "discrepancies" and the choice of the "best" value at length:

S1: I wonder... we didn't find the same values (all the groups). Why? ... err ... we used identical instruments (manometers) with the others ... also (identical) vessels...

S2: ... it is reasonable... they are similar (instruments), but not the same... it also depends on who measures ... the precision with which one observes ... the (digital) manometer is more reliable, while when using the ruler ... it is possible to have some error ...

T (teacher): Are these pressure values plausible? ...

S3: .... yes... they may deviate a little ... but...

S4 (2nd group): ... why do the others measure zero pressure units, while we measure 1 unit? (when they measure pressure at 0-cm depth)

S5: ...umm, it must be some uncontrolled effect... an error.

S4: ... Oh yes ... that's why our values are systematically higher than those of the other teams...

Then the discussion focussed on the values 30, 32 and 31 units that were found by the three teams for pressure at the level of 10 cm. The students argued at length about what should be reported as a result, namely "31", "31 ± 1" or "30 to 32". Regarding the last format, it was noticed that the students with a good knowledge of theory did not accept the use of the interval "30 - 32" to express their result. We infer that for these students, theory dominates the experiment with which they are not familiar, thus leading them to prefer more precise expressions of a measurement result. In contrast, the students with a poor knowledge of theory considered the interval expression as plausible, possibly because it incorporates the "errors of measurement". Our observations on the influence of students' views about the representation of measurements on the preferred format for reporting experimental results are similar to those encountered in the literature (see e.g. Evangelinos et al., Chapter 4 *of this volume*).

Another important aspect regarding measurements and errors was revealed when the students investigated the influence of density/type of fluid on pressure. It was observed that the students with good theoretical knowledge measured pressure in pure and salt water at only one depth (10 cm), finding a result of 30 and 35 units of pressure respectively. In contrast, the students with a limited knowledge of theory decided to take measurements at five different depths, justifying their choice as follows:

S: Since there are errors (in measurements), how can we tell if 31 is different from 33? ... If in all cases (at all depths) we measure a difference ... which is repeated ... this is more convincing... We just want to make sure...

From such results we may suggest that the students with poor theoretical knowledge understand more easily the implications of experimental errors, as compared to those with a better knowledge. It seems that the imbalance in the presentation of the theoretical and experimental aspects of scientific inquiry in textbooks may create obstacles in assimilating methodological aspects of measurements and data treatment.

Making inferences

Our students have difficulties in making inferences from the processed data and they need the teacher's help in most cases. In the following extract we have an indication about how the students understand the transmission of pressure (Pascal's principle), in accordance with the properties of pressure (2nd unit), as well as how the teacher scaffolds students' understanding:

T: ...What can you conclude from this data... (There is a table that contains the pressure' values in every manometer. After the creation of every single pressure difference, we have performed a set of measurements)
S1: ...The pressure increases with the depth...
T: ...Did the increase (in pressure) transmit to all the points (of the water)?
S2: ...Yes, but normally it should be the same at all points...
T: ...What do you mean? How did you note it?
S2: ...We observed it, the reading increases in all manometers...
T: ...Is this related to any property of pressure we have learned about in a previous lesson?
S3: ...Yes, hum... that it is not exerted at a specific point, err ... it just exists everywhere..."
S4: ... err ... it also has no direction...

This last comment was made by a student with a poor knowledge of theory, while the "better" students had more difficulty in understanding the scalar nature of pressure. Again, it seems that the students with a poorer knowledge of theory more readily assimilated the new knowledge presented to them through the experimental procedures used in the TLS than did those with a good knowledge.

Conclusions

In this work, we have described aspects of the design and theoretical assumptions for the development of an innovative, laboratory-based, teaching learning sequence, dealing with both the procedural and the conceptual knowledge concerning fluids and pressure. The results of this sequence application revealed significant conceptual rather than procedural student constructions. Most of our students achieved the conceptual target though they followed three (3) different pathways, two of which were not predicted by the design of the sequence. The students' constructions corresponded to their initial poor or good theoretical knowledge. Students with good initial knowledge improved their knowledge to an acceptable level or even higher. Students with poorer knowledge did not show the same evolution as the others, though many of them achieved the conceptual target.

It seems that this type of TLS may be able to help students achieve only some of all the possible aspects of procedural knowledge. These include, for example, the distinction and control of variables. Our students perhaps confronted severe difficulties in the performance of experiments due to their lack of experience. In the case of the representational part of the experimental approach, the problems are more complicated. Our students with good conceptual knowledge do not easily accept the experimental errors or other "discrepancies" of the experimentation. That is possibly why the students are led astray from their solid theoretical basis. In contrast, the students with less knowledge accept some of these problems more easily. It seems that adequate theoretical knowledge leads students to more difficulties regarding the understanding of experimental procedures than their lack of experience in handling apparatus and instruments, probably because theory is influencing and leading their observations.

Our attempt to overcome any gap that may exist between theory and practice by interlacing the content with methodology had significant conceptual results, such as the differentiation of pressure and force. The results were not so clear with respect to the procedural knowledge. In this case our students had difficulties in the interventional part of experimentation, e.g., performance of the experiments, using concepts as entities, although they did not have problems with the representational part of the experimentation. The constructions concerning errors depended on the initial level of theoretical knowledge; students with poor initial knowledge easily assimilated the knew knowledge about errors.

Recommendations

These results provide significant implications concerning the design of such a TLS and the teaching of fluids and pressure. The designer of a TLS should take into account students' initial level of knowledge, providing appropriate tasks and materials to scaffold the constructions of the students in each of the three groups. The three different levels in the understanding of the differentiation of pressure and force revealed in this work could offer useful insights into the learning of pressure and force.

The interlacing of theory and practice will be facilitated if conceptual achievements are associated with easy procedural tasks and vice versa. So the cognitive conflict approach is mediated by qualitative or semi-quantitative tasks, while the procedures of data treatment and measurements should enhance the students' conceptions, which are close to the scientific ones.

The interventional part of experimentation is more difficult for the students than the representational one. A familiarisation phase should give the students' the opportunity to handle the experiments and apparatus, applying the relevant scientific concepts and procedures and not just putting things to work.

The acceptance of experimental errors or "discrepancies" by students with a good theoretical knowledge but poor experimental practice could be facilitated by an epistemological discussion aimed at revealing aspects of the nature of Science.

References

Driver, R., Leach, J., Millar, R. and Scott, P. (1996). Young people's images of Science, Buckingham: OUP.

Duit, R., (1999). A model of educational reconstruction – A framework for the research and development in Science education. In P. Koumaras, P. Kariotoglou, V. Tselfes & D. Psillos (Eds.), *Proceedings of the 1st Panhellenic Conference on science education and Application of New Technologies in education* (pp. 30-34). Thessaloniki: Christodoulides (in Greek).

Engel-Clough, E. & Driver, R. (1985). What do children Understand about Pressure in Fluids? *Research in Science and Technological Education*, 3 (2), 133-144.

Gabel, D. L., (1994). *Handbook of research on science teaching and learning*. MacMillan Pub. Co. N. York.

Germann, P. J. & Aram, R. J., (1996). Student Performances on the Science Processes of Recording Data, Analyzing Data, Drawing Conclusions, and Providing Evidence. *Journal of research in science teaching*, Vol. 33, No. 7, pp. 773-798.

Howe, C. & Smith, P. (1998). Experimentation and conceptual understanding in school Science. In: Wellington, J. (Ed), *Practical work in school Science*. Pp. 220-236. Routledge, London and N. York.

Kariotoglou, P. and Psillos, D. (1993). Pupils' Pressure Models and their implications for Instruction. *Research in Science and Technological Education*, Vol. 11(1), 95-108.

Kariotoglou, P., Koumaras, P. and Psillos, D. (1993). A constructivist approach for teaching fluid phenomena. *Physics Education*, 28, 164-169.

Kariotoglou, P., Koumaras, P. and Psillos, D. (1995). Differentiation conceptuelle: un enseignement d' hydrostatique, fonde sur le development et la contradiction des conceptions des eleves. *Didaskalia*, 7, 63-90.

Kariotoglou, P., Koumaras, P., (1997). Using concept map for comparing pupils conceptions with physics concepts: the case of fluids. Paper presented in the First ESERA Conference, September, Rome, Italy.

Kariotoglou, P. (1998). Investigating aspects of an innovative experimental sequence: The case of fluids. Case Study final report, project *Labwork in Science Education*, European Commission DGXII, SOE2-CT95-2001.

Leach, J. & Paulsen, A.(Eds.). (1999). *Practical work in science education*. Roskilde University Press, Frederikseberg & Kluwer Dodrecht.

Linjse, P. L. (1995). "Developmental Research" As a Way to an Empirically Based "Didactical Structure" of Science, *Science Education*, 79(2), 189-199.

Millar, R. (1998). Rhetoric and Reality. In: Wellington, J. (Ed), *Practical work in school Science*. Pp. 16 – 31. Routledge, London and N. York.

Niedderer, H. (1997). Learning process studies in physics: A review of concepts and results. Paper presented in 1997 AERA Annual Meeting, Chicago.

Psillos, D., Kariotoglou, P. (1999). Teaching Fluids: Intended knowledge and students' actual conceptual evolution. *International Journal of Science Education* (special issue). Vol. 21, No. 1, pp. 17-38.

Smith, C., Carey, S. & Wiser, M. (1985). On differentiation: A case study of the development of the concepts of size, weight and density. *Cognition*, 21, 177-237.

Wellington, J. (Ed), (1998). *Practical work in school Science*. Routledge, London and N. York.

Development and Evaluation of a Laboratory Course in Physics for Medical Students

Heike Theyßen, Dieter Schumacher, University of Duesseldorf, Germany

Stefan von Aufschnaiter, University of Bremen, Germany

Abstract

This case study deals with the development and evaluation of a labwork course in physics for medical students using the model of educational reconstruction. Surveys among experts were used to determine the objectives and contents of the labwork course. By means of empirical studies the development of students' knowledge was analysed in detail. The case study describes the research methods used in the empirical studies and their results with regard to the objectives derived from the surveys. It gives a short description of the new labwork course and recommendations for the design of learning environments.

Introduction

Medical students in Germany have to attend a special physics lecture and to participate in a labwork course in physics, to learn about the physics foundations necessary for their further medical studies. The design of such labwork courses has to consider, that physics is a subsidiary subject for the students. This implies specific learning conditions and makes demands concerning the selection of content. Both aspects are necessary to investigate for this special combination of topics. Heise & Mittner (1975) found, that the mathematical knowledge of medical students usually is quite small, and many physicists will confirm (but only by experience), that the physics knowledge as well as the interest in physics is very small, too, compared to other groups studying physics as a subsidiary subject (e.g. engineers). Concerning the selection of contents Ucke (1977) states, that it has to be determined by their relevance with regard to the further studies and the subsequent professional life of the students. But the effect of such a choice of contents on the learning processes was not investigated. For other study paths the efficiency of physics labwork has been investigated (e.g. Toothacker 1983; Psillos, Niedderer & Séré 1998) but there is no statement concerning the transferability of the results to medical studies.

This case study deals with the development and evaluation of such a labwork course, according to the model of "Educational Reconstruction" (Theyßen 2000). This model, presented by Kattmann et al. (Kattmann, Duit, Gropengießer, & Komorek 1997; Kattmann, Duit & Gropengießer 1998), provides a theoretical framework for the iterative construction and improvement of a learning environment.

The constituent components of this iterative process are

- Scientific Clarification, which means the analysis of the scientific content aiming at its use in an educational context,
- Comprehension of the students' perspectives comprising their cognitive and motivational learning conditions and based on empirical studies,

D. Psillos and H. Niedderer (eds.). Teaching and Learning in the Science Laboratory. 91-104.

- Construction of instruction considering the results of the other two components equally and allowing them to be influenced by the application and evaluation of the "constructed" learning environment.

In this case study these components are realised as follows:

- A survey at and interviews with experts (physicians and lecturers in medicine) were used to clarify the objectives of physics education of medical students and to determine their relevant physics knowledge and abilities.
- Detailed analysis of the development of students' knowledge during labwork yielded information on the specific learning conditions.
- The results were the basis for the successive design of a modified labwork course, starting with one session on geometric optics and the function of the human eye.

An important aspect of Educational Reconstruction is that the three components influence each other and constitute an iterative process of analysis, development and evaluation. This study cannot describe this iterative process step by step. It is restricted to the description of the starting-point, the outcome of the development and the comparative empirical studies on the learning processes in the traditional and the modified learning environment.

Teaching approach

The starting-point for the development of the new labwork course was the existing, traditional labwork course and a number of theoretical assumptions concerning learning processes. The labwork course as well as the assumptions are described in this chapter. In addition, a brief characterisation of the new labwork course is given.

In the existing ("old") labwork course, the students carried out experiments without an obvious relation between the physics content and any medical context. In preparation for the experimental work they had to work out the theoretical physics background of each experiment. In the laboratory the students worked together in groups of two. During the labwork session they were supervised by tutors, who instructed them concerning the carrying-out of the experiments and the interpretation of the results (mainly using calculations).

The development of the new labwork course started with the clarification of the aims, experts (here physicians) contributed to it. The most important results were, that

- the physics content treated in the course should be restricted to those with a highly medical relevance and that the relations between physics content and medical contexts should become obvious,
- the focus should be on qualitative understanding of physic contexts rather than on the quantitative verification of physic laws and
- the students should work out and discuss the connection with the medical context during the labwork session.

According to a study within the EU-project Labwork in Science Education (Welzel, Haller, Bandiera, Hammelev, Kouramas, Niedderer, Paulsen, Robinault & Aufschnaiter 1998) the most important objectives physicists ascribe to labwork courses for physics students are to link theory and practice, to develop experimental skills and to develop scientific thinking. These aims significantly differ from those listed above: experimental skills are not in the focus of interest for the physicians who contributed to this survey, instead they favour students acquiring a qualitative understanding of relevant physics contexts and linking physics content and medical contexts. This comparison shows, that the objectives of a labwork course strongly depend on the group of learners it is addressed to. It confirms that, as demanded by the model of Educational Reconstruction, the scientific clarification starting from the clarification of the objectives has to be done anew for each specific constellation of learners and subjects.

The theoretical assumptions of learning underlying this study come from a (radical) constructivist description of cognitive processes (v. Aufschnaiter & Welzel 1999; v. Aufschnaiter & v. Aufschnaiter 2000), that is based on the results of empirical studies on the construction and development of students' meanings in labwork situations (e.g. v. Aufschnaiter 1999). According to this the situated construction of meanings always starts at lower levels of complexity (dealing with concrete objects and situations) from where it develops to higher levels (dealing with abstract scientific concepts). Since the understanding of abstract and "theoretical" descriptions of physics phenomena demands the construction of meanings on higher levels of complexity, it is necessary that in advance the students have the opportunity to gain a wide experimental experience with these phenomena. Furthermore the theoretical descriptions have to refer to this experimental experience. The results of Schoster (1998) suggest furthermore that the connection between experimental experience and theoretical description has to be initiated by the specific design of the learning environment.

The development of the new teaching approach was mainly influenced by the results of the clarification of the aims listed above and the empirical results of analyses of the students' learning processes. The outline of the new laboratory course is as follows:

- The laboratory course deals only with that physics content that has a high medical relevance.
- The physics content is embedded in the medical context: numerous hints, examples and questions throughout the laboratory guide, just as the design of the experimental setup itself, are used to demonstrate the relationship between the medical relevance of the physics content. It is not restricted to preparatory motivation.
- The students work with a laboratory guide, which contains a step by step description of the experimental setup and the measurements.
- Questions concerning the observations and the (qualitative) interpretation of measurements are embedded between the observational and experimental tasks, in order to initiate discussions concerning the physics content and its medical application.
- The experimental tasks start with the careful observation of phenomena and continue with systematic qualitative and quantitative investigations.
- The students acquire the theoretical background during and after the experimental phase, not beforehand. For that, the laboratory guide contains a description of the theoretical background that is closely related to the experiments and refers to the observations and experimental results.

Apart from these changes concerning content and educational concept, the size of groups, the number of identical experimental setups and the ratio of students to tutors is the same as for the "old" labwork course. Due to the new concept, the role of the tutors observably changed: Since instructions concerning the carrying-out of the experiments and the interpretation of the results are contained in the laboratory guide the tutors have more opportunity to support and enhance the discussion of physics and medical content.

To demonstrate the new teaching approach, the changes in the educational concept are illustrated as an example for one labwork context: Geometric optics and the function of the dioptric apparatus of the human eye.

In the traditional course different methods, one of them using the imaging equation, were introduced for measuring the focal length of one lens. In a second part of the labwork session the students worked with a model of a human eye. In this model two lenses with different focal lengths could be mounted behind the "cornea" in order to realise the extreme states of accommodation. The students measured the focal length of the lenses and of the "cornea"-lens-systems. The measurements were evaluated mathematically in order to verify physics laws, for example the additivity of refractive power.

According to the experts, a labwork session in geometric optics should focus on the process of accommodation and on the different forms of ametropia, like myopia and hyperopia. Thus the relevant physics content is optical mapping for a given image distance by means of variation of the focal distance. In order to understand the mechanism of accommodation the students have to be acquainted with the qualitative correlation between the focal length and the thickness of the lens. Performing and comparing different methods of quantitative measurement of the focal length of one lens has no medical relevance in contrast.

With regard to these major goals defined by the experts, a modified model was developed for the new labwork course, in which the focal length of the lens can be changed continuously as in the real human eye (Theyßen 2000). In this model the process of accommodation can be used to produce, for a given image distance (cornea-retina), a high-definition picture on the "retina". The sequence of tasks in the labguide starts with numerous qualitative tasks, preparing for the work with this model:

- The connections between object distance, image distance and the size of the image are examined and discussed for lenses with different focal lengths;
- the focal lengths are measured by means of a very simple and clear method (creating an image of the sun or an other far-off object)
- the correlation between the focal length and the thickness of a lens is examined qualitatively;
- the problem of optical mapping for a given image distance is studied and used to introduce the students to the model of the human eye.

With the model the students study the process of accommodation and its limitations, for the emmetropic eye as well as for different kinds of ametropia The correlation between the object distance and the focal length of the lens is examined qualitatively. Finally the students have to determine the suitable eyeglasses for each kind of ametropia and to study their effect. The experimental results for the different kinds of "eyes" have to be compared.

Due to the use of this enhanced model of the human eye the link between physics and medical content is quite easy for the students. In addition it is supported by numerous questions, for example whether a person with hyperopia needs eyeglasses for small and long distances or for small distances only. These questions can be discussed on the basis of the experimental results.

This example shows that the topic of the labwork session and the labwork tasks are designed according to the goals given by the experts: qualitative investigations dominate quantitative measurements and the link between physics and medical content is enhanced by means of the experimental setup as well as by means of the labwork tasks and questions.

Research Questions

The characteristics of the learning environment mentioned above yield several starting points for empirical studies comparing the old and the new labwork course. Questions of interest are for example

- How does the demonstration of the medical context influence the students' motivation and learning processes?
- How does the choice of the experimental experiences as the basis for the theoretical description affect the students' understanding of the physics?

The research questions this study deals with focus on the effects achieved by the implementation of questions with regard to the aim that the students should discuss their measurements and work out the correlation with the medical context:

(1): How does qualitative discussion and interpretation during experimental work influence the construction and development of students' meanings?

This question covers two aspects:

(1.1) First it has to be investigated, how frequently and in connection with which activities (measurement, interpretation etc.) the students verbalise physics or medical content.

(1.2) Furthermore the dynamic of the complexity ascribed to the students' meanings has to be analysed in detail, especially while they are dealing with the questions embedded between the experimental tasks.

A second question refers to the hypothesis, that the link between experiment (observations and measurements) and theory (qualitative and quantitative interpretation of observations and measurements) has to be initiated. Since only the predominantly qualitative interpretation during the labwork session is documented in the available data, the corresponding research question is:

(2): Is it necessary to initiate discussion of observations and interpretation of measurements by questions or does it occur "spontaneously"?

Research methods

In this case study the students' perspectives were investigated by means of empirical studies of learning processes. Videotaped laboratory situations were analysed with regard to the research questions mentioned above.

Data was gathered for one laboratory session of the old and one of the new labwork course. For the old labwork course data for 3 groups, for the new labwork course for 4 groups of two students were collected. The labwork sessions documented were those dealing with geometric optics and the function of the dioptric apparatus of the human eye. Due to the different designs of the labwork sessions described above, the data permit a comparative evaluation.

Two research methods complementing each other were used to investigate the students' learning processes: A category-based method suitable to analyse the complete data and a transcript-based method, which allows a very detailed analysis of the development of meanings with regard to content and complexity. The latter was applied only to short passages of the videos, which were selected with regard to the research questions and based on the results of the category-based analysis. Both methods are described in the following paragraphs.

The category-based analysis

The method used in this case study is derived from the CBAV-Method, which was developed in the EU-project Labwork in Science Education and is described in detail by Niedderer et. al. (Chapter 1 *of this volume*). It uses two types of categories: context-categories and knowledge-categories. The context-categories describe the contexts students' activities deal with and the knowledge-categories characterise the content of students' verbalisations.

The method itself was adopted from the CBAV-Method: while watching the videotapes every 30 seconds at least one context-category was marked for each student on a working-sheet. At the same time knowledge-categories were ascribed to the verbalisations of each student. In contrast to the CBAV-method, categories were ascribed to the activities and verbalisations of each student and not to the group of two students.

The context- and knowledge-categories were adapted to the research questions of the case study. Hence they are not identical with but quite similar to those of the CBAV-Method (see the contribution of Niedderer et. al., Chapter 1 *of this volume*):

Context-categories ("What context do the students' activities refer to?"):

- Other (O): students' activities are not referring to the labwork task
- Instructions (IN): students get instructions concerning the laboratory tasks (e.g. from the laboratory guide or from the tutor)
- Setup (SE): students carry out the experimental set up, prepare a measurement or take a test-measurement
- Measurement (ME): students take measurements
- Interpretation (IP): students evaluate and interpret their observations and measurements (including calculations)

Knowledge-categories ("What content do the students' verbalisations refer to?")

- technical (t): verbalisations referring to the handling of the apparatus, the technical organisation of the measurement and the reading of data
- physic-technical (pt): verbalisations referring to the experimental setup or the reading of data with a physics background or formulations of observations directly referring to the experiments
- physics (p): physics comments on observations or on data and their physics interpretation
- physics-medical (pm): verbalisations combining physics and medical content, e.g. medical conclusions drawn from the experimental results

- medical (m): verbalisations concerning the medical context but without immediate relation to the current experiment

A consistent attribution of these categories demands more detailed definitions, including examples, and training. Unlike in the CBAV-method, two categories were marked "half", if the context-category within a 30 second period was not clear. Therefore the sum over all entries for all context-categories multiplied by 30 s yields the total time the student spent on labwork. For a statistical interpretation it is possible to determine the relative portion of time for each context-category. In contrast to this (and like in the CBAV-method), between zero and five knowledge-categories were ascribed to one period and full and half (or fifths) entries were not distinguished between. Since one verbalisation usually takes less than 30 seconds, they were treated as "events" and only the frequency of verbalisations in the different knowledge-categories was calculated and compared. "Frequency of verbalisations" here means the number of verbalisation-"events" related to the number of 30 second periods analysed (for the particular student). From these data the portion of labtime spent on verbalisations cannot be derived.

The transcript-based analysis

To describe the development of students' meanings with regard to content and complexity in detail, an analysis of time scales of a few seconds is necessary. The method used in this case study was developed and tested in the Institute of Physics Education at the University of Bremen (group of Prof. S. von Aufschnaiter). It comprises the following steps (see C. v. Aufschnaiter & S. v. Aufschnaiter 2000):

1) With regard to the research questions, especially (1.2), and based on the results of the category-based analysis, relevant passages of the videotaped data are selected. In this case the focus was set to the passages containing the context-category "Interpretation".
2) The selected passages of the videos are transcribed in detail (including students' relevant actions).
3) The students' meanings are reconstructed from their actions and verbalisations. This procedure is based on the radical constructivist point of view that individuals' meanings are not directly accessible to any observer. In order to differentiate between the students' meanings and observers' reconstructions, the latter are called "ideas". As an indicator of the accordance with the students' meanings, the viability of a sequence of ideas ascribed to the passage is used. Based on the sequences of ideas, the content of the students' development of meanings can be analysed qualitatively .
4) Levels of complexity are ascribed to the ideas (reconstructed meanings). The levels belong to a model for the quantitative description of the complexity of situated meanings (see Table 1; C. v. Aufschnaiter & S. v. Aufschnaiter 2000). In this model, ten discrete levels of complexity are used to characterise the complexity of students' meanings. On the lower levels "Objects", "Aspects", and "Operations" the meanings deal with concrete objects or situations and their interdependence. On higher levels classes of objects or situations are identified on the basis of (stable) properties and (within "Programmes" and "Principles") meanings related to variable properties as well as to covariations between at least two variable properties are constructed.

Table 1: Levels of complexity [1]

Systems	Construction of stable networks of variable principles
Networks	Systematic variation of a principle according to other principles
Connections	Links between several principles with the same or different variable properties
Principles	Construction of stable co-variations of pairs of properties
Programmes	Systematic variation of a property according to other stable properties
Events	Links between some stable properties of the same or of different class(es) of objects
Properties	Construction of classes of objects on the basis of common or different aspects
Operations	Systematic variation of objects according to their aspects
Aspects	Links between objects and/or identification of specific features
Objects	Construction of stable figure-ground-distinctions

The development of complexity during the different parts of labwork is visualised in diagrams such as shown in Figure 1: with the time on the x-axis and the levels of complexity (as far as they occur in this passage) on the y-axis. For each idea the ascribed level of complexity is marked as a dot in the diagram. The vertical lines mark episodes of coherent occupation with a narrow area of content (E1 to E7).

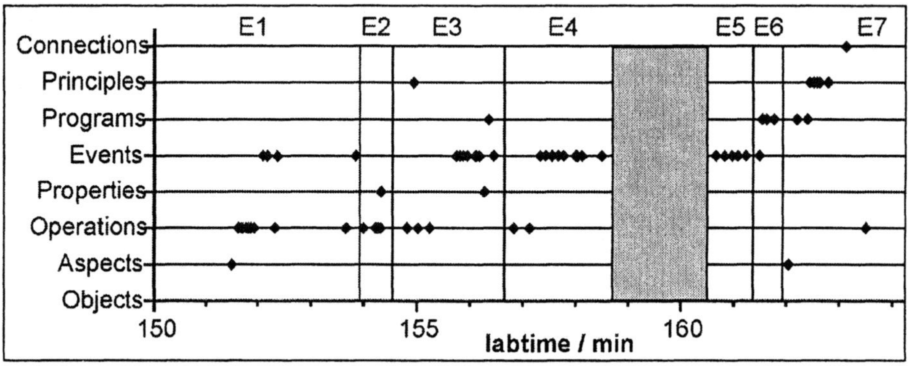

Figure 1: Development of complexity during the interpretation of measurements (new labwork course)

The diagram shows, that in almost every episode the development of complexity starts anew at a very low level. But during the interpretation of the measurement it reaches increasingly higher levels of complexity and it reaches these levels within shorter time.

Research Results

The working-sheets of the category-based analysis were statistically analysed first: the relative portion of time spent on each context-category and the frequency of verbalisations in each knowledge-category, which is the portion of the analysed 30 s-periods, in which the category could be ascribed to at least one verbalisation,

[1] According to C. v. Aufschnaiter & S. v. Aufschnaiter; 2000

were determined for each student. The results were averaged over all videodata for the old and the new laboratory course and compared with regard to the research questions, especially (1.1).

The results of this analysis are:

- In the new labwork course (compared with the old one) the portion of labtime devoted to instructions considerably decreased in favour of the occupation with the measurements (see Figure 2, left diagram).
- Although the portion of verbalisations with technical content predominates in both labwork courses, the proportion of verbalisations with physic or medical content is considerably larger in the new labwork course (see Figure 2, right diagram).

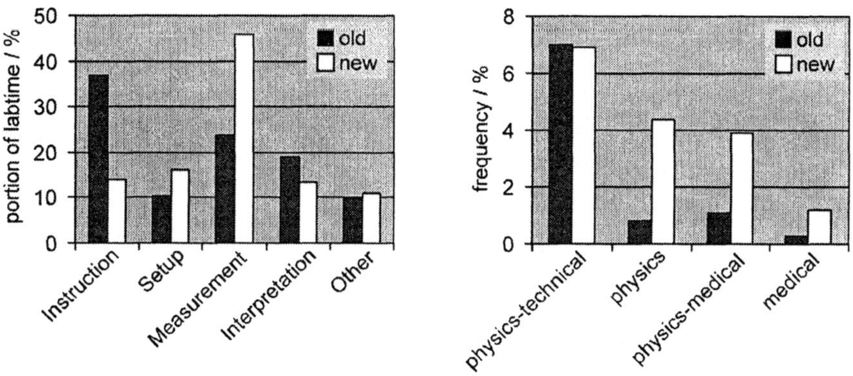

Figure 2: Distribution of labtime on the context-categories (left diagram) and frequency of verbalisations in different knowledge-categories (right diagram) [1]

Figure 2 shows that "Interpretation" takes about 15% of the total labtime in both courses and is rather lower in the new course. But the videodata and the distribution of verbalisations show, that the way of "interpreting" has distinctly changed: whereas in the old course the interpretation almost only comprises calculations and statistical analysis of the results, in the new course it consists of the formulation and discussion of observations. Furthermore, as will be discussed below, the temporal distribution of the periods of interpretation is different.

The results of the statistical analysis show the desired increase in the discussion of physics and medical contents, but they are not sufficient to ascribe this increase to the implementation of the questions, since the correlation between activities (context-categories) and verbalisations (knowledge-categories) can not be seen from Figure 2. In order to analyse this correlation and the temporal distribution of activities and verbalisations for single students the data were illustrated in diagrams as presented in Figure 3: with the time on the x-axis and the context-categories on the y-axis. Every (full or half) entry for a context-category is marked as a black dot in the diagram. White dots indicate that parallel to the activities (context-category) a

[1] 100 % means: in each analysed 30 s-period the category could be ascribed to at least one verbalisation

verbalisation concerning "physics", "physics-medical" or "medical" knowledge is documented.

The diagrams presented in Figure 3 show patterns of the temporal distribution of activities and verbalisations and correlations between activities and verbalisations, that are typical for the old and the new labwork course. Obviously in the new labwork course the interpretation of observation and measurements is well embedded in the experimental work. Whereas in the old labwork course interpretation almost only occurs after all (or at least half of the) measurements are done, here each measurement is usually immediately followed by its (qualitative) interpretation, initiated by the respective question in the labguide. The white dots indicating verbalisations concerning physics or medical contents are predominantly found in the context-category "Interpretation", i.e. in connection with the answering of the questions in the labguide.

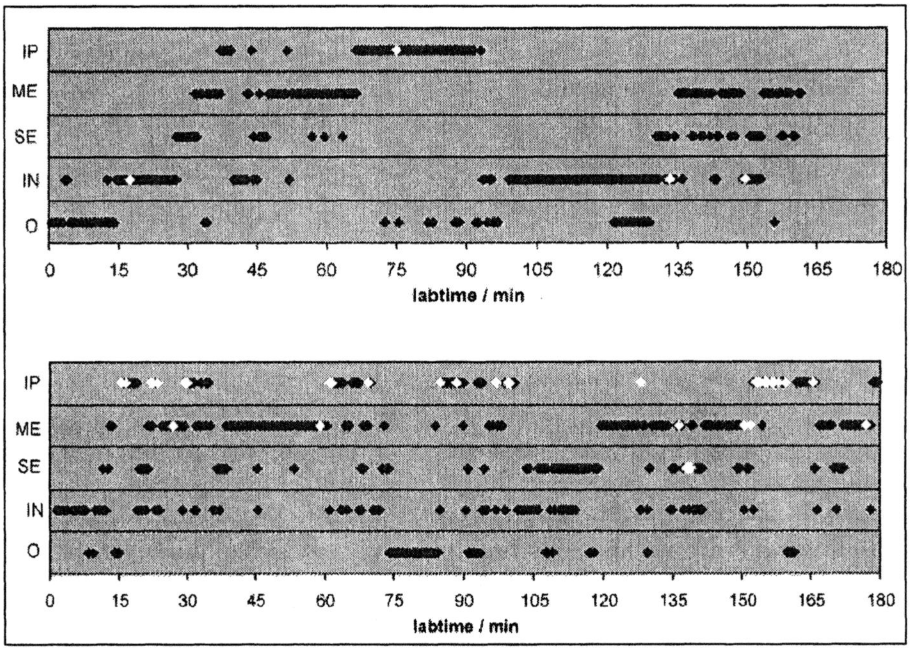

Figure 3: Patterns of context-categories during labwork
(Upper diagram: old course; lower diagram: new course)

Thus the analysis of the diagrams shows that the (immediate) discussion of observations and measurements indeed has to be initiated and that the questions in the labguide of the new course are suitable to perform this task. Recommendations concerning the formulation of questions, especially with regard to the complexity of meanings, cannot be derived without a detailed analysis of the development of meanings.

The analysis of the development of meanings with respect to content and complexity demands the transcript-based method described above. This detailed analysis on short time scales allows the comparison between the structure of the tasks (questions) on the one hand and of the development of meanings on the other hand.

This is illustrated here for one example: the sequence of ideas, which the diagram shown in Figure 1 is based on. The sequence belongs to the interpretation of the measurements for the emmetropic and myopic eye and the task is: "Compare the results for the emmtropic and the myopic eye, with special regard to the point of close and to the volume of liquid in the lens for large object distances. Which problems may occur for the myopic eye, when it is looking at distant objects?"

This task is structured into two parts: the first part demands the comparison of data, which is possible at the level of complexity called "operations". The second part demands the extrapolation of the measurement and a conclusion concerning the medical context. To answer this part of the question, the construction of meanings at higher levels of complexity ("events" up to "principles") is necessary. The student, who the data in Figure 1 belong to, answered the questions in 7 episodes. During the first two episodes (for about 3 minutes) he dealt with the first part: he compared the measured data and, beyond the task, generalised his results. According to the requirements of the task the complexity of the ideas steadily reached the level "operations" (comparison of data) and sporadically reached the levels "properties" and "events" (generalisation of results).

In the following 5 episodes (for about 7 minutes) he dealt with the second part: he started with a (correct) hypotheses concerning the covariation of two variables ("principles" in episode 3). Due to an interaction with his partner, he established a new (wrong) hypothesis, the verification of his own hypothesis with the data ("operations") was disturbed. After the discussion of the new hypothesis ("events") he did not succeed in verifying this one by means of the data. Episode 4, which comprised another unsuccessful try of verification, was followed by an interaction with the tutor, that was not documented in the videodata. During episode 5 the student reconstructed the explanations given by the tutor in detail ("events"). Episode 6 and 7 comprised the verification of these explanations by means of the data and their repeated and increasingly elaborated formulation. Thereby he constructed variable properties ("programs") and their covariation ("principles").

As already mentioned in the previous chapter, in the course of time the development of complexity on the one hand reaches higher levels and on the other hand it reaches these higher levels faster. The analysis of the sequence of ideas yields that this development is supported by the structured formulation of the task. The recommendation that is derived from these results is, that the tasks should

- contain several steps, that can be treated separately,
- start with "simple" questions, that can be answered at a low level of complexity, and
- progress with questions demanding answers at increasingly higher levels of complexity.

Conclusions

The aim of the "Educational Reconstruction" of this labwork course for medical students was to design a learning environment, in which the students are able to acquire the physics foundations they need for their further medical studies (and the following professional life).

According to the scientific clarification and to the results of previous empirical studies, important features of this labwork course are:

- to make the correlations between the physics content and the medical context obvious,
- to focus on a qualitative understanding of physics contexts and
- to demand the discussion of physics and medical contents during labwork.

With regard to these aims, the contents of the labwork tasks were chosen and detailed and precise questions concerning the interpretation of the observations and measurements were integrated into the labguide. This design of a labwork course is not only innovative compared to the old one, but to traditional physic labwork as subsidiary subject in general, since these courses usually are "reduced versions" of labwork courses for physics students and specific for the students main subject and learning conditions.

The research questions discussed in this study focussed on the correlation between the integration of these questions and the discussion of physics and medical contents by the students.

The results of the category-based analysis of videotaped labwork sessions show that in the new laboratory course the time students spent on getting instructions was noticeably reduced, whereas the frequency of verbalisations concerning physics and medical contents substantially increased, these verbalisations predominantly occur during the interpretation of observations and measurement, initiated by the questions in the labguide, and without this initialisation the discussion of physics and medical contents hardly occurs.

The transcript-based analysis of passages, where students interpret their measurements in order to answer the questions, shows, that:

- during the discussion of physics and medical contents students' meanings sporadic reach rather high levels of complexity ("programs" and "principles") and
- the development of meanings with regard to content and complexity is "guided" by the structuring of the questions. A well structured task, with slowly increasing demands concerning the complexity of the answers, can support the development of meanings.

Recommendations

This study shows, that the model of Educational Reconstruction is a suitable theoretical framework for the development of a new learning environment. Both scientific clarification and investigations of the students' learning conditions are necessary contributions and have to be done with careful regard to the specific constellation of learners and subject. The significant differences between the objectives physicists attribute to a labwork course for physics students (Welzel et al. 1998) and those found in this study emphasise the limited possibilities of transfer of results. Consequently the design of labwork courses in physics for students' studying physics as a subsidiary subject has to start from the clarification of the aims, calling on scientists of the students' main subject as experts. The significant changes in the design of the labwork course compared to a traditional one, show that it is just as necessary to investigate the students' special learning conditions, which may be different for each study path, and to consider them in the construction of the learning environment.

For the design of a labwork course in physics addressed to medical students the following recommendations can be derived from the results of this study:

- The contents should be selected with regard to their medical relevance. This relevance should be emphasised and visualised by the sequence of the laboratory tasks as well as by the apparatus itself.
- Apart from the contents and the demonstration of the medical relevance, the sequence of tasks should guide the students step by step, starting from operations with concrete objects, going on with careful observations of phenomena and finally demanding systematic qualitative and quantitative investigations. This structure should follow the specific way of learning of this group of students which does not necessarily correspond with the technical structure of the physic topic.
- Questions concerning the interpretation of measurements should be embedded between the experimental tasks, in order to initiate reflections on the physic content and its medical application.

References

Aufschnaiter, C. v. (1999). Bedeutungsentwicklungen, Interaktionen und Situatives Erleben beim Bearbeiten physikalischer Aufgaben. Fallstudien zu Bedeutungsentwicklungsprozessen von Studierenden und Schüler(inne)n in einer Feld- und einer Laboruntersuchung zum Themengebiet Elektrostatik und Elektrodynamik. Dissertation im Fachbereich I (Physik / Elektrotechnik) der Universität Bremen. In: Niedderer, H. & Fischler, H. (Hrsg.). *Studien zum Physiklernen*, Bd. 3, Berlin: Logos.

Aufschnaiter, C. v. & Aufschnaiter, S. v. (2000). Students' Cognitive Development While Working on Physics Tasks. Submitted to 'Journal of Research in Science Teaching', Feb. 2000.

Aufschnaiter, S. v. & Welzel, M. (1999). Individual Learning Processes - a Research Program with Focus on the Complexity of Situated Cognition. In: Bandiera, M.; Caravita, S.; Torracca, E. & Vicentini, M. (Eds.). Research in Science Education in Europe. Dordrecht: Kluwer, 209-215.

Heise, H. & Mittner, J. (1975). Über die elementarmathematischen Kenntnisse von Medizinstudenten der Anfangssemester und die Konsequenzen. Der mathematische und naturwissenschaftliche Unterricht, 28 (7), 436-438.

Kattmann, U.; Duit, R. & Gropengießer, H. (1998). The Model of Educational Reconstruction. Bringing together issues of scientific clarification and students' conceptions. In: Bayrhuber, H. & Brinkman, F. (eds.). What - Why - How? Research in Didactic of Biology. Kiel: IPN, 273-282.

Kattmann, U.; Duit, R.; Gropengießer, H. & Komorek, M. (1997). Das Modell der Didaktischen Rekonstruktion. - Ein theoretischer Rahmen für naturwissenschaftsdidaktische Forschung und Entwicklung. *Zeitschrift für Didaktik der Naturwissenschaften*, 3 (3), 3-18.

Psillos, D., Niedderer, H. & Séré, M.-G. (1998). The main results of case studies: About the effectiveness of different types of labwork. WORKING PAPER 8 from the European project LABWORK IN SCIENCE EDUCATION (Targeted Socio-Economic Research Programme, Project PL 95-2005).

Schoster, A. (1998). Bedeutungsentwicklungsprozesse beim Lösen algorithmischer Physikaufgaben. Eine Fallstudie zu Lernprozessen von Schülern im Physiknachhilfeunterricht während der Bearbeitung algorithmischer Physikaufgaben. Dissertation am Fachbereich I (Physik / Elektrotechnik) der Universität Bremen. In: Niedderer, H. & Fischler, H. (Hrsg.). Studien zum Physiklernen, Bd. 2, Berlin: Logos.

Theyßen, H. (2000). Ein Physikpraktikum für Studierende der Medizin. Darstellung der Entwicklung und Evaluation eines adressatenspezifischen Praktikums nach dem Modell der Didaktischen Rekonstruktion. Dissertation im Fachbereich I (Physik / Elektrotechnik) der Universität Bremen. In: Niedderer, H. & Fischler, H. (Hrsg.). Studien zum Physiklernen, Bd. 9, Berlin: Logos.

Toothacker, W. S. (1983). A critical look at introductory laboratory introduction. *American Journal of Physics*, 51 (6), 516-520.

Ucke, C. (1977). Didaktik naturwissenschaftlicher Praktika: Beispiel Physik für Mediziner. Hamburg: AHD.

Welzel, M; Haller, K.; Bandiera, M; Hammelev, D.; Kouramas, P.; Niedderer, H.; Paulsen, A.; Robinault, K. & Aufschnaiter, S. v. (1998). Teachers Objektives for Labwork. WORKING PAPER 6 from the European project LABWORK IN SCIENCE EDUCATION (Targeted Socio-Economic Research Programme, Project PL 95-2005).

The Biology Textbook as a Source of Ideas about Scientific Knowledge and Experimental Activity

Milena Bandiera, University Roma Tre, Italy

Abstract

References, indications and proposals that can be seen as experimental activities supporting the presentation of biology topics in Italian secondary school textbooks and aspects of communication aspects that can affect the image of scientific knowledge are the focus of this study. From lower secondary school to the threshold of the university a poor and decreasing demand for participation in a concrete, direct or experimental approach to natural objects and events has been noted. The prominence of assertions and definitions, the presentation of a certain number of unrealisable and "wrong" experiments, and experimental evidence restricted to the subject matter of Ethology (a young discipline, not yet sufficiently stated epistemologically) reinforce the image of scientific knowledge as a well-established and unquestionable set of information, which perhaps has never had (and does not need) any empirical support.

Introduction

Science textbooks have been analysed and compared in terms of the development of their presentation style, beginning in the early part of the 20th century (Skoog 1979), mainly with reference to their treatment of relevant themes – principally, in biology textbooks, that of evolution (Jiménez Aleixandre 1994; Swarts, Anderson & Swetz 1994; Moody 1996) – and their readability (Wright 1982). These same points were particularly taken into consideration in the 1980s by Italian researchers with regard to Italian textbooks (Lumbelli 1984; Rigutti & Santaniello 1985).

Even though data varied considerably from country to country and from school level to school level, what emerged in all cases was the importance of the textbook in determining the scope and organisation of classroom instruction (Helgeson, Blosser & Howe 1977; Stake & Easley 1978; Woodward & Elliot 1990a).

Issues relating to the epistemological approach of science teaching and the educational role of practical work in school science have also been an object of investigation since the beginning of the 20th century (Armstrong 1903); this has intensified over recent years, revealing a variety of problems. Examples include the evidence that experiments and theory are somewhat disconnected in pupils' thinking (Duveen, Scott & Solomon 1993) and that the different kinds and purposes of practical work are often indistinct and indistinguishable (Gott & Duggan 1996; Nott 1996).

Occasional studies combine those two issues in teaching/learning and look at the influence of the layout of the textbook on the student's prefiguration of an image of science, while putting in evidence the tendency of texts to concentrate only on the products of science, rather than on its nature and processes, and to have activities limited to a cookbook approach (Glynn, Yeany, & Britton 1991; Woodward & Elliot 1990b). With regard to Italian education, this combining seems crucial. Since Italian science teachers rarely engage in practical work, it is quite obvious that students – at school – get their image of science and their representation of experimental activity

D. Psillos and H. Niedderer (eds.). Teaching and Learning in the Science Laboratory. 105-118.

mainly from textbook readings (chiefly from historical anecdotes and from described processes and experiments).

Therefore the original design of this investigation was centred on the analysis and categorisation of the "experimental activities" presented in the textbooks, looking for traces of the epistemological status of Biology and potential impact on the students' image of scientific research.

Preliminary analysis revealed a very poor repertory of true experiments (as opposed to the – in any case few – experimental activities), and consequently it became necessary to take into consideration other elements of a textbook that could be concerned with the foundation of an experimental attitude and the development of the image of science: verbal (Sutton 1996) and iconographical (Mathewson 1999) languages first of all. It was substantially assumed that the image of science does not result from the information that is given, but rather from the way in which it is presented. The relative quantity and the quality of examples and experimental activities that are implied, shown or suggested are among the most relevant contributions.

All the considered elements concern, at a general level, the definition of the nature of scientific knowledge, and, at a personal level, the stimulation of curiosity toward natural objects and events, the triggering of motivation towards engagement in a scientific activity, either research or applied, consciousness of one's own expectations and the evaluation of the chances of success. Within a constructivist perspective (Novack 1987) these elements represent the environment from which the learners take information and start to construct personal interpretations and meanings about science. This is a spontaneous process in which the learners are well trained for achieving common knowledge and which operates outside the possibility of teacher control.

It also seemed reasonable to be alert to the possibility that the observations, manipulations and experiments proposed in the textbooks do not correlate with the foundation of investigative methods, but rather support rote learning instead. Textbooks could, in fact, be built around different aims with respect to the structuring of a meaningful supply of information about objects and events, and the laying of the basis for investigative attitudes, e.g. fostering teacher competence and habits and pursuing commercial success.

Attention to these collateral aims is undoubtedly encouraged by the presence on the market of a very large number of textbooks (78 for lower and 101 for upper secondary school in Rome and its province). This unexpected finding could be relevant to the quality of scientific education as a whole and to the availability of a plurality of tools for teachers.

The considerations on which this study is based can be put explicitly in the form of three crucial research questions concerning the responsibility of textbooks for epistemological education:

- how and to what extent does the verbal presentation of disciplinary topics account for the experimental basis of scientific knowledge?
- how and to what extent does the presentation highlight the fact that scientific knowledge is characterised by consensus and temporariness?

- how and to what extent do the text-aids (figures, experiment descriptions, experimental procedures, and so on) contribute to the foundation of a scientific attitude?

Sample and Categories

Sample

The investigation was restricted to Biology topics (and Biology-related Chemistry topics) in the most frequently used textbooks in the school year 1996/1997 in all the schools in Rome and its province. Therefore analysis was performed on a selected sample of lower (Appendix 1) and upper (Appendix 2) secondary school textbooks. The representativeness of the sample can be derived from data reported in Table 1.

Table 1: Science teaching in Rome and its province: lower (LSS) and upper (USS) secondary school sections, textbooks adopted and consistency of the analysed sample

Subject	Science (LSS)	Biology (USS)
school sections in Rome and province (total no.)	1.019	888
textbooks on the market (total no.)	78	101
textbook sample (no.)	10	10
school sections using the textbooks of the sample	394	470
% of school sections involved	38.7	52.9

The eventuality of taking whole books into consideration was rejected due to the concrete impossibility of carrying out an in-depth work on a meaningful sample, to the different total length of the books constituting the sample, and to the different length of complementary sections in individual books. Rather it seemed reasonable that the analysis be carried out with reference to themes spanning different disciplinary contexts. Themes were chosen (bearing in mind the weight of Chemistry/Biochemistry in the overall status of Biology and the absence of a separated teaching of Biology in lower secondary school) for their relevance with respect to the curriculum, length of treatment (similar, as far as possible, in all the analysed books) and, preferably, the organic unity of the presentation. Still, in a few cases some minor adaptations integrating material from separate sections had to be made. The chosen themes are shown in Table 2.

Table 2: Themes that have been chosen for analysis in lower (LSS) and upper (USS) secondary school textbooks

	LSS	USS
Biology	breathing, sense organs	breathing, symbiosis and parasitism
Biochemistry		enzymes, cell permeability
Chemistry	changes of phase, chemical reactions	
Interdisciplinary	pollution	animal behaviour

Categories

In keeping with the premises and declared aims, the analysis focused on three aspects of presentation, which refer respectively to verbal communication, to iconographic communication and to the quoted, suggested and prefigured experimental activities.

Since some already adopted descriptors and procedures seemed not to fit with the specific aims of the analysis, new ones were devised, which were validated (both in terms of definition and attribution to the categories) in a textbook not included in the analysed sample, by comparison of the classifications independently formulated by three judges. 90% agreement was considered satisfactory.

Categories for paragraphs

As far as the verbal text is concerned, each paragraph in selected sections was classified according to the typology of its essential content as:

a. assertion/definition, when it is reduced to one piece - or to a collection - of information ("Then, temperature measures the mean speed of each molecule, considered separately: heat, on the contrary, measures the energy received by all the molecules, considered as a whole." Gori Giorgi 1994);
b. example, when it pivots upon references to concrete objects or events (Conditioning is learning through association. ... Does this not recall your cat's behaviour when it hears the rustle of a paper which could wrap some food?" Curtis & Barnes 1996);
c. experimental practice/literature, when it concerns experimentally substantiated contents or concepts ("Lorentz spent a lot of time in a close contact with geese ... new-born geese were "imprinted" with his person, and specifically with his hand." Zullini & Sparvoli 1994).

Categories for figures

All the figures that illustrate the textbook excerpts dealing with the selected subjects were categorised with respect to their main requisites:

a. pertinence (the figure is substantially and formally in keeping with the verbal text, connecting logically, directly or easily with it);
b. self-consistency (all the information potentially included in the plate is readable as a result of the availability of all needed specifications: size scale, location, manipulation and treatments in figures; variables, scale values in graphs);
c. meaningfulness (the figure is complementary to the verbal text, renders it clearer, integrates or probes it).

In addition, the same figures were screened in order to quantify the consistency of sub-samples with different functions, or – in other words – whose main objectives corresponded to the following three connotations:

a. decorative element/decorativeness (the "pretty" figure is superfluous; the relationship between figure and text is weak or absent);
b. support for attention or motivation (the figure, superfluous from an informational point of view, shows a pertinent object or event which is at the same time, familiar and emblematic);
c. substitute suggestion or introduction for direct observation and lab work (figure shows a pertinent object or event, which is not familiar in itself or due to some feature of presentation: enlargement, sectioning, schematisation, data).

Categories for experimental activities

In the same paragraphs of the above mentioned verbal text the form of the verbal presentation of the "experimental activities" – the proposed interactions with objects and events – were classified as:

a. assertion/definition, when the "interaction" is simply cited or summarily described ("Moreover cholesterol helps to maintain free and separate the phospholipid fatty acid tails. Cells experimentally cultured without cholesterol fail to adhere to each other since the plasmatic membrane is weakened", Curtis & Barnes 1996, see Appendix 1);
b. explanation, when it is punctually described and exhaustively analysed ("Finely chop the moth-balls (naphthalene), weigh carefully, scatter on the bottom of a large box and leave uncovered, exposed to the air for several days. Then weigh again. ... A quantity of the moth-balls passed into the gaseous state", Durante et al., see Appendix 1);
c. problematisation, when the reader is requested to get involved, to take a position on it ("How long can one last without breathing? Breathe deeply three times, then stop breathing for as long as possible, measuring time in seconds. Repeat the experiment three times and at different hours of the day and write down *the highest value you reach.",* Giorgi 1994, see Appendix 1).

In addition, the same "experimental activities" were analysed taking into consideration the typology of implied objects and events. They were identified as:

a. those from everyday life (see the examples cited in "verbal presentation", classes b and c);
b. those specifically used in learning and training activities, in the laboratory (see the example cited in "verbal presentation", class a);
c. those represented by models and schemes ("Using elastics, attach the rubber balloons to the ends of the plastic tube ... your simple model of the breathing apparatus is now complete." Bargellini 1992, see Appendix 1).

Categories for interactions

Finally, the typology of interactions with the above-mentioned objects and events were taken into consideration and classified as:

a. observation/recognition: description, identification, citation of concrete, usually well-known or easily imaginable objects or events ("The wet laundry on the line will dry in the sun and air after a short time." Fiaccavento & Romano 1996);
b. manipulation: alterations/changes of objects or events, which make observation easier or deeper (see the examples cited in "verbal presentation", classes b and c);
c. experiments: manipulation of objects or events planned on the basis of a hypothesis - in fact the core of scientific activity ("During the mating season the male stickleback's abdomen is tinged with red: the involvement of this colouring in inciting the aggressive reaction of the other males was demonstrated by their same effective reaction to simple shapes with red painted on their lower sections", Alberghina & Tonini 1994).

Results

Paragraphs: the forms of the verbal communication of scientific information

The classification of a total of381 paragraphs in lower and 337 paragraphs in upper secondary school textbooks clearly showed the overwhelming prevalence of assertions and definitions (Figure 1).

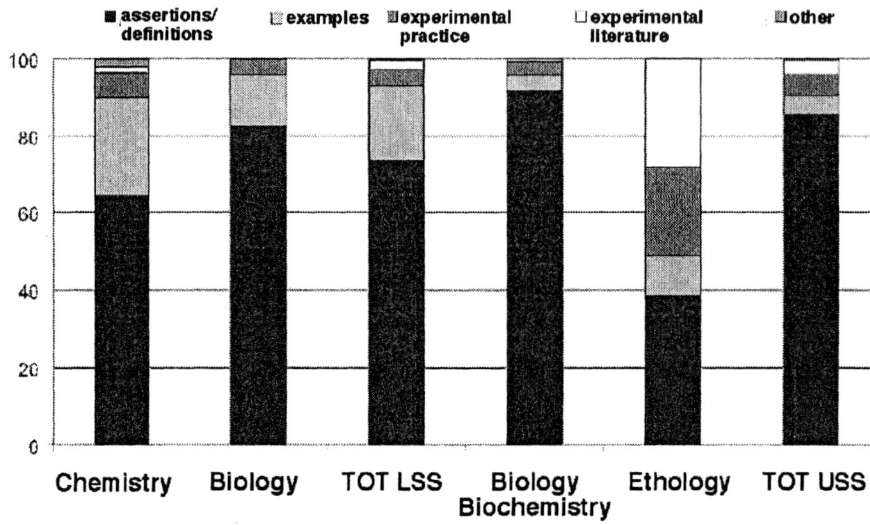

Figure 1: Classification of paragraphs in books according to their essential content [1]

The weight of paragraphs where examples are presented spans from 25% in the sub-sample concerning Chemistry in lower secondary school textbooks down to 5% in the sub-sample concerning both Biology and Biochemistry in upper secondary school textbooks, while that of paragraphs presenting experimental practices stays at around 5%. The relevance of experimental literature (historical or effected observations and experiments) is scanty, with only around 2 - 3% of paragraphs.

A notable deviation from the average represented by the entire sample is shown in the sub-sample concerning animal behaviour (Ethology); in fact, nearly every concept elaborated in the textbooks where this topic is present is substantiated by an example or, more often, by a detailed description of an emblematic or historical experiment, such as those by Lorentz, Spencer, Tinbergen, or Thorndike.

[1] Lower secondary school textbooks (LSS) and upper secondary school textbooks (USS) "TOT" refers to the entire sample (books and subjects), "Chemistry" to the treatment of changes of phase and chemical reactions (in all the LSS books), "Biology" to the treatment of breathing and sense organs (in all the LSS books), "Biology, Biochemistry" to the treatment of all the subjects except animal behaviour (in all the USS books), "Ethology" to the treatment of animal behaviour (in all the USS books). Data are given in percentages.

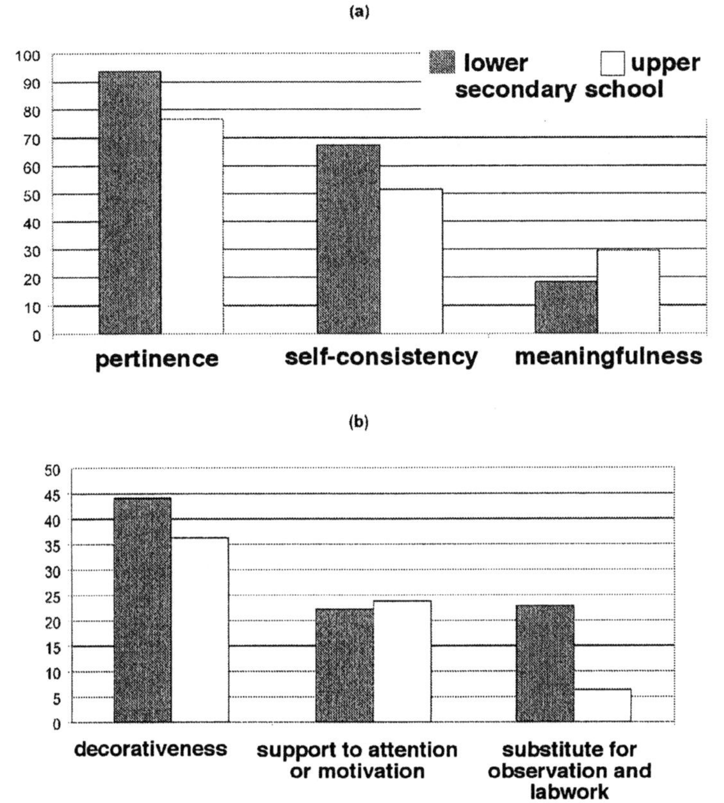

Figure 2: Categorisation of illustrations [2] (a) according to their main requisites (b) according to their main objective

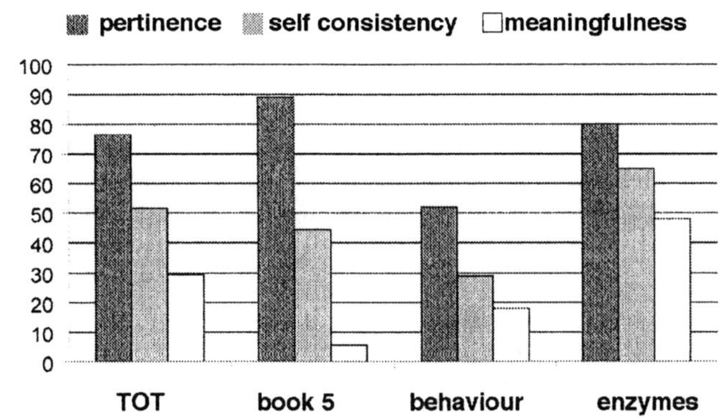

Figure 3: Categorisation of illustration in USS textbooks [3]

[2] Categorisation of illustrations in the selected excerpts of lower (559 illustrations) and upper (308 illustrations) secondary school textbooks. Data are given in percentages.

Figures: the function of illustration

559 plates for LSS and 302 for USS were inspected. The high percentage of pertinent plates must be set off against the lower percentage of self-consistent ones (67 - 52%) accounting for their lower usability, and with a very low percentage of meaningful ones, demonstrating their scanty rigour - and therefore their scant utility - on the level of information (18 - 29%) (fig. 2a). On both school levels nearly two figures out of five appear essentially decorative, and nearly one out of five supports attention without providing any further information regarding the presumed knowledge and experience of the reader (fig. 2b). Very rarely are figures aimed at substituting for observation or lab work or at supporting them.

Figure 3 demonstrates the range of variation in terms of the three parameters cited, comparing the general mean values with those of an individual textbook (no. 5 in Appendix 2) and those of two specific subject matters (enzymes and behaviour), which are characterised by the extremes in the range.

"Experimental activities"

The analysis concerns the presumable "experimental activities" that are presented the selected portions of textbooks (387 activities from LSS and 277 from USS textbooks). On both school levels verbal presentation of the interactions with objects and events is for the most part composed of assertions and definitions (63 - 71%) (fig. 4a). It should be pointed out that the majority of the residual type of presentation (explanation and problematisation) is given in lower secondary school only in an open form, i.e. within the framework of a dialogue with the reader: questions are asked and answers are delayed or not quite given. At the upper secondary level the disappearance of any kind of open exchange is accompanied by the extreme reduction in the number of objects and events related to everyday life (fig. 4b). Attention is paid and treatment refers mainly to "cultural" objects and events, which characterise or emerge from learning and training activities. The models and schemes, which are just as frequent, belong to this same category, but are unambiguously aimed at encouraging learning and at lowering the rate of abstraction and extraneousness.

The emerging type of interaction is predominantly that of identification and recognition (54.8% in LSS and 77.3% in USS; fig. 4c). The stimulus to deepen personal knowledge of objects and events through manipulation activities or experiments, or else through exploration of the applicative side of scientific knowledge, is drastically lowered in the passage from LSS to USS (from 45% to 22%).

Worth noting is the singling out of a number of incorrect experiments, whose description is misleading both from the methodological and the conceptual points of view (Bandiera 1999).

[3] According to their main requisites: comparison among the entire sample (total), one single book (no. 5 in Appendix 2) examined for all five selected subjects, and two exemplary subjects (enzymes and animal behaviour) examined in all the books. Data are given in percentages

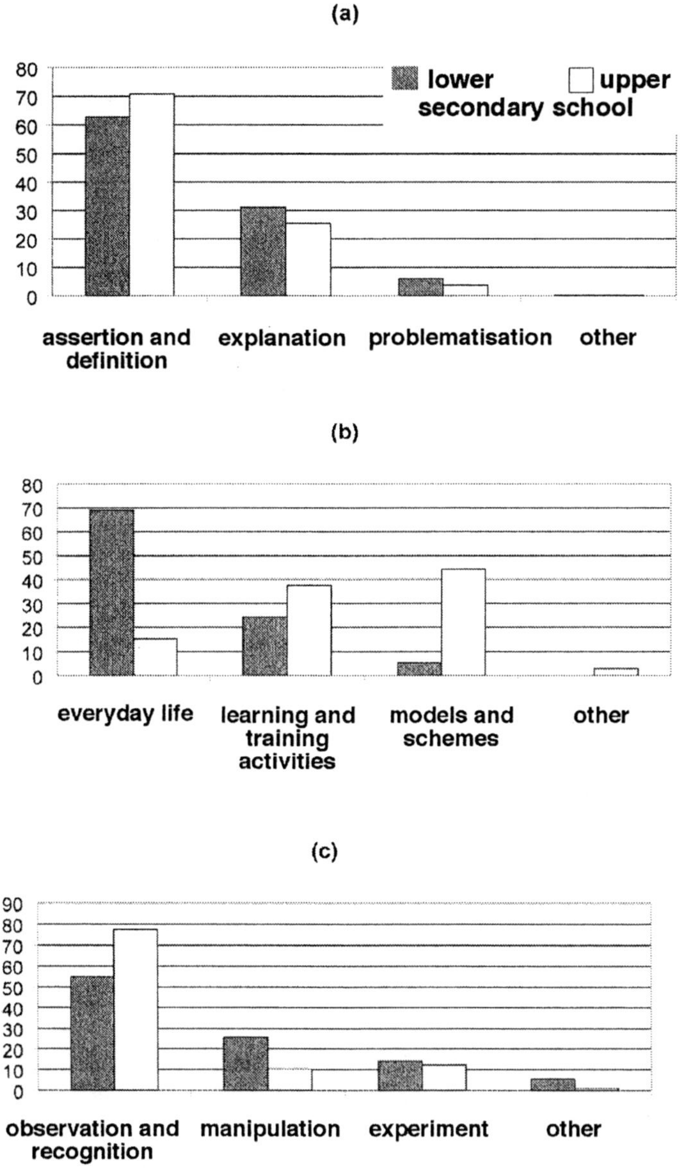

Figure 4: Classification of "experimental activities" [4] (a) Verbal presentation (b) considered objects and events (c) interaction with objects and events

---

[4] Presented in the selected excerpts of LSS and USS textbooks. Analysed according to the categories listed in "Sample and instruments". Data are given in percentages.

Conclusions

The discussion of the data presented cannot but begin with a comment on the extraordinarily wide availability of textbooks. These do not seem to differ in their more or less advanced didactic/methodological approach, or in their greater or lesser suitability to different curricular streams. Some differences were found with respect to the presence and length of chapters dedicated to particular subjects. But the aspect apparently treated with the most care and creativity is surely that concerning the editorial layout, which has already been noted and analysed (Westbury 1990; Woodward 1987): from type of articulation to multiple appendices (glossary, abstract, test, conceptual map), from relevance to visual impact of iconography, from paper quality to originality of jacket design. These features influence particularly the price, but do allow teachers to give students pretty textbooks, gauging their choice on their type and level of competence and on their willingness to commit themselves to didactic elaboration of the contents (verification of intelligibility, depth of investigation, synthesis, execution of laboratory sessions).

As far as science is concerned, textbooks have been charged – already and in eminent context (AAAS 1989) – with actually impeding progress toward scientific literacy, emphasising "the learning of answers more than the exploration of questions, memory at the expense of critical thought, bits and pieces of information instead of understandings in context, recitation over argument, reading instead of doing". They have been shown to be used by students mainly like a dictionary, to look up definitions that must be memorised for written and oral tests (Driscoll, Moallem, Dick & Kirby 1994). This perception agrees with the outcome of the presented analysis.

The Italian textbooks most used (not considering correctness and currentness of contents) share certain features, which are listed below, together with observations about the scientific knowledge and experimental work they induce:

1. the prominence of assertions and definitions (fig.1 and 4a), which would weigh on the image of scientific knowledge as a well-established and unquestionable set of information;
2. correlatively, the underestimation of doubts, open questions, problems in past and present scientific research processes, as well as in science learning, which would discourage a speculative attitude and personal involvement in all processes and, at the same time, indicate the prevalence of information and facts over ideas;
3. the very limited attention paid to the experimental and epistemological dimension (see examples, experimental practice, and experimental literature, fig. 1, 4a and 4c), especially in upper secondary school compared to lower secondary school, which would strengthen the above-cited image, missing the opportunity for an explicit reflection on science;
4. as far as illustrations are concerned, the choice of trivial and superfluous subjects, the careless explanations and the lack of the data necessary for understanding (fig. 2 and 3), which would actively oppose the structuring of a proper critical and investigative attitude;
5. the presentation of natural objects and/or events as a (pleasant or dutiful) side of the informative apparatus, due to the shortage of involvement strategies (fig. 4a and 4c), which would end up sustaining rote learning rather than illustrating

the role of experimental activities (observations and experiments) in constructing scientific knowledge.

Emblematic of and encapsulating this state of affairs is the poor store of historical, classical, crucial experiments. The surprising exception in the ethological field supports the hypothesis that in educational contexts experiments are held to be useful or necessary only where, as in the case of a "young" discipline such as Ethology, factual knowledge is not consolidated and generalisations are not sufficiently legitimised. On the other hand, "sure" knowledge may simply be communicated (along with examples), which accounts for the limited capacity for abstraction of lower secondary school students.

Finally, descriptors and categories which have been devised in order to analyse textbooks, and data which have been collected through them, permit the answering of the research questions with reasonable certainty, albeit with undeniable discouragement. In fact, the above listed points and respective comments clearly indicate that verbal presentation denies the experimental basis of scientific knowledge while avoiding frequent correlation between pieces of knowledge and concrete objects/events, tactics/strategies/plans of productive experiments, and promoting rote instead of reflective-learning. They suggest that presentation in general ignores the intersubjective nature of scientific knowledge and the lack of durability of science content, while widely preferring assertions/definitions over data/theory-based interpretations; that text-aids paradoxically oppose the foundation of a scientific attitude while excluding specificity and individuality of involvement, completeness and rigour of information, and relevance of contributions from the history of science.

Recommendations

Italian Biology textbooks seem to ignore the widely claimed assumption that scientific knowledge about natural objects and events should be marked by personal or reported experience of observations, experimental manipulations and experiments carried out with technical and methodological rigour, systematic reflection on the data, within an appropriate and explicit theoretical framework.

Although it seems quite impossible to couple such an experience with each piece of *information* communicated at school, the relevance of direct experience, of experimental rigour, of hypothesis and data discussion and of theoretical framing seem unquestionable. This appeals to the most specific features of scientific knowledge and is directly concerned with the promotion of attention to epistemological issues. This, in the absence of a specific education, is the least attainable goal and the one with which teachers should be most deeply involved. This should therefore represent the basic guideline for (planning and) choosing a textbook. Teachers should be mindful that textbooks do convey an image of science through indirect and direct signals, principally through coherence of language and reasoning, where involvement in observation and manipulation activities is, or is not, encouraged, and by means of the function given to experimental activity directly performed or quoted from different sources.

Consequently, some recommendations should be addressed to textbook writers as well. They should begin by adopting simple practical strategies. Some of these, mainly at the level of readability, have already been clear for some time now: these include the emphasis on the relationship between ideas (Meyer & Freedle 1979), on the linking words which help point out the logical structure of treatment (Meyer, Brandt & Bluth 1980), on open questions addressed to the reader (LaZansky, Spencer & Johnston 1987). As far as illustration is concerned they should take care to use an adequate iconographic language to communicate relevant and meaningful information (Kearsey & Turner 1999) and to assure the presence of systematic and appropriate links between pictures and written text (Harrison 1980; Wright 1981; Levie & Lentz 1882; Reid & Beveridge 1986; Reid 1990). At the same time teachers should be concerned with students' gradual and careful training in interpretation (Bluth 1981; Mathewson 1999).

Finally textbooks should favour presenting pieces of scientific knowledge (in particular those present within the current debate) as the best approach to the explanation of the objects and events involved, to connecting pieces of present knowledge with experiments and data on which they are based, to exploiting the educational resources of scientific practical work as defined by Kerr way back in 1963. In brief, readers should be urged to put themselves to the test, in the certainty that learning scientific knowledge today is a necessary requirement for getting involved in using and building scientific knowledge tomorrow.

References

American Association for the Advancement of Science (1989). *Science for all Americans.* AAAS, Washington, DC

Armstrong, H.E. (1903). *The teaching of Scientific Method.* Macmillan and Co Ltd, London.

Bandiera, M. (1999). Bone composition: laboratory mistakes, presumptions and events. In K. Nielsen. & A.C. Paulsen (Eds.) Practical Work in Science education: the Face of Science in Schools, RDSES, Copenhagen, 61-75

Bluth, L.F. (1981). The illustration-comprehension relationships: a review of the literature. *Reading and Improvement, 18*, 251-25

Driscoll, M.P., Moallem, M., Dick, W., & Kirby, E. (1994). How does the textbook contribute to learning in a middle school class? *Contemporary Educational Psychology, 19*, 79-100

Duveen, J., Scott, L. and Solomon, J. (1993). Pupils' understanding of science: description of experiments or a passion to explain? *School Science Review, 75*, 19-27

Glynn, S.M., Yeany, R.H., & Britton, B.K. (1991). A constructive view of learning science. In S.M. Glynn, R.H. Yeany, & B.K. Britton (Eds.) *The psychology of learning science*, Erlbaum, Hillsdale, NJ, 3-20

Gott, R., & Duggan, S. (1996) Practical work: its role in the understanding of evidence in science. *Int. J. Sci. Educ., 18*, 791-806

Harrison, C. (1980), *Readability in the classroom.* Cambridge University Press, Cambridge

Helgeson, S.L., Blosser, P.E., & Howe, R.W. (1977). *The status of pre-college science, mathematics, and social science education.* The Ohio State University, Center for Science and Mathematics Education, Columbus, OH

Jenkins, E.W. (1999). Practical work in School Science – some questions to be answered. In J. Leach & A.C. Paulsen (Eds.) *Practical Work in Science Education – Recent Research Studies*. Roskilde University Press, Frederiksberg, 19-32

Jiménez Aleixandre, M.P. (1994). Teaching Evolution and Natural Selection: A Look at Textbooks and Teachers. *Journal of Research in Science Teaching, 31*, 519-535.

Kearsey, J., Turner, S. (1999). How useful are the figures in school biology textbooks? *Journal of Biological Education, 33*, 87-94

Kerr, J. (1963). *Practical Work in School Science.* Leicester University Press. Leicester

LaZansky, J., Spencer, F., & Johnston, M. (1987). Learning to read: Setting students up. In R.J. Tierney, P.L. Anders, & J.N. Mitchell (Eds.) *Understanding readers' understanding: Theory and practice.* Erlbaum, Hillsdale NJ, 255-282

Leach, J. (1999). Learning Science in the laboratory. In J. Leach & A.C. Paulsen (Eds.) *Practical Work in Science Education – Recent Research Studies*. Roskilde University Press, Frederiksberg, 134-147

Levie, W.H. and Lentz, R. (1982). Effect of text illustrations: a review of the literature. *Educational Communication and Technology*, *30*, 195-232

Lumbelli, L. (1984). Per la diagnosi della comprensibilità. *Riforma della Scuola*, *5*, 23-34

Mathewson, J.H. (1999). Visual-Spatial Thinking: An Aspect of Science Overlooked by Educators. *Sci. Ed.*, *83*, 33-54

Meyer, B.J.F., Brandt, D.M., & Bluth, G.S. (1980). Use of top-level structure in text: Key for reading comprehension of 9th grade students. *Reading Research Quarterly*, *1*, 72-103

Meyer, B.J.F., & Freedle, R.O. (1979). The effect of different discourse types on recall. (*Tech. Rep. No. 6*). Department of Educational Psychology, Arizona State University

Moody, D.E. (1996). Evolution and the Textbook Structure of Biology. *Science Education*, *80*, 395-418

Musheno, B.V., Lawson, A.E. (1999). Effects of Learning Cycle and Traditional Text on Comprehension of Science Concepts by Students at Differing Reasoning Levels. *J. Res, Sci. Teach.*, *36*, 23-37

Nott, M. (1996). When the black box springs open: practical work in schools and the nature of science. *Int. J. Sci. Educ.*, *18*, 807-818

Novak, J. (1987). Human constructivism: toward a unity of psychological and epistemological meaning making. *Second International Conference on Misconceptions and Teaching Strategies in Science and Mathematics.* Cornell University Press, Ithaca, NY

Reid, D. (1990) The role of pictures in learning biology: part 2, picture-text processing. *Journal of Biological Education*, *24*, 251-258

Reid, D., and Beveridge, M. (1986). Effects of text illustration on children's learning of a school science topic. *British Journal of Educational Psychology*, *56*, 294-303

Rigutti, M. & Santaniello, M.A. (1985). Esperimenti come ricette da cucina. *Riforma della Scuola*, 9/10, 57-63

Skoog, G. (1979). Topic of evolution in secondary school biology textbooks: 1900-1977. *Science Education*, *63*, 621-640

Stake, R.E., & Easley, J. (1978). *Case studies in science education.* Vols I and II. Center for Instructional Research and Curriculum Evaluation, University of Illinois at Urbana-Champaign, Urbana, IL

Sutton, C. (1996). Beliefs about science and beliefs about language. *Int. J. Sci. Educ.*, *18*, 1-18

Swarts, F.A., Anderson, O.R., Swetz, F.J. (1994). Evolution in Secondary School Biology Textbooks of the PRC, the USA, and the Latter Stages of the USSR. *Journal of Research in Science Teaching*, *31*, 475-505

Westbury, I. (1990). Textbooks, textbook publishers, and the quality of schooling. In D.L. Elliot, & A. Woodward (Eds.) *Textbooks and schooling in the United States: 89th yearbook of the National Society for Education*, University of Chicago Press, Chicago, 1-22

Woodward, I. (1987). Textbooks: Less than meets the eye. *Journal of Curriculum Studies*, *19*, 511-526

Woodward, A., & Elliot, D.L. (1990a). Textbooks use and teacher professionalism. In D.L. Elliot, & A. Woodward (Eds.) *Textbooks and schooling in the United States: 89th yearbook of the National Society for Education* (Part I), University of Chicago Press, Chicago

Woodward, A., & Elliot, D.L. (1990b). Textbooks: Consensus and controversy. In D.L. Elliot, & A. Woodward (Eds.) *Textbooks and schooling in the United States: 89th yearbook of the National Society for Education*, University of Chicago Press, Chicago, 146-161

Wright, P. (1981). Tables in text: the subskills needed for reading formatted information. In L.J. Chapman (Ed.), *The reader and the text*, Heinemann Educational Books/United Kingdom Reading Association, London, 60-69

Wright, J.D. (1982). The effect of reduced readability text materials on comprehension and biology achievement. *Science Education*, *66*, 3-13.

Appendix 1 - Lower secondary school textbook sample

1. Fiaccavento, G., Romano, N.: "Dal perchè alla legge" Fabbri, Milano, 1996	(79)
2. Gori Giorgi, C.: "Corso di Scienze" Zanichelli, Bologna, 1994	(48)
3. Bertini&Danise&Franchini: "Dire, fare, conoscere le scienze" Mursia, Milano, 1995	(44)
4. Longoni, E.: "Dagli atomi all'uomo" La Nuova Italia, Firenze, 1994	(43)
5. Bargellini, A.: "Le vie della Scienza" Signorelli, Milano, 1992	(39)
6. Vantaggio, D., Febbraro, F.: "1,2,3... Le Scienze" Signorelli, Roma, 1996	(32)
7. Alfani, Di Bernardo, Palumbo: "Corso di Scienze" B.Mondadori, Torino, 1992	(29)
8. Durante et al.: "Introduzione alle scienze sperimentali" Le Monnier, Firenze, 1980	(27)
9. Rovelli, E., De Capitani, R.: "L'ARCA" Principato, Milano, 1995	(27)
10. Colombi et al.: "I grandi temi delle scienze naturali" Il Capitello, Torino, 1996	(26)

Appendix 2 - Upper secondary school Biology textbook sample

1. Postlethwait&Hopson&Veres: "Biologia" McGraw-Hill, Milano, 1991, tr. 1992	(113)
2. Curtis, H., Barnes, N.S. : "Invito alla Biologia" Zanichelli, Bologna, 1996	(101)
3. Biggs et al.: "Biologia: la dinamica della vita" Zanichelli, Bologna, 1991, tr. 1994	(61)
4. Casagrande et al.: "15 moduli per lo studio delle scienze della natura" Bovolenta, Ferrara, 1996	(48)
5. Gainotti, A., Modelli, A.: "La biologia: diversità e unità della vita" Zanichelli, Bologna, 1995	(35)
6. Zullini, A., Sparvoli, F.: "Biologia: dalle molecole all'ecosistema" Atlas, Bergamo, 1994	(29)
7. Biggs et al.: "Lineamenti di Biologia" Zanichelli, Bologna, 1991, tr. 1995	(23)
8. Purves, W.K., Orians, G.H.: "Corso di Biologia" Zanichelli, Bologna, 1992, tr. 1995	(21)
9. Mazzoni&Pirone&Cerofolini: "Biologia e Scienze della Terra" Archimede, Cuneo, 1995	(20)
10. Alberghina, L., Tonini, F. : "Scienze della Natura: l'ambiente e i viventi nel sistema Terra" A. Mondadori Scuola, Milano, 1994	(19)

- The number of school sections adopting each textbook is given in parenthesis.
- The list of textbooks on the market and their frequency of adoption have been supplied by the "Organizzazione Provinciale Confesercenti".

Ten of the textbooks rated at the same time on the supplied list had the same authors and publishers as textbooks listed in higher positions and resembled them closely, and were, therefore, not included in the analysed sample. These textbooks were used in a total of 371 cases and substantially raised the representativeness of the sample: to 56.6% (lower) and 74.9% (upper secondary school).

Note: Italian students make use of one inclusive book ("Physical, Chemical and Natural Science") in lower secondary schools and single-discipline books in upper secondary schools. National science curricula recommend experimental activities, direct observations, experiments. Nevertheless, practical work is, in fact, totally optional except for laboratory courses in vocational schools. Non-vocational schools seldom have a laboratory; the existing ones are mainly equipped for physics and sometimes chemistry experiments; in old, *historical* Institutes collections of biological specimens are kept.

Chapter 3

OPEN-ENDED LABWORK

Introduction

This chapter includes studies on open-ended labwork in which students are required to make some decisions for themselves as to how to act in various types of projects. In order to carry out an open project in an autonomous way, students have to draw upon knowledge and understanding of scientific content and experimental procedures and, frequently, exhibit sophisticated positions regarding the relationship between knowledge claims and experimental data. The presented studies focus mainly on aspects of the understanding and use of scientific procedures on the part of the students, as well as on the improvement of their epistemological knowledge when engaged in investigative work. As a matter of fact, epistemological issues related to labwork have only recently become an object of investigation by researchers. In the present chapter special conceptions of developing epistemological reflections through various forms of projects at university level are discussed, from various perspectives, in all three studies.

We may note that, as in other chapters, all three studies are parts of wider projects, which have been running for years. One study (Guillon and Séré) concerns research-based innovative labwork, while the other two (Lewis and Ryder respectively) investigate existing laboratory courses. The effectiveness of these courses with regard to students' acquisitions after labwork is the main focus of these three studies, rather than what they actually do during labwork, as in Chapter 2.

Guillon and Séré attempt to model the procedures physicists use in investigative work, conceived as a confrontation between a variety of theoretical models and experimental data. The epistemological analysis of scientific procedures carried out by the authors adds to our knowledge of the nature of the procedures from a didactical perspective. The study involves an explicit presentation, to first year physics undergraduates, of epistemological information related to investigation processes and strategies, followed by sequences of labwork to make students familiar with different procedures and a number of open-ended projects within a two-year course. The effectiveness of this strategy is discussed by the authors who,

D. Psillos and H. Niedderer (eds.). Teaching and Learning in the Science Laboratory. 119-120.

among other findings, point out that students who were familiar with "one experimental method" to a certain extent found strange the use of a variety of models and strategies during investigative work. In their recommendations the authors stress the need for a combination of conventional and open–ended work, with clearly set objectives, as a means of improving student autonomy.

The second study (Lewis) concerns a specific type of open-ended labwork: the mini projects which are sometimes included in undergraduate courses with the expectation that they will help students make the transition from set practicals to open-ended investigative work. In effect, this Lewis raises the issue of the transition from conventional labwork to open-ended investigations, which seems to be important, as the study by Guillon and Séré suggests and which the next study (Ryder) discusses in a different context. Despite the noted success in developing students' understanding of the nature and processes of scientific research, the mini-projects left many students feeling demoralised and largely unaware of the learning which had taken place. The need for clear objectives clearly set by the tutors is stressed by the author in his recommendation for improving the effectiveness of the mini projects.

In the last study (Ryder), students are taken out of the laboratory to carry out residential field work in geology, at the intention being to develop students' ability to interpret geological data. The learning aims of such a field course included sophisticated epistemological activities such as developing alternative interpretations of a single data set and comparing and evaluating multiple interpretations. The author argues that the field course and particularly its residential nature were effective in getting students to engage with the intended epistemological issues and develop personal interpretations of data instead of simply looking for the correct answer, which is considered as a manifestation of naïve realism.

The Role of Epistemological Information in open-ended Investigative Labwork

Alain Guillon, University of Cergy – Pontoise, France

Marie-Geneviève Séré, University Paris-Sud XI, France

Abstract

At university level, to become a physicist, students have to learn not only about the content of physics but also about strategies of investigation. We developed a two year sequence of labwork to introduce epistemological knowledge, and to allow students to go through an entire investigation strategy in open-ended projects. This study describes the epistemological knowledge passed on to students at undergraduate level, followed by sequences of labwork to make students familiar with different procedures, and open-ended projects. Questionnaires at different times within the sequence assess this teaching sequence. We also analysed the two reports made by each pair of students involved in a project.

Introduction

Teaching epistemological knowledge about physicists' processes through labwork

The aim of this study was to teach epistemological knowledge at undergraduate level, through a teaching sequence involving labwork. 'To help students to get to know scientific processes' is frequently stated as the main aim of labwork. As Hodson (1992) wrote:

> *Though necessary, conceptual knowledge and knowledge about procedures that scientists can adopt (and may have adopted in particular circumstances in the past) are insufficient in themselves to enable students to engage successfully in scientific inquiry. That ability is only developed through hands – on experience of doing science in a critical and supportive environment*

In this study we use the term 'scientific processes' to describe the different strategies of investigation used to answer questions about phenomena. How do physicists use models and theories? How do they carry out experiments? How are data collected and interpreted? How do physicists judge the fit between experimental data and models? We include all these aspects in the following expression: the processes and strategies of investigation of the physicists.

This type of knowledge has an epistemological dimension and, as such, has been conceptualised by several authors. For instance, in line with a tradition from Francis Bacon & Claude Bernard, Develay (1989) describes the experimental process as follows:

> *To word a question, to set out hypotheses, to test hypotheses by designing experiment, to carry out experiment, to analyse the results, to give an interpretation.*

However, if this sequence is considered as chronological, it does not take into account the frequent switches back and forth between theory and practice.

Another conceptualisation is proposed by Gott & Murphy (1987) who consider science as a problem solving activity, but say little about the variety of scientific processes and the roles of models. Vicentini also recognises the need for metareflection about processes and proposes an organisation of an experimental

---

D. Psillos and H. Niedderer (eds.). Teaching and Learning in the Science Laboratory. 121-138.

science starting from 'The world of events and phenomena' and implying loops linking empirical referents, empirical laws, primary models, secondary models and theories (see Vicentini 1994 and Buty, Chapter 5 *of this volume*).

We first developed an original conceptualisation of physicists' work, and, second, we realised a complete teaching sequence based on it, including tools of assessment. The first step was based on a study of physicists' scientific publications as well as some interviews with engineers and researchers (Guillon 1996). This resulted in a general framework, destined to be taught (see paragraph 2), which identifies four different 'ideal' basic processes represented in Figure 1. For each of them, the type of model employed is different.

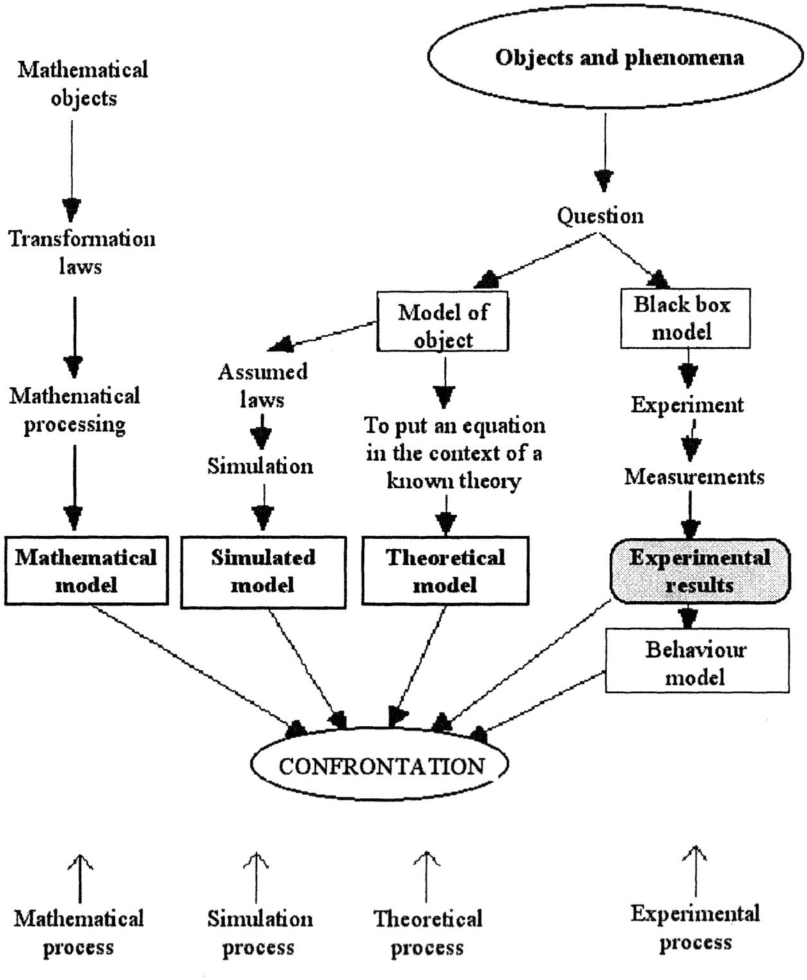

Figure 1: Four 'ideal' processes existing in physicists' activity. Real investigation strategies are combinations of them

This schema excludes some existing approaches (analogy, trial and error, etc.), since it selects the investigation strategies that we wanted to teach because they are combinations of basic processes. It means that we distinguish between *Behaviour Model, Model of Object, Simulated Model, Black Box Model and Theoretical Model*[1] (the fourth 'ideal' process, on the left, implies purely mathematical models and has not been explained in depth during the teaching sequence).

For instance, in an *Experimental process*, data are measurements carried out to answer a question. A part of the experimental device can be considered as a black box, the choice of it being a construction. For this reason, the term *Black Box Model* is used. The mathematical relation between output quantities and (controlled) input quantities is called the *Behaviour Model* (Trigeassou 1988) (Walter & Pronzato 1994). An example has been chosen for the sake of teaching: students were given the chronophotography of a golf ball thrown upwards in a vertical plane. Here, the Black Box is the golf ball, the input quantity is time and the output quantities are x and y co-ordinates of the ball. The experimental dots [(x,t) , y(t)] are obtained from the chronophotography. The linear function giving account of x(t) and the binomial function giving account of y(t) are the *Behaviour Model.*

A *Theoretical process* is described in the following way by Bunge (1973):

> *A theoretical model always includes a schematic representation of the real system under consideration; this representation is often called model of object.*

In the example of a tennis ball, the *Model of Object* of the ball (real object) is a material point under the influence of two forces: its own weight and air friction force, given for instance by f = -kv_ (alternative expressions exist). The *Theoretical Model* is made of two 2 differential equations obtained by applying Newton's theory:

$$\ddot{x}=-\frac{k}{m}\dot{x}\sqrt{\dot{x}2+\dot{y}2} \quad ; \quad \ddot{y}=-g-\frac{k}{m}\dot{y}\sqrt{\dot{x}2+\dot{y}2}$$

The *Theoretical model* can be either the equations themselves or the analytical solutions when they exist or the two graphs x = f(t) and y = g(t) obtained by numerical solving, when they do not exist.

The *Simulation process* is more and more frequent nowadays, mainly in cases when there is no coherent or complete theory, or when there are too many parameters in the system under study. For instance, a physicist studying interactions between atoms or molecules has to choose between different potentials of interaction. Starting from a theory, he/she makes successive approximations and obtains a system of equations solved numerically by a computer. *The Simulated Model* is the result of this calculation, or the corresponding graphs.

Any strategy represented in this schema passes through a central step of *confrontation* between experimental data and models, or between models. Several confrontations may be necessary in a real strategy of investigation. Their role is central in the sense that it determines the following steps of the work: new experiment, theoretical study, choice of another model or another process, etc. We established a typology of confrontation functions (with a reference value, playing

[1] In this study the different types of processes and models as used in the teaching will be written with capital letters at the beginning.

the role of validation, resulting in a model) and we listed all the tools used for confrontations (Guillon 1996).

This is the epistemological analysis of the physicists' investigation approaches that we wanted to teach at undergraduate level. This new curriculum was expected to have positive effects on students' comprehension of the links between the different steps in a complete investigation strategy, particularly between theoretical elements and experimental results. The curriculum also provides specific teaching on different methods used in physics concerning measurement and data processing.

Teaching approach

The teaching sequence was aimed at teaching epistemological knowledge as analysed in the introduction, teaching the corresponding tools through labwork, and finally applying it to open-ended labwork, namely projects, in order to enable students to use epistemological knowledge when necessary. Learning epistemological knowledge cannot be instantaneous. It requires time to be understood, experienced and used. It requires internalisation. For this reason the teaching sequence was spread over two years[2].

It was implemented, observed and assessed at Undergraduate level at the University of Cergy – Pontoise (students aged from 18 to 22 years participating in the course: Science of matter). The mark obtained for the whole sequence, represents 27% of the global mark of physics in the first year and 33% in the second year.

According to our basic assumption concerning the role of epistemological knowledge in labwork, the whole sequence[3] comprised various types of teaching, in three sections.

During the first semester, a general epistemological framework is provided. The aim is to make students aware of the different investigation strategies undertaken by physicists and of the different models that they use. During the first two sessions (experiments in mechanics) students carry out different strategies involving various models and processes. The introduction of epistemological knowledge takes place at the end of the second session, during one hour. At the beginning, the teacher discusses the structure of the first two sessions with students. The notion of 'strategy of investigation' is introduced from the comparison between the steps of the previous experiments. This gives the students insights into processes and models using examples. A text with proper vocabulary, definitions and explanations is given to students: every model and process is introduced and the general framework of Figure 1 is presented, discussed and applied. Finally, some insights are provided into the historical and social aspects of physicists' work.

The second section comprises eighteen labwork sessions, spread over the next two semesters and the beginning of the fourth semester. They are intended to teach the different tools associated with the confrontation step. In doing so, the

---

[2] In France, students are admitted to University when having passed the baccalauréat, at the end of Upper secondary school. After two years they aim at the diploma called DEUG (Diplôme d'Études Universitaires Générales). A proportion of students need a third year in order to graduate.

[3] The whole sequence that we describe was specifically organised as part of the research study. Years later, it is still being taught, with some adjustments as a result of our research.

epistemological structure presented in the first section is constantly referred to. Computers are used in nine sessions to collect experimental data, to analyse data, to search for an appropriate model, to simulate phenomena and for statistical data processing. The labwork sessions last four and half hours during the first year, and four hours during the second year. Students have no preparation at home; the text-book is given at the beginning of each session and all students work on the same experiment during the same session. They work in pairs within a group of ten to twelve pairs, a professor or a lecturer being in charge of the whole group. They have to write a report during the session. The experiments are related to subject matter dealt with in the courses and lectures: mechanics and electricity during the second semester, waves and optics during the third semester.

The sequence ends with open-ended projects undertaken within seven sessions over a period of three months. Ten to twelve pairs of students work under the responsibility of one professor or lecturer in a laboratory dedicated to teaching, rather than in a research laboratory. At the beginning of the fourth semester, a list of physics problems to be solved by experiments is given to students. For about half of the problems a computer is necessary to collect and analyse data. Having chosen one problem, each student pair writes a report and gives a short talk on the planned experiments: we call it *'The first version v1 of the protocol'*. During the next five sessions students carry out the experiments. During the sessions, teachers endeavour to answer students' questions by other questions, to help them to go further by themselves as much as possible. Finally, at the end of the semester, students write a report and present it orally. *'The final version of the protocol'* leads to a discussion with teachers.

Research questions

Consistently with what has been said, the main aim we assigned to this research concerned teaching scientific processes and offering students the opportunity to undertake such activities. We wanted to measure the effectiveness of the epistemological analysis taught during the first period, associated with the different sections implemented during the first two years of university. We were faced with the problem of assessing the impact of epistemological teaching. In this aim, we distinguished two aspects, as set out below.

Students were required to learn declarative knowledge. It was possible to assess what they have learned through written questions concerning the notions themselves (the various processes, the various models, etc.), but also the application of these notions to particular experiments (as carried out during labwork sessions). A part of the assessment was to check the correct use of the specific vocabulary, though we attached no intrinsic value to the vocabulary acquisition.

Students were also required to engage in project work. We wanted to check if the taught processes and strategies are relevant to addressing the physics problems we proposed. This was not done a priori, but by considering students' practice. We observed which processes they carried out and if they are described by the epistemological knowledge we gave. If it is the case, it demonstrates that this knowledge is useful for students and that projects are really opportunities to use it and internalise it. A further question was to assess if students used the epistemological knowledge provided several months before. With this question, we encountered the difficulty of assessing epistemological knowledge used (rather than articulated) to make choices, decisions and judgements about methods, confrontation, conclusions, and so forth. Nowadays, the need for an understanding of epistemology amongst students is recognised, named alternatively as acquiring a proper 'image of science' (Driver, Leach, Millar & Scott 1996; Leach, Millar, Ryder, Séré, Hammelev, Niedderer & Tselfes 1998) or an 'experience of science' (Woolnough 1989), or 'scientific methods' which are sometimes considered by authors as impossible to be taught (Ntombela 1999). The challenge of this study is to teach and to assess epistemological knowledge, compared with what has been taught, with different tools. In previously published work, the tools of assessment generally do not address the effectiveness of a given teaching sequence. Rather, they intend to assess epistemological knowledge independently from teaching.

In order to assess the two year teaching sequence, we addressed two sets of research questions:

The first set concerns *learning* physicists' processes:

1.1 How did students learn the epistemological knowledge of the first sessions?

1.2 Are they able to recognise and give the correct name to the different phases of their own project as they conducted it?

The second set concerns *using* physicists' processes during projects:

2.1 Are the taught processes and strategies put into operation? How did students make choices of the experiment, setting-up, measuring instrument?

2.2 How did they carry out judgements at the step of confrontation and how did they conclude their study?

2.3 Which are their difficulties? Do they stem from a lack of epistemological knowledge?

The first set of research questions concern the way students articulate epistemological knowledge, at the beginning and at the end of the sequence. By contrast, the second set of research questions concern their action and personal decisions, as well as the link between knowledge and action, at the end of the sequence. What happens in students' mind in the meanwhile is not possible to describe. The learning processes themselves are not accessible. Nevertheless, we expect these two sets of research questions to provide insights into the effectiveness of this framework for students, as well as identifying areas for possible improvement.

Research methods

Our study functioned as 'action research' implying feedback and interactions between research issues and decisions in teaching approaches. It also functioned as a 'case study' referring to a real context, namely the existing labwork curriculum set up in the University of Cergy-Pontoise. Consequently the data were collected from all students and not only from a restrictive 'experimental' group.

According to research question 1.1, epistemological knowledge is checked immediately after having provided the framework of physicists' strategies (short-term assessment). For the other research questions, the three assessment tools focus on the projects, i.e. the last teaching section. These two assessment activities were separated by 18 months.

Short-term assessment

A written questionnaire (denoted Q, and presented in Appendix 1) was implemented at the end of the third session. The questions concern the specific situation studied during this third session: large oscillations of a heavy pendulum with air friction. Students have to identify their own processes during the session, the different models, the input and output quantities and the parameters of the models. The questionnaire is not aimed at testing memory, since students may consult the textbook, particularly the text defining processes. Analysis of the 224 questionnaires Q collected involved the categorisation of 4 (out of 5) open questions.

Medium-term assessment

Three sources of written information were available: *'the first version v1 of the experimental protocol'* (written report during the course of the project), *'the final report'*, and a post-questionnaire (denoted PQ, and presented in Appendix 2). Each pair of students writes a report corresponding to a single project. The questionnaire PQ was completed by individual students.

The three sets of data provide information, to a certain extent, on what has been done by each student pair. The two versions of the reports have been analysed using a common analysis grid. The different items are as follows.: What is the strategy? What processes are used? How is the report structured? Is the step of confrontation expected, and what is its nature? What are the tools used? Are the links between several steps explicit? We focused mainly on the two stages of protocol development and confrontation; stages particularly fruitful in learning about the processes and strategies of investigation.

The three sets of data provide information on the relevant use of vocabulary introduced in the first teaching section. In the reports, we can see students' own use of vocabulary. Questionnaire PQ also gives insights into how students match their actions to specific terms presented to them (See PQ1, PQ6 and PQ10 in Appendix 2).

In fact the short-term assessment and the medium-term assessment were undertaken in the same year. This means that the two samples were made of totally different students (first year university students responded to questionnaire Q, and second year students responded to questionnaire PQ). Second year students are less numerous, because every student follows the first teaching section, whereas only students majoring in physics perform the totality of the curriculum and specifically projects. Thus, there were only 63 students involved in 32 projects. We received 30 final written reports instead of 32. The PQ questionnaire was given at the very end of the year and students were asked to mail it back. We only received 38 post-questionnaires representing 30 projects. The small size of the sample means that our study is a case-study providing careful observations which permitted progressive improvement of research questions as well as of teaching of physicists' methods.

Results

The first section concerns the first set of research questions about learning epistemological knowledge. As said before, it utilises data at different times: Q, PQ and the two successive written reports.

The second section concerns the second set of research questions, namely how students manage during projects, in order to check which processes, strategies, models they use. It gives some indication about what epistemological knowledge has been internalised by students, in other words, used as a metaknowledge. It employs the three types of data arising from the projects.

Assessment of epistemological knowledge learning

We present a selection of results concerning three aspects of epistemological knowledge: processes and strategies, models, confrontation.

About processes and strategies

At the end of the first semester, just after having presented the epistemological framework, we posed a question about a process, which has just been carried out:

Q1 – Which adjective do you associate with the process followed during the last laboratory work session?

Theoretical – experimental – mathematical – simulated

[The correct answer is *Experimental Process*]

- A small majority of students (58%) identify correctly an Experimental Process.
- Other answers involved all the other processes: Simulated 19%; Theoretical 9%, Mathematical 7%.
- [Multiple answers 5%, no answer 2%]

At the end of the sequence, 18 months later, we posed a rather differently formulated question in the post questionnaire. The question PQ6 (See Appendix 2) is supposed to provide information about students' consciousness of the different steps of the physicists' processes. In this question, a list of 30 'actions' involved in any scientific process is given in no logical order. This means that students are not tested on memory of the specific terms. They are required to select which of these actions they think they performed and to classify them in chronological order. Of course, each project is original. No comparison is made between students.

The action *'to do a theoretical study'* is selected 32 times out of 38.[4] Thus students are highly conscious that the theoretical elements have a critical role in all experimental work. The selection of other items shows that few of them are able to distinguish the various roles of theory, as taught. Moreover, for some of them, 'theoretical study' has the superficial meaning of calculations, formulae, mathematical demonstration, etc. This explains the discrepancies with what students really achieved: 27 student pairs put into operation a theoretical process, associated or not with an experimental process.

[4] The action 'documentary research' is selected 35 times out of 38. 27 students found an experiment, 8 a measurement method, and 23 reference values. Projects gave the opportunity of a step of research, unusual in labwork sessions, namely documentary research.

From these two results, we conclude that the importance of theory (the word being possibly interpreted inappropriately) is recognised. The exact distinction between the different processes seems to be acquired by a small majority only.

About models

In our description of physicists' processes, a particular type of model is associated with each process. We attempted to understand how students distinguish the different models *at the beginning.*

Q5 - Concerning the system you just studied, state the Model of Object you used to support the theoretical study.

[The correct answer is: *a material point subject to the forces of gravity and friction*]

- Here, the rate of 'no answer' is higher than in other questions (33%), which probably means difficulties in understanding.
- Moreover the rate of correct answer is very low (11%), among a wide variety of answers.
- The main confusion (33%) is to give the name of model to the object itself (the pendulum). For 20%, a harmonic oscillator is a Model of Object. For them, probably, it is no longer a model if a friction force is added.
- 10% attribute the same type of model to objects which can be modelled mathematically by the same differential equations (for the pendulum, the 'model' is said to be a spring or a RLC circuit).
- 7% use an analogy by quoting an everyday object like a clock. To these students the word 'model' appears to be understood as a copy; an idea far from what has been taught. The pendulum in the laboratory appears as a simplified reproduction of more complex objects from everyday life.
- 7% give a formula or an equation. They probably have in mind a Theoretical or a Behaviour Model but not a Model of Object.

Turning to the 30 *final reports* of projects, the words 'model' or 'modelling' occur 6 times only. This is disappointing, since it shows that students did not apply spontaneously the vocabulary they had been taught.

Question PQ1 of the post questionnaire provided similar results, in a less spontaneous context.

PQ1 - Put a cross in the box corresponding to the type of model you used during your work. Which was your model?
The proposed models are:
Theoretical Model, Behaviour Model, Model of Object
[the exact formulation is provided in Appendix 2].

Frequently, the descriptions of the models given by students are inadequate. For instance, 13 students thought 'they had used a Behaviour Model' but 10 of them identify the model as the object itself, such as in the following answer: *'the model I used was a square coil'.*

- 16 students believed *'they used a Theoretical Model'*, but only 7 did so.

The question PQ6 addresses the same problem by giving even more hints (the proper vocabulary as described in paragraph 4) to make students establish a link between what they did and the epistemological information provided. When responding to PQ6, all students recognise that they had used a model. A majority (32/38) specified the model they used as Theoretical.

From these three types of data, it can be concluded that many students are unable to match each model to each type of process appropriately.

About the step of confrontation

At the beginning of the curriculum, identifying the nature of the confrontation appeared difficult for students. This is shown by the high rate (22%) of no response to question Q4:

Q4 – During the session today, a confrontation between theory and experimental results has been completed.
In which of the two parts of the session, has the confrontation occurred?
What is the relation associated with this confrontation?

- 42% gave a correct answer to the second part of the question.
- Many students (26%) identified the relation with the mathematical relation of the Behaviour Model.

Turning now to the projects, several months later, the words 'confrontation' and 'to confront' appeared explicitly in the final reports, in the titles of paragraphs or in comments. In question PQ1, a majority of students thought they had performed a confrontation.

- 15 between a Theoretical Model and experimental results;
- 4 between a Behaviour Model and experimental results;
- 13 between a Behaviour Model and a Theoretical Model.

Though PQ1 shows that students are highly conscious of having performed a confrontation, PQ6 shows that it is difficult for them to identify the sort of confrontation.

Taking into account the possibility of multiple answers in PQ1, these two questions imply that 32/38 students were conscious of having performed a confrontation, even though all of them did. But only 17/38 students were able to characterise it.

These results show that students encounter difficulties in identifying models, in recognising the nature of a confrontation, and to a less extent in identifying processes. They are puzzled by the fact that words like 'experimental' or 'object' or 'behaviour' may be associated with the word 'model'.

Our first results, namely difficulties encountered by students to put in words epistemological knowledge, are below complemented by what students do when they have to undertake an investigation. By the three sets of data concerning the open activity which closed the sequence, our intention was to assess if

They know more than they are able to say (Woolnough & Allsop 1985)

The achievement of projects

This paragraph addresses the second set of research questions, and utilises data from the projects only. The aim is to make clear to what extent the description of processes and strategies we taught, fits with students' practice during the projects. In other words, we wondered if projects were good opportunities to apply and recognise epistemological knowledge. The first section reports how students started and developed their strategy. The second section addresses the step of confrontation, which leads to conclusions. Quite frequently, we will describe students' difficulties.

This is because these negative aspects are more evident, and are also relevant to the further improvement of teaching.

Positive aspects are shown by students' satisfaction at the end of the two years studying epistemology and projects. Answering PQ8, only two students stated that they learnt 'nothing new'. Responses to question PQ10 (open question, as shown in Appendix 2) show that the sequence was highly appreciated. Some are 'reconciled' with physics. Some evoke a personal attitude towards physics. The aspect of a personal involvement is the most frequent idea expressed.

The development of the experimental set up

The data are derived from the final reports. Generally speaking, when students develop an experimental set up, they have many choices to make: a strategy, the quantities to be measured, the experiment itself, the measurement method, the type of data processing. All these choices have an influence on one another and also on the step of confrontation. The first choices must be studied in order to understand which epistemological resources and knowledge would be necessary.

a. Choice of the first process and strategy

In fact, when starting the work, students are strongly influenced by the availability of theoretical results on one hand, and the familiarity they have or they have not, with the objects presented in the problem on the other hand.

The different strategies chosen in the 30 projects were the following:

- Theoretical Process - experiments - confrontation (22)
- Experimental Process - Theoretical Process - confrontation (5)
- Experimental Process - confrontation (2)
- Simulation Process - theory - simulation - experiments - confrontation (1)

This list shows that all the strategies described during the first sessions have been put into operation, in at least one project. A large majority of the strategies start with a Theoretical Process. This is not surprising because searching for information often begins with academic textbooks.

For instance in project XVII (How do the different parameters influence waves in the water of an aquarium?) the objects were familiar and the associated conceptual knowledge, learned in the previous semester, not very difficult. Thus students could start with an experimental process after having identified the two parameters to be studied. As early as the introduction itself, they wrote:

The formation and progress of these waves can be modified by various physical parameters:

- *the depth of water in the aquarium*
- *the pulsation of the source originating the waves*

In this project, students completed a Theoretical Process rather late, when they came to interpret their results. A preliminary overall view of the strategy they adopted would probably have allowed them to go straight on to a more relevant data analysis. However, this was not considered because of the apparent familiarity of the problem.

Some choices seem to be made at random or under the pressure of purely practical constraints.

b. Choice of the quantities to be measured and of the corresponding experiment

- Difficulty in using the Black Box model.

We expected that, when the quantities involved were not given explicitly in the wording of the problem, this would be the opportunity to use the Black Box Model as well as the definition of input and output quantities. In fact the choice of the quantities to be studied was guided by what they have learnt previously, or by a text book, or by a documentary paper on an apparatus, or by a preliminary theoretical study.

For instance in project I about the Holweck-Lejay pendulum, students found a formula in a textbook:

$$g = (2kd^2 T^2 - 4\pi^2 MD^2)/MDT^2$$

(k, d, D, M are characteristics of the pendulum. T is the period.)

In the report v1, we found the following:

To be able to calculate g, it is necessary to know k, M, D and d, together with their uncertainties.
The measurement of the pendulum oscillation period, also necessary to calculate g, will be done...

It seems again that students rely solely on formulae.

- Difficulty in separating variables to be kept constant and quantities to be measured.

We have noticed that students were seldom able to vary more than one quantity. Often, the choice of the values of the constant quantities was not explicit. Most students did not realise at the beginning that the values of the constant quantities could have a strong impact on the accuracy of the measurements and also on the quality of the confrontation.

In project V about light polarisation, the polarising angle had to be studied in relation to the length of a small tank, to the concentration C of glucose in it and to the wavelength. The quantities to be kept constant change within the same report v1:

We got new measurements modifying the length of the tank. Cglucose = 0.4 g/ml.
The solution of glucose is lighted up by the D ray (λ = 589 nm) of a sodium lamp.
Further:
We use a length of the tank l = 0.8 dm constant.
We fix l = 0.8 dm and C = 0.4 g/ml.

In fact, the choice of the length in the second and third experiments was not appropriate because a tank of 1.6 dm was available and would have enabled a doubling of accuracy.

c. Choice of the measurement method

This choice implies consciousness of the problem of uncertainties. Many students did not worry about uncertainties at the beginning of the work. However they became progressively concerned with them. This was the case in project VII. In the first version v1, we found:

We measure R directly with an ohmmeter at a fixed temperature.

and in the final report:

We connect an operational amplifier. This circuit enables us to limit the uncertainties.

Finally it was the calculation of uncertainties which influenced them in choosing the apparatus and the method of data collection.

d. In all choices, the importance of being able to evaluate orders of magnitude

An example is project XIII, which is the influence of the bouncing of a ping-pong ball upon the rotation of the ball:

> *The video camera taking only 25 images per second, the ball must have less than 25 rounds per second to allow the study of the rotating movement.*

In the report, a brief theoretical study follows concerning the distance along which the ball has to roll on an inclined plane.

Other examples show that orders of magnitude, for quantities and uncertainties, have a special role and are useful for a range of decisions. Obviously, it is a very different situation to what is encountered in conventional labwork sessions during which the uncertainties and the orders of magnitude are required after the measurement in order to verify the plausibility of the final value and to have arguments for a possible confrontation.

The step of confrontation

The main feature is the obvious difference between students' intentions at the beginning of the project and at the end.

a. Students' intentions from the first version v1 of the protocol

Confrontation and uncertainties were not a major preoccupation for students developing their first version of the experimental protocol. To find a 'good' experiment seems the most important preoccupation for them.

- Confrontation was explicitly mentioned in very few v1 reports: 13 reports out of 30.
- In 5 of these reports, statistical data processing is planned in a relevant way. For instance, in project VII (v1).

We will set up the Cavendish experiment several times in order to have several measurements of a and then of g (gravity constant). Thus a statistical study of g will be possible.

- In 5 other projects, a visual fit between a calculated line and the experimental points is implicitly planned.
- The use of quantitative fitting criteria is planned only once.
- Project XXVIII (v1), is an example of the lack of quantitative criteria to judge if a quantity is constant or not:

  For every tension, the mass of the string is calculated, if it is constant then the relation is correct.

b. The final type of confrontation

All final reports give account of a step of confrontation. This confrontation is generally correct. The more students proceed with the experiments, the more they are pushed to envisage confrontation, and the more they use the resources they have been taught previously. The main difficulties encountered are that students often missed clear criteria to validate a confrontation theory-experiment. In this case, the personal conviction has a pre-eminent effect upon the conclusion. Here are some examples of such difficulties:

- The confrontation with theory was only qualitative.
  Even if, because of our modest experiment, we have only made a qualitative observation, we have understood the possible applications of the phenomena.
- Students were conscious that their confrontation was qualitative and insufficient.
- By contrast, in project X, 'What are the characteristics of the objective lens of a camera?', the comparison was quantitative in relation to the uncertainties. All along the work, the students evaluated rather carefully what they called 'errors' (uncertainties). Whatever the value obtained, they were content with them. For instance:
  We made an error of 2.1%, which seems acceptable.
  We made an error of 0.5%, which is a very good value.
  We made an error of 16.6%, which is reasonable according to the dimensions of the objectives.

One can wonder if the succession of the different steps above is problematic. In fact, what is problematic is the co-ordination of the steps. From our observations, we can claim that students do need a framework organising their decisions and actions. Ideally the epistemological framework we tried to teach should work as a 'frame of knowledge' defined by cognitive psychology (Richard, Bonnet & Ghiglione 1990) as:

> *... both a manner to represent the organisation of knowledge inside memory and a manner to tell how this knowledge is used in order to understand, to memorise, to make inferences and also to perform.*

In addition Richard et al. emphasise that learning such a framework, should involve repeated opportunities to *use* the framework. Students have difficulty in *using* epistemological knowledge to plan an investigation. However, the more they proceed with the investigation, the more they do.

Conclusion

The basis of our work was the desire to define as fully as possible physicists' processes and strategies to help students when faced with real experimental investigation situations. By doing so, we took into account that, at undergraduate level, professional training is more and more relevant. Though simplified for the sake of teaching, the description respects the variety of models that physicists use , the central role of confrontation between experimental data and models, as well as the complexity of the real approaches.

The description we elaborated, structured in an epistemological framework, was taught in a two years sequence during which students experienced a variety of complementary activities from guided work to open-ended labwork.

The analysis of what students did when addressing physics problems during the open-ended projects, confirmed that all processes, strategies, confrontation types and tools have been used at least once. Most of our students showed great ability in handling sophisticated tools of confrontation. This suggests that all of these issues should be included in teaching.

Written questionnaires about the taught epistemological knowledge gave additional information and revealed difficulties concerning the effectiveness of teaching such material. Short-term assessment demonstrates that the notions and words are difficult for students. Medium-term assessment, 18 months later, at the end of the sequence, demonstrated that students have difficulty in matching what they did with the taught models and types of confrontation. The recognition of processes and strategies is better after the teaching, though not acquired by all students.

These difficulties show that the variety of strategies and models appears strange and new to students. We interpret this feeling as arising from habits acquired during several years of lectures that communicate the idea that there is only one experimental approach and that physics always deliver the truth, supposedly unique.

Our study points at some limitations of conventional labwork. For instance, we show that the ability to evaluate orders of magnitude and uncertainties before designing an experiment is a great help. It allows students a degree of autonomy in their work, which is not possible in guided labwork. Other aspects of our teaching enhance students' autonomy, by giving a structure to the different steps of physics processes.

The difficulties encountered by students show that this teaching is only a start in making them autonomous in addressing physics questions, designing experiments and taking advantage of physicists' experience. Most students expressed how they felt more personally involved in physics tasks, as a result of this sequence. It can be expected that continuing open investigative tasks will enhance their epistemological metaknowledge and consequently their autonomy.

Recommendations

As in any action-research, lessons have been learnt from the work. What we have continued and improved in our university is the following:

- Ensure continuous two-way exchange between practice in labwork and epistemological knowledge, in order to distinguish the different types of models. For instance, presently in our university, the method of commenting on and considering what has just been done in a labwork session has been enhanced. This is an activity which is original and does not exist in conventional labwork.
- Provide students with two types of 'tools', one indispensable to carry out proper data processing and the other for confrontation; both deserve to be taught in depth and the corresponding learning not left to chance.

This implies a variety of teaching contexts, conventional sessions having their own justification as an effective preparation to open – ended activities, and projects being highly adapted to the application.

The following improvements could be made to the use of projects:

- In our university, projects are now presented more clearly as an application and result of the preceding phases of teaching. This has consequences on the wording of the physics problems proposed for projects. Our experience is that they must be rather short and they should allow students to take responsibility for the choice and the definition of the quantities and models to be handled. This justifies new research.
- It is relevant to ask students to write an initial short report supported by a short oral presentation after a few sessions of personal work (what we have called the 'first version of protocol'). The presentation should provide the opportunity to make students conscious of the approach in which they are engaged, knowing

their difficulty to engage in epistemological reflection. Such a presentation should also provide the opportunity of encouraging a discussion between students and the group of teachers. This is intended to allow an authentic scientific debate.

References

Bunge, M. (1973). *Philosophy of Physics*. Dordrecht: Reidel Publishing Company.

Develay, M. (1989). Sur la méthode expérimentale. *Aster*, n°8, pp.3-15.

Driver, R., Leach, J., Millar, R., and Scott, P. (1996) *Young people's images of science*. Buckingham: Open University Press.

Gott, R. and Murphy, P. (1987). *Assessing Investigations at Ages 13 and 15. Assessment of performance Unit Science report for Teachers: 9.* London, Department of Education and Science / Welsh Office / Department of Education for Northern Ireland.

Guillon, A. (1996). Étude épistémologique et didactique de l'activité expérimentale en vue d'un enseignement et de l'apprentissage des démarches du physicien, dans le cadre des travaux pratiques de première et deuxième années d'université. PhD dissertation (Paris. Université Paris XI).

Hodson, D. (1992). Assessment of practical work: some considerations in philosophy of science, *Science and Education,* 1, 115-144.

Leach, J., Millar , R., Ryder, J., Séré, M.G., Hammelev, D., Niedderer, H. and Tselfes, V. (1998). *Survey 2: students's images of science as they relate to labwork learning* Working paper 4 'Labwork in Science Education'.

Ntombela, G.M. (1999). A marriage of inconvenience? School science practical work and the nature of science. In J. Leach and Albert Paulsen (Eds) *Practical Work in Science Education: Recent Research Studies.* Roskilde University Press. Kluwer Academic Publishers.

Richard J.F., Bonnet C. and Ghiglione R. (1990). *Traité de psychologie cognitive.* Vol.2. Paris: Dunod.

Trigeassou, J.C. (1988). Recherche de modèles expérimentaux assistée par ordinateur. Paris, Tec and Doc Lavoisier.

Vicentini M. (1994). Thinking physics for teaching in European Research in D. Psillos (Ed.) *Science Education II: Proceedings of the Second Ph. D. Summer School.* Leptokaria (Greece).

Walter, E. and Pronzato, L. (1994). *Identification de modèles paramétriques à partir de données expérimentales*. Paris, Masson.

Woolnough, B.E. (1989). Towards a holistic view of processes in science education. In Wellington, J.J. (Ed), *Skills and processes in Science Education: a critical analysis.* London Routledge.

Woolnough, B.E. and Allsop, T. (1985). *Practical work in science*. Cambridge: Cambridge University press.

APPENDIX 1: the questionnaire Q

Short-term assessment after the epistemological information after the third labwork session: LARGE OSCILLATIONS OF A PENDULUM [5]

Q1 Which adjective do you associate with the process followed during the last laboratory work session?

theoretical - experimental - mathematical - simulated

Q2 Concerning the studied system, which are

THE INPUT-QUANTITY(IES) ?

THE OUTPUT-QUANTITY(IES) ?

Q3 Which are the parameters associated with the studied system?

Q4 During the session to-day, a confrontation between theory and experimental results has been completed.

In which of the two parts of the session, has the confrontation occurred?

What is the relation associated with this confrontation?

Q5 Concerning the system you just studied, state the Model of Object you used to support the theoretical study.

[5] Without the boxes to answer

APPENDIX 2: the post questionnaire PQ

Medium-term assessment at the end of the teaching sequence, about the project of each student [5]

PQ1 - During your work, you have:

- defined a Model of Object? which one ?
- used a Theoretical Model ? which one ?
- established an Behaviour Model ? which one ?
- made a confrontation between a Theoretical Model and experimental results ? how ?
- made a confrontation between a Behaviour Model and experimental results ? how ?
- made a confrontation between a Behaviour Model and a Theoretical Model? how?

PQ6 - During your work, you have:

1- to choose an apparatus for experiments	2- to do a theoretical study	3- to look for causes of error
4 - to choose a model of object	5- to identify which quantities to measure	6- to collect measurements
7- to evaluate uncertainties	8- to choose a measurement apparatus	9- to do a simulation
10- to calculate the order of magnitude of the results	11- to draw the graph of a theoretical model	12- to search for input and output quantities
13- to evaluate the precision of measurement instruments	14- to choose which parameters will be constant and which will be variable	15- to calibrate a measurement instrument
16- to choose the measurement method	17- to plot experimental points	18- to calculate the standard deviation between experimental points and the model
19- to verify if an apparatus is in good working order	20- to decide to repeat measurements N times	21- to look for the behaviour model, by plotting a known function
22- to look for an experiment in a book or in a journal	23- to choose the quantities to put on the axes of an experimental graph	24- to look for a measurement method in a book or in a journal
25- to draw a histogram	26- to calculate the relative deviation between a measured value and a reference value	27- to carry out experiments
28- to search for the parameters of the system	29- to look for reference values in a book or in a journal	30- to draw on the same axes a calculated curve and the experimental points

In the list above, tick the actions you carried out. Report below the corresponding numbers in the chronological order of your successive actions [5]

PQ10 - In a general or a particular manner, indicate below all points you think related to:

what you have appreciated	what to improve

The Effectiveness of Mini-Projects as a Preparation for Open-ended Investigations

Jenny Lewis, University of Leeds, UK

Abstract

Mini-projects are sometimes included in undergraduate courses with the expectation that they will help students to make the transition from set practicals to open ended investigative work. This case study assesses the extent to which one particular use of mini-projects was able to effect this. Despite their success in developing the students' understandings of the nature and processes of scientific research, the mini-projects left many students feeling demoralised and largely unaware of the learning which had taken place. Factors contributing to this outcome were identified and their implications for the design and management of such projects are considered.

Introduction

Practical work in British universities can take a number of different forms. Traditionally, labwork during the first two years of a degree course has focused on set practicals. Typically, students are given detailed instructions for a task which uses established techniques to achieve an expected outcome in a limited time - usually 1 or 2 sessions. One major objective of such set practicals is to develop students' technical skills. In the third year the focus changes. Labwork then takes the form of a small research project. These projects are open-ended in that the final direction and outcome are unknown at the start. The student is responsible for developing and planning the work and an extended amount of time is made available - 1 or 2 terms, as many hours as the student chooses. The purpose of such projects is to develop both the students' understanding of the processes of scientific enquiry and their awareness of the nature and practices of scientific research within their particular discipline. It is assumed that they will do this through a process of enculturation (Brown et al. 1989). That is, they will learn the 'craft knowledge' of their scientific discipline by working alongside more experienced scientists as an 'apprentice'.

While enculturation is an important mechanism by which students learn how to work as scientists it is not, within the timescale of an undergraduate project, sufficient. Many students find the transition from traditional practical work to open ended projects difficult and they need some preparation and support if they are to make the most of the experience. Recent research supports this view (Driver, Leach & Ryder 131 1996). Most students are unused to taking personal responsibility for their labwork and have little experience of thinking through the theoretical or organisational demands of a project. They are often unprepared for the difficulty of collecting reliable data and interpreting results and largely unaware of the relationship between theory, data and practice - that theory can be used to inform initial planning of a project and to interpret results; that initial results, combined with theory, can be used to inform further planning. As a consequence they have unrealistic expectations as to what can be achieved and problems in interpreting the

D. Psillos and H. Niedderer (eds.). Teaching and Learning in the Science Laboratory. 139-150.

results which they do get. These difficulties lead some students to feel very demoralised.

In an attempt to overcome some of these difficulties mini-projects are sometimes included in second year courses. These are short projects of limited scope which are intended to teach students about the process of designing and conducting open-ended inquiries. The expectation is that students, through their work on these mini-projects, will develop more realistic expectations of, and greater confidence in, open-ended investigative work. While the need for a transitional form of labwork has been recognised there appears to be no explicit rationale, based on a consideration of effective strategies for achieving the desired learning outcomes, for the design of such mini-projects.

This chapter reports an evaluation of one particular use of mini-projects and considers obstacles and affordances to the effective use of mini-projects as a preparation for open ended laboratory work.

Research Methodology

This case study reports on the use of mini-projects within a second year undergraduate biology module on gene structure. It identifies the learning aims and objectives of both the lecturer and the students and considers the following questions:

- to what extent did these mini-projects succeed in achieving the learning aims and objectives?
- what were the obstacles to achieving the learning aims and objectives?

It focused on the process rather than the outcome and the researcher acted as participant observer, working alongside one group of students throughout the sessions. In addition to observation notes, data were collected from course documentation, end of project presentations and student diaries in which students in the observation group recorded their personal view of the progress that they were making – the difficulties, frustrations and successes. Semi-structured interviews were conducted with the technician (a single individual interview), the lecturer (a pre-interview to ascertain aims and objectives; a post interview, after project reports had been marked, to discuss outcomes) and 6 second year students (a series of 3 paired interviews at different stages of the mini-projects). 8 third year students who had previously experienced the mini-projects were also interviewed (single interviews in two groups of 4), providing an opportunity to assess the longer-term impact of the mini-projects. All interviews were audiotaped and transcribed for later analysis.

Analysis of the data was qualitative and these multiple sources of data were used to validate the findings through triangulation. The learning aims and objectives were identified first, then each source of data was analysed to identify statements or information relating to each of these objectives. The results of this analysis were then used to inform a consideration of the two key research questions noted above.

The Teaching Approach

The Context

All students in the class were given the same basic task - to map uncharacterised mutations of a microscopic worm, *C. elegans*. A number of worm stocks, carrying different mutations, were made available and information on the reproductive habits and genetics of *C. elegans* was provided, together with a description of the process of gene mapping, in the course manual. Working in groups of 6, students had to decide which mutations they wished to map and to develop (and implement) a strategy for doing this within the time available - 9 two hour sessions over a period of 5 weeks, plus any additional time they wished to spend in the laboratory. *C. elegans* has just 6 chromosomes so one possible strategy for distributing the work within the group was for each student to collect the data for just one of the chromosomes. Each group was expected to present their results to the rest of the class during the last session. Just one lecturer and one technician were responsible for supervising and supporting 66 students organised into 12 groups. There was no demonstrator. Assessment of the mini-projects was based on the final written report and 17% of the marks for this module were allocated to this.

The Demands of the Task

This task made a number of demands on the students. They needed a certain amount of practical expertise to work effectively with the worms. Worms needed to be picked up individually and transferred between plates without damage. The worms also needed to be sexed accurately, to ensure that the correct matings were set up. In addition, the students needed to become familiar with the physical and behavioural characteristics of different mutations in order to identify the offspring of the matings correctly.

The students needed to think the project through before starting and develop a strategy for achieving their practical aims. This planning needed to be based on a sound understanding of genetic concepts if it was to be effective, including gene structure, linkage, Mendelian genetics and the application of all this to *C. elegans* (which is usually hermaphrodite). If their plans were to succeed they also needed to use these concepts in the development of a hypothesis which would enable them to interpret their results and so inform the next step in the process.

There were certain personal skills and attitudes, which the students also needed. Having developed a plan, they needed to distribute the workload effectively amongst the group members. To ensure that findings were collated, interpreted and used effectively to inform the next stage, they needed to develop effective lines of communications between the whole group and the individual members. Finally, they were expected to recognise and accept that educational benefits could be as important and as valuable as marks. This extended piece of work was very demanding in terms of time and effort and the students needed determination and resilience to keep at it. Despite this, the marks allocated to the projects were relatively low.

Results

The Lecturer's and the Students' Aims and Objectives

The aims and objectives identified by the lecturer and the students are summarised in Tables 1 and 2 respectively. These aims fitted into one of three categories:

- A. practical and scientific aims, which focused on extending the students' understanding and application of the science;
- B. educational aims, which focused on developing the students' understanding of the nature and practices of science, and the skills which they might need when working in the laboratory;
- C. other aims, which seemed to relate to personal needs.

Table 1: The Lecturer's Aims and Objectives	Table 2: The Students' Aims and Objectives
A - practical and scientific aims (i) mapping of 1 or more mutations. (ii) development of understanding of Mendelian genetics.	A - practical and scientific aims (i) determine linkage for a number of mutations. (ii) experience working with the whole genome.
B - educational aims (i) preparation for Year 3 projects. 1. recognising the difficulty of achieving results: developing realistic expectations; setting achievable goals; learning to be flexible. 2. thinking for themselves: deciding the research questions; solving problems; interpreting results; deciding the next step. 3. developing organisational skills: planning; time management.	B - educational aims (i) prepare for work as a scientist. 1. develop team working skills; communication, organisation, leadership. 2. learn to think for themselves: decision making; explanation, justification, evaluation. 3. developing organisational skills: planning; preparation, record keeping.
C - other aims (i) reduce students' negative feelings. (ii) differentiate between problems related to *C. elegans* and problems intrinsic to research.	C - other aims (i) get results. (ii) get marks.

While the lecturer and the students identified some similar aims and objectives, they each had a number of objectives, which were not shared by the other. For the lecturer the practical aim was to map uncharacterised mutations of *C. elegans.* Although this was made quite explicit to the students, most of the students said that the aim of the practical was to determine the linkage of uncharacterised mutations of *C. elegans* - a less demanding task. They seemed to be unaware that there were two distinct steps in the process of gene mapping - the identification of the chromosome on which the mutation was located (and hence the linkage group) *and then* the location of the uncharacterised mutation in relation to that known linkage group - and unaware of the discrepancy between their own and the lecturer's perception of the task. The lecturer's scientific aim was to develop the students' understanding of Mendelian genetics. The students did not recognise this as one of the aims of the mini-project. What some of them did identify and value was an opportunity to work with the whole genome. The lecturer recognised that this might be particularly pertinent to students specialising in biochemistry, but providing this opportunity was not one of the aims that he identified.

Both the lecturer and the students identified a number of similar educational aims and objectives, but the focus was slightly different in each case. While the students recognised the lecturer's specific and explicit aim of preparing them for third year projects, their aim was broader - to prepare for work as a scientist. Within these aims both the lecturer and the students identified 'learning to think for themselves' and 'developing organisational skills' as important objectives, but they had different views as to what these might mean - the lecturer focused on specific aspects of the methodology while the students focused on general skills. The students also included 'development of team working skills'. For the lecturer, group working was a pragmatic response to the limitations of time, resources and student numbers rather than a key objective. What the lecturer hoped to develop was the students' understanding of the difficulty of achieving the desired results. The students remained unaware of this, despite the lecturer's efforts to prepare them for it during the introductory session. The students' prior experiences of practical work led them to believe that achieving the necessary results would be unproblematic. The absence of 'managing workloads' as an organisational objective for the students may have been related to this.

The lecturer identified 'preparation for Year 3 projects' and 'development of the students' understanding of Mendelian genetics' as the main aims of the mini-projects, and tended to see the practical task as a vehicle for achieving these. In contrast, the students tended to see the successful achievement of the practical task as the main aim and this was reflected in their 'other aims' - to get results and to get marks. The lecturer's introduction to the mini-projects explicitly highlighted the issue of results and also emphasised the educational rather than assessment purposes of the labwork. While students appeared to recognise and accept that results were not the only benchmark, they still wanted results (and needed results if they were to achieve their main aim - to determine linkage). Similarly, while they appeared to recognise the educational value of the project, they found it hard to accept this as a substitute for marks.

The lecturer also had two objectives, which were not shared by the students. He was concerned about the very strong negative feelings that some students developed in response to the mini-projects and one objective was to reduce these, if possible. He was also concerned about the long-term effect of these negative feelings on his own area of research. On the basis of their experience of the mini-projects, many students were reluctant to work with *C. elegans* again - either as a third year student or as a postgraduate. He felt that the students, instead of recognising that practical difficulties are an intrinsic part of any research, blamed *C. elegans* for the problems which they experienced. One of his objectives was to clarify the distinction between problems which were an intrinsic part of research and problems which were specific to *C. elegans.*

The Learning Outcomes

The learning outcomes for each set of aims and objectives are reported below under three separate subheadings.

The practical and scientific aims and objectives

Only 2 groups succeeded in identifying the linkage group for their chosen mutations, and none succeeded in mapping their mutations. It was clear from the group presentations at the end of the mini-projects that the discrepancy between the lecturer's practical aim (mapping) and the students' practical aim (linkage) was never really resolved. This difference appeared to arise from the students' lack of understanding of the difference between the two processes and it seems likely that this lack of understanding contributed to the inability of students to achieve even their more limited aim of identifying the linkage groups.

The lecturer's scientific aim - that students should develop their understanding of the theory through using it to inform their planning of the project - was never made explicit to the students and there was little evidence of this aim being achieved. While most students recognised (often with hindsight) the importance of planning, few groups saw the need for a rational plan based on theory. This was highlighted in the observation group. Several weeks into the project they were having difficulty in interpreting their results. In passing they mentioned that they still didn't know how they were going to work out linkage. They seemed unconcerned about this and unaware of the need to base practice on theory - that they needed an understanding of linkage in order to inform the design of their practical activities

The students' written reports, produced after the group presentations, showed that while some students did eventually develop a good understanding of Mendelian genetics and how to apply it to *C. elegans,* many did not.

The educational aims and objectives

At the start of the mini-projects students had little understanding of the nature of open-ended investigations or the culture of scientific practice. Consequently they had unrealistic expectations of what could be achieved, based on their perception of results and their experience of set practicals:

> *At the time we thought it would be easy, well not easy but we thought we'd be able to do it, we didn't think it would be so hard and we expected everything to work.*

As a first step in developing a better understanding of the nature of scientific research the students needed to recognise that there were differences between open ended investigations and set practicals. In particular they needed to recognise the uncertainty of open-ended investigations (the technical problems and the difficulties of collecting the required data) and the need to take personal responsibility for the work (to plan experiments, keep records, interpret results, and to revise or develop plans in response to findings). Evidence from a number of sources suggested that students did begin to develop such awareness through their work on the mini-projects.

They began to recognise the different nature of open-ended labwork:

> *The closed practicals you have clear instructions, clear aims to what you're meant to achieve and you come in, follow what they say, get results and write it up. It's much more open-ended in the worm project. So you do have to learn to ask your own questions and work out your end results. There's no straight path. And of course all closed practicals have been tried and tested and if you do them right they work.*

They also came to recognise that the outcome is not always known in advance - that there may not be one 'right' result:

> *...For this project there wasn't really a list of answers. It was so open. I don't think [the lecturer] could really have gone through it all [and said] this is what the answer was for this part. I don't think there are definite answers for some things.*

In addition they began to identify the personal characteristics that a research scientist might need:

> *As far as I can understand research, people doing it would get disappointments like we did. But they'd have more time to plan it and think it through. Keep going at it over and over again.*

Some also began to reflect on their own attitudes towards research:

> *In some ways it's possibly put me off doing research because I like to see something happening when I'm doing work. If I can't see any progress then I get fed up with it. I'm probably not cut out for research.*

A few students began to reflect on their changing understanding of the nature and practices of scientific research:

> *You read a research paper, these research papers that get published are really good ... It's easy to think that science has been 100% successful all the time ... but when something fails like this [their project] you can imagine there's a lot of hard work and a lot of failed attempts and what you see in the publications are probably just the top 0.01%.*

While the evidence appeared to suggest that these mini-projects did succeed in the educational aims identified in Tables 1 and 2, students often remained unaware of this success. This was evident in interviews with second year students at the end of their mini-projects:

> *Student: [using those criteria] I would say it's a success because it did teach us a lot about research and what we can do in the time allowed and why things go wrong. If that was the aim I think it's a more worthwhile experiment than mapping (....)*
>
> *Interviewer: But you didn't actually see that as being the aim of the project while you were doing it?*
>
> *Student: I don't think so, no.*

It was also apparent from interviews with third year students. When asked if any previous work had prepared them for their third year projects very few spontaneously mentioned the mini-projects. When a number of these students were asked explicitly *'Do you think that the mini-projects prepared you for your third projects, in any way?'* three students said 'yes', three said 'no' and one still felt so negative about the whole mini-project experience that she found it difficult to give an objective answer.

Other aims and objectives

Despite the lecturer's wish to reduce negative feelings about the project, the students clearly continued to develop them. There were a number of reasons for this, two of which related directly to the students' additional objectives - obtaining desired results and collecting marks. Students were anxious about their lack of results:

> *It's natural to assume that you've got to get results. We always got results for our biochemistry experiments. Not to get results was awful.*

This anxiety was compounded by their continuing belief, despite a growing awareness of the importance of the educational aims, that the main aim of the project was the successful completion of the practical activity.

They also felt angry that the marks for the work in no way reflected the amount of work or the stress involved:

We did a multiple-choice paper in the other practical, which literally took 15 minutes and was worth 10%. That was worth twice as much as a 6 week project. It doesn't seem right. It was really easy, I got 70% in it or something like that without any work. Then you're doing all this which is only worth 5% and it's a lot of hard work.

Even those students who recognised the advantages of so few marks being allocated to the project - that if they didn't do very well it wouldn't matter very much - felt that the weighting of marks was unfair and demotivating. This sense of unfairness was so strong that it was still being expressed by third year students. Surprisingly, given the anger which they felt about marks, few students seemed to know exactly what percentage of the module marks had been allocated to the mini-projects. Assumptions ranged from 5 - 15% but the figure was actually 17%.

The Process

An analysis of the mini-project process identified a number of factors likely to have influenced these outcomes, particularly the strong negative feelings expressed by some students.

Conflicts and contradictions

As already noted, one of the main aims of the mini-projects, from the lecturer's point of view, was to prepare the students for their third year projects. In particular he wanted to develop their awareness of the difficulty of getting results and of the need to set realistic targets. For most students, the main aim of the mini-projects appeared to be the successful completion of the practical task. Consequently there was a tension between the practical aims and the educational aims of these mini-projects. Students were expected to achieve the lecturer's educational aims through their experience of, and reflections on, the difficulties of achieving their practical aims. As a result, most students were acutely aware of the extent to which they had failed in their practical aim. This sense of failure was, to an extent, an inevitable part of the learning process.

Expectations of what could reasonably be achieved within the time available were over-optimistic and misleading. Information presented to the students implied that 6 weeks were available for the mini-projects and that 10 matings could be achieved in this time. In practice only 5 weeks were available and only 7 - 8 matings were possible. This was just enough time to map a mutation, assuming that this number of matings could be achieved, and that each mating was successful. However these assumptions contradicted the initial premiss - that students would experience difficulties and problems in trying to achieve the practical and scientific aims. This tension between achieving the educational aims and succeeding at the practical task were not helped by the lecturer's own ambivalence. While he said that the main aims of the mini-project were to prepare students for their third year projects and to develop their understanding of Mendelian genetics he very much wanted students to get results and to succeed at the practical task. Aware of the very tight time constraints he strongly encouraged the students to set up their first matings during the first session. This pressure to get started on the practical work was in direct conflict with the students' need to think through their plans and to develop their technical skills:

> *Yeah, 'cos we thought the first week was going to be planning, whereas (the lecturer) sat us down with a box of plates and said make up your stocks - like, we don't know what we're doing yet (.....) we needed to get more organised really.*

The difference between what the students thought that they were doing and what the lecturer intended them to do led to a number of problems and was a major source of the negative feelings which many students experienced.

Problematic issues

Based on their prior experiences of set practicals, students expected practical work to be largely unproblematic and to lead to predictable and achievable results. The lecturer tried to change their expectations by warning them explicitly of the danger of being over ambitious in the mini-projects. He stressed the difficulty of getting results and emphasised the need take this into account in the planning. The students, with no clear understanding of the nature of the differences between set practicals and open ended investigations, were unable to assimilate this information and continued to believe that achieving results would be unproblematic. Only after successive set backs and difficulties did they begin to appreciate what might be meant by 'a realistic plan'.

During the projects the students seemed largely unaware of the extent to which they were succeeding in the educational aims. Recognition of what they had learnt appeared to come with reflection, but the point at which this reflection took place, if at all, seemed very variable. While for some students the stimulus appeared to be the need to make a presentation, for others it was the act of writing up the work. From interviews it was apparent that in some cases reflection didn't come until a year later, at the end of their third year project. For some students reflection only took place in response to a direct stimulus, such as that provided by the interviewer.

There were also technical problems. Many of the students had no experience of working with C. elegans and found the worms very difficult to handle and to sex. While the lecturer recognised this:

> *Once you've seen the difference in the sexes it's so obvious. Yet always there are a significant proportion of students that have set up the first crosses and clearly couldn't see the difference.*

He seemed less aware of the consequences, and their effect on the students. This lack of skill seriously limited the students' ability to make progress with the mini-projects - either because the damaged worms failed to produce many offspring or because wrong assumptions about parental types led to misinterpretations of the results. Frustration was inevitable, given the students main aim (to succeed in the practical task). Such experiences also led students to blame *C. elegans* for the practical difficulties which they experienced rather than consider the possibility that practical problems and difficulties are an intrinsic part of scientific research.

Omissions

A number of features not included in the mini-project sessions might have enhanced their effectiveness in achieving the educational aims and objectives.

If students are to develop an understanding of the nature and practices of science through enculturation they need to develop some awareness of the processes by which knowledge is socially constructed and agreed within the laboratory. This requires some understanding of the relationship between co-workers within a laboratory. Prior to their third year projects, the relationship which most students experience with practising scientists is very hierarchical and authoritarian - they go to their demonstrator[1] or lecturer to check that they have the right result or to be told what to do. With the third year project this relationship changes. The students are expected to integrate into the life of the laboratory and to develop a more equitable relationship with other researchers, sharing expertise and discussing ideas. Many students find this transition difficult. Initially they may not be aware of the changed expectations or the role this new relationship plays in the development of their own thinking and learning. In addition, many lack the confidence to participate in this way even when they are aware of the expectation. With hindsight a number of the third year students felt that the mini-projects would have been improved by the presence of a demonstrator:

> *You need to confront somebody who's like a demonstrator or whatever and set up some sort of rapport between your demonstrator and your group a bit better. (...) because I think when I came to do my proper project when I got introduced to my demonstrator I (...) assumed that it was kind of like you're on your own here and you don't go and see him unless something terrible's gone wrong (...).*

While the lecturer had tried to fulfill this role, there were too many other demands on his very limited time. In any case, most students needed someone closer to themselves in terms of experience and status if they were to gain sufficient confidence to try to develop this new type of relationship.

Integration into the life of the laboratory could also be enhanced by the development of good group working skills. The students rightly recognised the importance of such skills but the mini-projects did not include any interventions or activities, which might help the students to develop such skills.

Drawing on their combined experience of mini-projects and open ended investigations, third year students also felt there was a need to re-structure the start of the mini-projects to include activities which would encourage students to think the projects through more carefully, for example, students to produce a written plan before starting the work. The issue was not so much that second year students were unaware of the need to plan rather that they were unsure of how to plan. Although the practical manual provided detailed information on the genetics of *C. elegans* and described procedures for determining linkage and for mapping, the importance of using this information to inform their planning or interpret their results was never made explicit and few students made use of this information at the start of their project.

Conclusions

There is a clear need for some form of labwork which can help undergraduate students to make the transition from set practicals which are designed to develop their technical skills to open ended investigations which are designed to develop

[1] Demonstrators are postgraduate students, actively engaged in research, who help and support undergraduate students during practical sessions.

their research skills. Findings from this study show that mini-projects can be a very effective form of transitional labwork. However, an inevitable part of the learning process is a certain sense of failure as students come to realise the difficulties of obtaining the necessary data and learn to set themselves more realistic goals. This study identified a number of factors, which influenced both the effectiveness of the mini-projects in preparing students for open ended project work and the extent to which students were left feeling demoralised by the experience. These factors, and the pedagogical implications, are identified below.

Recommendations

The factors which were identified in this study as influencing the effectiveness of mini-projects as a preparation for open ended labwork can be loosely grouped under 5 inter-related recommendations (see below). While these 5 recommendations might appear to reflect commonly agreed good practice in school teaching, the need to apply such principles to university teaching is less well documented. These recommendations could be used to guide the design of transitional labwork in order to maximise its effectiveness as a preparation for open-ended investigations while minimising the negative feelings which such projects often generate.

Ensure that Aims and Objectives are Unambiguous, Achievable and Explicit

Within this case study there was a need to identify and resolve ambiguous, contradictory and conflicting objectives. For example, as the practical aim was less important than the educational aims it would have been more effective to reduce the pressure on students to achieve the maximum number of matings and instead encourage the students to spend more time on planning and reflection.

Recognise the Nature and Difficulty of the Demands

In the process of explicitly identifying and prioritising the objectives, the demands which the project is likely to make upon the students become apparent and can be analysed (Leach & Scott 1995). Within this case study such an analysis would have identified the need to make explicit the importance of understanding the theory and of using theory to interpret data and inform subsequent planning. The significance of the students' difficulty in sexing the worms, and the need to develop a strategy for checking this, would also have become more apparent.

Devise Strategies to Address Specific Learning Demands

Once the learning objectives and demands have been identified it is possible to develop strategies to address these. Within this case study the students might have felt less frustrated if they had understood what 'planning' entailed at an earlier stage. One possible strategy for developing this understanding would have been to set aside time during the introductory sessions to explain the process of planning - making explicit the importance of using theory to inform the planning and allowing the members of each group to work together to prepare and agree a plan and to establish an effective mechanism for checking those plans (to ensure that theory was being used, correctly). In the process students would have recognised the need to develop their understanding of Mendelian genetics and perhaps been encouraged to use the practical manual more effectively. *This approach might also have encouraged students to articulate their thinking, justify their reasoning and (possibly) helped them to develop group working skills too.*

In planning such strategies it is important to recognise that students find it difficult to respond to information and advice which is beyond their experience. Within this case study the lecturer told the students at the start that collecting the required data and achieving the practical aims would be problematic. Despite this the students continued to believe, until they found otherwise from personal experience, that it would not be problematic. Some way of illustrating the problem, in a context that they could relate to, might have helped. For example, it might have been more effective to ask a postgraduate student to give a brief account of their experiences of research - their expectations and how these had changed, common problems and frustrations and how they tried to overcome them, and the occasional pleasures which made it all worthwhile.

Provide Sufficient Time and Appropriate Support and Guidance

If students are to gain maximum benefit from this type of labwork they will need support, guidance and sufficient time to develop the required skills and understanding. For example, in this study students needed to spend more time planning before beginning their project. They would also have benefited from opportunities to discuss results and justify their ongoing plans with a more experienced researcher. It is not possible for one lecturer to provide this level of support unaided, nor would it be desirable. The students found the social and intellectual gap between themselves and the lecturer intimidating. If they were to gain confidence in talking about their work they needed someone that was closer to them in the hierarchy, such as a demonstrator.

Devise Strategies to Foster the Development of Metacognition

Within this case study problems at the metacognitive level, with differing views as to the main purpose of the mini-projects, meant that students were unaware of the learning which they had achieved. In turn, this led to difficulties in contextualising that learning and to the development of strong negative feelings. What was needed were strategies which would:

- help them to develop shared understandings of the aims and objectives (within their working groups; between themselves and the more experienced researchers);
- provide opportunities to assess their progress against these explicit objectives;
- encourage them to reflect on the development of their own understanding.

One possible strategy for doing this would be to expect each group to hold regular but brief meetings within class time to collectively discuss findings and consider the next step. This would encourage individuals to reflect on and articulate their own understandings, help the group to develop a shared understanding of the project and its purposes and provide the group with opportunities to monitor progress.

References

Driver, R. Leach, J. and Ryder, J. (1996). Working Papers numbers 1-7. ULISP, University of Leeds.

Leach, J. and Scott, P. (1995). The demands of learning science concepts: issues of theory and practice. *School Science Review*, 76, (277), pp.47-52.

Seely Brown, J. Collins, A. and Duguid, P. (1989). Situated cognition and the culture of learning. *Educational Researcher*, 18(1), 32-42.

Data Interpretation Activities and Students' Views of the Epistemology of Science during a University Earth Sciences Field Study Course

Jim Ryder, University of Leeds, UK

Abstract

This case study follows the activities of first year university earth sciences students and tutors during the final days of a two week residential field study course in Northwest Scotland. A central purpose of this field course is to develop students' ability to interpret geological data. The majority of students were able to generate and justify interpretations using available evidence. Furthermore, students also tended to recognise that a single field data set could be interpreted in a variety of ways: a sophisticated epistemological position. However, students were less successful in generating a geological history of the field site and surrounding area, tending to focus on local interpretations of rock geometry at specific field site locations.

Introduction

Students' views about the ways in which knowledge claims in science are developed and justified (i.e. their views of the epistemology of science) have been identified as playing an important role in their learning of subject matter knowledge and scientific enquiry processes. For example, Tiberghien & Megalakaki (1995) identify students' views about the ways in which phenomena can be interpreted using theoretical ideas as an important feature in the development of understanding of energy transfer in electrical circuits. Cartier & Stewart (2000) emphasise the relationship between the development of genetics understanding in a high school classroom and teaching and learning about the epistemology of science. In a scientific enquiry context Ryder & Leach (1999) show how views about the relationship between data and theory can influence students' activities during open-ended science investigations. However, despite the important role played by epistemology in science education, several studies have shown that students of all ages often exhibit naïve views of the nature of scientific knowledge (Lederman 1992; Ryan & Aikenhead 1992; Ryder, Leach & Driver 1999). For example, in interviews with 15 - 18 year old students, Larochelle & Désautels (1991) identify strongly empiricist views about the nature of scientific knowledge, with students showing limited recognition of the role of theoretical ideas in the design and interpretation of investigations.

Given such concerns about the impact of students' views about the epistemology of science, the study reported here examines an undergraduate field course in earth sciences in which the epistemology of science was a central feature of students' experiences. The study considers learning aims for the field course, students' attempts to justify knowledge claims in the field, and features of the field course that prompted students to engage with aspects of the epistemology of science. Whilst not an archetypal 'laboratory work' teaching context, the earth sciences lecturers involved saw the field course as an integrated part of laboratory work provision in the university teaching programme. In addition, this is a novel case study since the context of undergraduate fieldwork in the earth sciences has rarely been the focus of

D. Psillos and H. Niedderer (eds.). Teaching and Learning in the Science Laboratory. 151-162.

educational research (Mason 1980). Indeed, to the author's knowledge, there has been limited research into the impact of university fieldwork in any of the science disciplines.

This study was influenced by a number of commitments concerning students' views about the epistemology of science. Science is a multifaceted activity. For example, the ways in which palaeontologists gather and interpret data are significantly different from those of condensed matter physicists. As a result, no single view of the relationship between knowledge claims and data is applicable in all science contexts. Thus, it is appropriate that students' views of the epistemology of science are probed in a specific science context. In this study students are encouraged to draw upon specific field course activities as they talk about the nature of scientific knowledge. This study also recognises the distinction between epistemological commitments informing *action* (e.g. views about scientific knowledge implicit in students' actions during school science investigation activities, Millar, Lubben, Gott, & Duggan 1994) and what students *say* about the epistemology of science. In this study students are provided with many opportunities to talk about their views of the relationship between knowledge claims and data in earth sciences fieldwork. However, to supplement these interviews, field observations were used to gain some insight into students' activities with data and data interpretation in the field.

Teaching context

This case study examines a field course organised by a university in the North of England. Degree courses in earth sciences at this university are composed of discrete course modules. In the first year of the course half of the modules cover aspects of earth sciences, with the remainder being optional modules in chemistry, physics, mathematics, computing or biology. In the second year of the course modules focus entirely on earth sciences. In the final year students choose from a wide range of specialist earth sciences modules. Prior to beginning their degree studies the students on this course will have had a variety of educational experiences. Most students will have studied 3 'A' level subjects (two-year courses in the UK usually taken by 16 - 18 year old students) typically including one or two of chemistry, geography, geology, mathematics or physics.

The study focuses on a two-week residential field course in the Northwest of Scotland held in the summer between students' first and second year at university. The field course is part of a compulsory second year module and is assessed. During the field course students develop their field measurement skills and prepare a detailed geological map. Activities include planning a data gathering exercise, looking for patterns and links in data, generating and evaluating hypotheses and interpreting findings in terms of established theoretical models. In the first nine days of the course students practice and develop specific mapping techniques and learn some advanced mapping skills. During this part of the course student activities are strongly guided by the tutors. The main focus of this case study is on the final five days at Mill na Claise, a remote field site in Northwest Scotland. During this part of the course students are given much less guidance. Over this period students work in groups to produce a *geological map* of a $2km^2$ area showing rock types and bedding angles. In addition students are expected to generate a *geological history* of the area, i.e. a description of the geological processes which have led to the current rock structure. This requires students to make predictions about the underlying rock types in the areas with no rock outcrops, where direct measurements cannot be made. Students also need to interpret their data, suggesting mechanisms and processes that might have caused the current geology. During these final five days students work with field data, established scientific knowledge and tentative knowledge claims of their own. Whilst students have been taught the individual skills of generating a geological history during the early part of the field course, and also during university-based activities, the Mill na Claise activity is the first time that these students have been asked to use all of these skills together to generate a geological history of their own.

Details of typical student activities over the final five days

The field study area at Mill na Claise is enclosed to the north by a stream and road, and to the southwest by a broad ridge rising to higher mountains. This southwestern section of the site is flat and boggy. To the northeast of this boggy section rises a horseshoe-shaped ridge system enclosing another boggy area to the east. Vegetation over the site as a whole is mainly tussocky grassland, with stunted tree growth on the steeper slopes. The area has a large number of exposed rocky outcrops making it an ideal field study site.

Students worked in the field all day. In groups of 3 - 7 the students walked over the site visiting individual rock outcrops. At each outcrop they would identify the rock type by considering the colour, lichen cover and mineralogy of the rock. In some cases students found a section of rock which showed the edges of the beds of rock laid down over time. They would use this cross-section of the rock beds to measure the orientation of bedding. This involved measuring the 'dip and strike' of the bedding planes using a compass clinometer. Students would then mark these measurements of rock type and dip and strike in pencil on their maps. To do this they needed to establish their position on the ground to within 10 metres using a compass and nearby landmarks. Students were expected to use the data they had gathered to make strategic decisions about where to make further measurements. During the field observations some students made measurements of 'palaeocurrents' which enabled them to establish the direction of flow of rivers in which the rock beds were laid down. In some cases students also searched for faults in which rock beds had been displaced rather than folded. In the evenings students would look over the pencilled results on their maps and 'ink in' their maps. Different colours are used to indicate the rock type at each outcrop.

Geological history of the Mill na Claise area.

The horseshoe-shaped ridge to the northeast is comprised of two types of Lewisian Gneiss: acidic and basic. These were originally laid down horizontally on sea beds but have been deformed into a synform: a fold structure with a U-shaped cross-section. A possible deformation considered by some students was an antiform: an inverted U-shaped fold. The wide, flat, boggy section to the south-west contains Torridonian sandstone laid down on a bed of Lewisian Gneiss. Torridonian sandstone is a much younger rock than Lewisian Gneiss. In the past a large river flowed over the Lewisian Gneiss coming down from the mountains to the south-east. This carried eroded rock from nearby mountains and deposited this rock within the broad valley: a fluvial plain. Over time this became sandstone. One of the key findings of the field study was that the Lewisian Gneiss was deformed before the Torridonian sandstone was deposited. Other student activities involved establishing the boundary between acidic and basic Lewisian Gneiss, and between Lewisian Gneiss and Torridonian sandstone.

Research design

Given the lack of previous research into university earth sciences field work, and the author's unfamiliarity with the field work context, this study aimed to provide a preliminary exploration of a range of issues. For a full analysis of the case study covering a broad range of research themes see Leach, Lewis & Ryder (1998). The focus of this report is on the field course as an opportunity for students to gather data in the field and interpret this data by drawing upon their existing knowledge in earth sciences. Specific issues addressed are outlined below.

a. What does the course organiser intend students to learn during the field course about the relationship between data and knowledge claims in earth sciences? (Section "Learning aims").
b. In what ways do students work with data during their field course? (Section "Students working with data in the field").
c. What activities on the field course resulted in students engaging with epistemological issues? (Section "Teaching / learning activities")

The research methodology used in this exploratory study has much in common with the research technique of *ethnography* in which the researcher attempts to interpret and understand the context under research from each participant's perspective. In order to establish the details of the context in which students were working the author attended the field course, working with staff and students during days 12 and 13. This provided an opportunity to follow up unexpected issues raised by the study participants. Informal discussions and semi-structured interviews with participants on the field course were an important feature of the study. An outline of the wide range of data sources is provided below. Full details of the research instruments are given in Leach, Lewis & Ryder (1998). All interviews were audio recorded and transcribed for further analysis.

Interviews with the course organiser

The course organiser was interviewed before the start of the field course and on the evening of day 13. During these semi-structured interviews the course organiser's perceptions of the aims of the field course, and the nature of the knowledge used or generated by the students, were discussed. The second interview also enabled the course organiser to clarify some of the technical earth sciences issues and respond to preliminary research findings.

Written survey given to tutors

This survey was distributed to all 9 tutors on the field course on the morning of day 12. The survey included questions about their perceptions of the aims of the field course and the nature of the knowledge used or generated by the students during the field course.

Field observations

The author shadowed three student groups as they performed field work on days 12 and 13. This involved the author following each group of 3 - 7 students as they collected field data, recorded their findings, discussed interpretations and decided what to do next. Overall, 14 students were involved in field observations.

Written survey given to students

A survey was given to all 55 students attending the course on the morning of day 12, resulting in 37 returned surveys. Students were asked to describe their experiences of working with scientific knowledge and data during the field course. The survey also probed students' views about the relationship between knowledge and field data in earth sciences, using questions derived from a larger study of students' ideas about the nature of science (Leach, Millar, Ryder, Séré, Hammelev, Niedderer & Tselfes 1998).

Interviews with students

Once the written survey and field observations had been completed interviews were held with students from two of the three student groups observed in the field. These students were interviewed in groups of 2 - 3. Overall, 7 students were interviewed. Issues discussed included students' experiences of working with knowledge and data during the field course, and key episodes from the author's field observations.

Research findings

Learning aims

In this section the course organiser's perceptions of the aims for the course concerning the collection and interpretation of data are examined in detail. These show the variety of ways in which students were intended to work with data, and provide a basis for evaluating the experiences of students reported in later sections. Relevant statements from the first interview with the course organiser were collated and grouped into six main categories. These are described below together with illustrative quotes from the interview. Some of the categories are further exemplified using open response written statements from the written survey given to tutors.

Observations and measurements in the field

Students are expected to make observations and measurements to establish rock types, bedding plane angles and other geological factors. Many of these observations are routine classification procedures, as described by the course organiser:

> *There are observations that need to be made which are simple observations (...) it's like a botanist looking at leaf shapes (...) there is a set of routine bits of data you collect about a rock.*

Generating an interpretation

The course organiser intended students to use data gathered in the field to generate an interpretation of local rock geometry, e.g. using data to suggest how rock beds had folded over the field site and predicting the location of geological faults. In the context of the interview 'generating an interpretation' also meant developing a geological history of the field site on the basis of data collected and students' own understandings of geological processes. A geological history is a chronology over geological time of the physical processes that have occurred at a field site:

> *If for example a rock was made up of granular shaped crystals which they were able to identify as dominantly quartz, and these were arranged in plane beds which showed some features of differing grain size, that told them based on what they know already that this is a sandstone and it was deposited in a river environment characterised by high weathering rates.*

Using an interpretation to guide subsequent data collection

The course organiser stated that he expected students to use preliminary interpretations to inform subsequent data collection:

> *What we try to teach is that understanding the history informs the map making and the map making informs the history and you have to go along thinking of both of those things in parallel (...). If you understand [an] historical process you're better able to predict where you should go look for the next key outcrop in your mapping process. So both of them inform each other in that way.*

Students were also expected to use preliminary interpretations to make *predictions* about the characteristics of rocks at another outcrop in the field, and then collect data at that outcrop to test their prediction.

Dealing with discrepant evidence

In the context of the interview 'discrepant evidence' refers to data that contradicts a student's preliminary interpretation of the geological history in the field study area. At two points in the interview the course organiser suggested that students should be able to deal with discrepant evidence. In particular, the course organiser expected students to collect more data to resolve any discrepancy.

Considering alternative interpretations.

Having made an initial interpretation, the course organiser expected students to consider other interpretations of the data that they had collected. Tutors made similar comments:

> *[I ask students] how do [you] think the rock has been deposited? Go [and] look for evidence to prove/disprove/make another hypothesis.*
>
> *[I try to encourage students to] generate possibilities, [for example] folding, and explore the consequences. For instance the sandstone should be deformed, but it isn't.*

The absence of definitive interpretations

The course organiser also described how he felt that students on the field course needed to accept that in some cases they may not be able to develop a definitive interpretation of their data. Rather, students needed to develop 'the most probable interpretation':

> *It's quite often a case of taking strands of evidence and producing a most probable outcome or most probable interpretation. So we try and teach the attitude of this is going to be the most probable outcome.*

One of the tutors on the field course made a similar point:

[Students need to learn] how to cope with situations where there is no unique answer.

Students working with data in the field.

This section summarises students' written and verbal descriptions of data interpretation activities during the field course. A comparison between these reported activities and the organiser's intended learning outcomes provides a tentative evaluation of the effectiveness of the field course.

Data gathered by students

Students reported gathering a wide range of data in the field. By far the most common activity mentioned was measurement of 'strike and dip': establishing the orientation of rock beds using a compass clinometer. Students also mentioned gathering mineralogical data, and establishing geographical position using a compass, map and nearby landmarks. Some reference was made to gathering 'palaeocurrent data': using features on the rock surface to establish the direction of flow of rivers in which rock beds were laid down. The variety of data gathered gave students a rich resource with which to generate and evaluate data interpretations: a central aim of the field course.

Interpretations made by students

Table 1 summarises the interpretation activities mentioned by students in response to a written question asking them to describe interpretations that they had made during the field course. The table gives the number of statements made about each interpretation. Responses to questions about predictions and hypotheses made in the field were broadly similar to those summarised in Table 1.

Table 1: Interpretations made by students during the field course

Interpretation	Number of statements [1]
Cross-sectional data [2]	
general statement	7
fold geometry	12
fault location	2
tilting of bedding planes	2
history of the area over geological time	8
historical direction of river flow	2
Other	3
unclear/vague	9
no response	1
TOTAL	46

The following student describes predictions about the location of rock boundaries:

You're making predictions all the time. Little things like you're walking along and you see a bit of the boundary and you think 'It's either there or there, so I'll walk there' and you do it and it's there, and you're doing it all the time. (student interview, group A1)

Table 1 suggests that the course organiser's aim of getting students to generate interpretations in the field was broadly met. Subsequent interviews with groups of

[1] from 37 returned surveys

[2] i.e. interpretations concerning rock structures at the field site

students show that, at least in some cases, the aim of getting students to evaluate alternative interpretations was also achieved. The following discussion amongst a group of students describes a disagreement about the nature of the fold geometry at a particular location in the field study area:

S1: We had quite a heated discussion [in our group] about whether there's an antiform [at a particular location at Mill na Claise] (...)

S2: The evidence is sketchy but we need to go back there again really.

S1: We all agreed that it could be [an antiform] but we can't actually find anything to support it; or not to support it, because lack of evidence doesn't say it's not there. (...)

S2: This idea we've had has been floating around for a couple of days. (...)

S1: Yes. We actually went looking for it this time. We were looking to see if we could find any evidence. It's like "we think it could be here". It helps explain things. We went looking whether we could find any support for it at all or whether we couldn't find any, and we found one little outcrop which isn't enough to say it is there. (student interview, group A2)

Whilst interpretations, predictions or hypotheses are not mentioned explicitly here, this interview sequence hints at the ways in which students discussed ideas and related them to evidence. However, students appeared to be less confident at generating interpretations of the geological history of the field site and surrounding area; a key learning aim identified by the course organiser. Table 1 shows that the majority of student statements referred to local interpretations of rock type and structure. Such 'local' data collection and interpretation is clearly an important aspect of the field work in this context. However, many tutors were concerned that a large proportion of students did not look beyond these local issues to provide a time-sequenced view of the geological processes occurring at the field site and surrounding area.

Teaching/learning activities

Field observations, interviews, and informal discussions with the author were used to identify activities on the field course that enabled students to collect and interpret data. Furthermore, field observations provide a probe of student activities that does not rely on students' descriptions of their activities; a weakness inherent in the findings reported earlier. Two key features of the course appeared to promote engagement with epistemological issues: working in student groups and the intensive, residential nature of the field course.

Working in student groups

Each group member was required to generate their own geological map of the field site. However, in most cases students pooled their efforts and shared their data. Discussing how to organise the time available, making group decisions about where to go next, and discussing their understanding of geological processes, were key activities in all three of the groups observed in the field.

A major aspect of student group work was how students worked with interpretations:

Interviewer: What do you reckon is the most important thing that you've learnt in Scotland?

S1: Believe what you see, believe your own results and interpretations.

S2: Just be your own person and basically say "everyone else is wrong".

S1: Yes go with what you think.
S2: (...) discuss it with people, but you've got to make sure you've got your own ideas and you're going to stick with them.
S1: But also if somebody else throws in an idea you have to take heed of that
S2: It's the first time I've ever realised that I'm actually as right as the next person. There's no-one else on there who knows any more than I know (...) even the lecturers. (student interview, group A1)

Field observations showed examples of individual students arriving at an interpretation of the geological history that was inconsistent with an interpretation from another group member. This provided students with an opportunity to justify their own views, debate the validity of the various interpretations, and possibly come to an agreement. Student groups worked without support from tutors for most the time in the field. As a result there was no 'authority figure' to which students could refer in order to resolve their debate. Indeed, many students felt that as long as they could justify their own interpretation using the data that they collected, then they were 'as right as the next person':

It sounds kind of weird but I think if in any time when you go out into the field everybody can add their own interpretations. We've all been told that basically there is never really an exact right answer and everybody else can go out there and get a slightly different answer. (student interview, group B1)

Students also talked informally about inter-group rivalry, and heated discussions about various interpretations between groups. However, one student suggested that in his experience the fact that the field course was assessed had limited the amount of inter-group discussion, with some students feeling reluctant to share their ideas.

Residential teaching and learning activity

The staff are brilliant. You really can't beat them. The thing is you see them nearly 24 hours a day so basically you get to know them so well and you get to know the people who you can talk to. So if you've got a problem you go to them, talk to them. But having said that there's not a single member of staff here who I wouldn't approach with a question. (student interview, group A1)

Students emphasised repeatedly the importance of the social element of their teaching/learning environment. Students and tutors worked and lived together for a period of 2 weeks. Many students stated that as a result of this familiarity with the tutors they were more likely to ask questions, and express their own opinions, about the geology of the field area. Furthermore, discussions with students suggest that students' ideas about right and wrong in earth sciences are strongly influenced by their interactions with professional earth scientists during the field course:

S1: We expect them [tutors] to know everything. But the thing is they don't know everything and that's what we can't get used to. They don't actually know what happened there [the geological history at Mill na Claise]. They have their own theories, which is fair enough, and they can probably prove them with their data, but nobody knows for definite, which is really hard to get used to because we're always used to a right answer.
S2: Even up to the end of 'A' Level [two-year courses in the UK usually taken by 16 - 18 year old students] there was a right answer (...) you were either wrong or right, but now you're almost making the answers yourself. (student interview, group A1)

Students and tutors were immersed in the culture of geological field study for the whole of the two week residential course. Many students felt that this intensity of activity focused and motivated them:

> *It was quite bizarre the other day because we were all in the pub the other evening (...) and all around you you could just hear people talking about what they'd been doing in the field and talking about their different views (...) and you were like "Oh God, it's eight o'clock in the evening and people are still talking about geology".* (student interview, group B1)

Discussion

The course organiser's learning aims for the field course include many sophisticated epistemological activities: developing alternative interpretations of a single data set; comparing and evaluating multiple interpretations; developing and justifying a personal interpretation of geological history by drawing upon local field data and 'text book' knowledge of geological processes. In this sense the field course contrasts with other laboratory work in university science education in which a principal aim might be to acquire competence in using a technical procedure (e.g. the preparation of potassium wires in an argon-filled environmental chamber), or to establish the value of a known physical quantity (e.g. the electrical resistivity of potassium). Whilst such aims may be entirely legitimate, typically they do not provide contexts in which students are required to exhibit sophisticated positions concerning the relationship between knowledge claims and data. From an epistemological perspective, the field course examined here is a distinctive teaching and learning activity, at least compared with 'typical' university laboratory work activities in the early years of university education in the UK, and perhaps in most European countries.

This study has shown that the earth sciences field course was broadly successful in getting students to engage with epistemological issues in the context of their field studies. Students gathered a wide range of data, generated their own interpretations, justified their interpretations using available evidence, and compared and evaluated competing interpretations. Many students were very clear that their goal was to provide a 'personal interpretation' that they could justify using the available evidence. This contrasts with the goal of 'finding the correct answer'; a position characterised by Nadeau & Désautels (1984) as 'naïve realism'. Students on the course appeared to appreciate that in the context of their field work no definitive answer was available. Many students stated that they had initially found this lack of a correct answer unusual and a little unsettling. Comments in the student interviews showed that the tutors had repeatedly emphasised that a single, correct answer was unavailable. Indeed the favourable ratio of students to tutors on this course (6:1), and its intensive residential nature, appear to have helped in reinforcing this message. Many of these issues are illustrated in the following quote taken from students' discussions with the author concerning their written responses to questions about the nature of science (reported in detail in Leach, Lewis & Ryder 1998):

> *S1: There's never a right or wrong [answer] (....) It's basically your opinion.*
>
> *S2: If you can prove what you've said with your data then it's right (...) No-one's ever lived long enough to see it happening so if you can say with your data that's what's going on and you can prove all the pros and cons for your theory then.*

S1: ..then you can accept that. That's what they [tutors] have been telling us.
S2: It takes some getting used to (...) it has only started to sink in with me [when I ask] demonstrators and lecturers "Is this right?" and they say "I don't know, how would I know if it's right?" (student interview, group A1)

Such an emphasis on 'personal interpretation' contrast sharply with the views of many students examined in previous studies, for example the view that knowledge claims in science are solely the result of careful observations and the meticulous collection of data (Nadeau & Désautels 1984; Ryder & Leach 2000).

There have been many calls for courses about the history and philosophy of science (inevitably including aspects of the epistemology of science) to be incorporated into science courses (Giere 1991; Matthews 1994). These courses need to recognise that there is no single view of the epistemology of science that can be applied to all science contexts. Hence aspects of the epistemology of the science need to be exemplified by drawing upon a range of science disciplines. However, such 'generic' courses will not provide the close contact with university students' chosen subject discipline. One striking feature of the field course described here is the extent to which engagement with aspects of the epistemology of science is embedded firmly in the context of the activities of earth scientists engaged in small scale field study work. Informal discussions with students on the course suggest that such an approach provides a natural and motivating context in which students can explore aspects of the epistemology of their discipline; and one in which the relevance of such issues is easily seen.

Recommendations

This study provides a list of epistemological learning aims identified by the course organiser. The data collected within the case study does not provide details of the extent to which such aims are discussed amongst the course organiser and tutors. This list could be used as a focus for such discussions both prior and during the field course. In addition the list could be discussed with students to help clarify what they are expected to learn from the field course.

The field course appears to be a teaching/learning context that is effective in enabling students to exhibit sophisticated epistemological views. The intensive interaction with the tutors as professional earth sciences, and the use of student group work, are key features of the field course. It is recommended that these features of the course be recognised as valuable and unique by organisers of university earth sciences courses. This is particularly relevant for those courses in which time and financial constraints are leading to pressure to reduce the amount of field study.

When working within their peer groups students tend to collect data at particular sites in the field study area without interpreting their data in terms of what they know of the geological history of the site as a whole. We suggest that tutors consider both formal and informal activities in which students are encouraged to generate historical accounts of their data. Formal activities could include group presentations of students' interpretations of their data towards the end of the field study period. Informal activities could include discussions with tutors about these geological processes during data collection in the field. It is also suggested that the development of students' ability to provide historical interpretations be built into the sequence of field work activities over the three years of the earth sciences course, and also incorporated into university-based tutorials and problems classes.

Acknowledgements

The author thanks all staff and students involved in the field course for their co-operation and support.

References

Cartier, J. L. and Stewart, J. (2000). Teaching the nature of inquiry: further developments in a high school genetics curriculum. *Science and Education*, 9, 247-267.

Giere, R.N. (1991). *Understanding Scientific Reasoning (3rd ed.)*. New York: Holt, Rinehart and Winston.

Larochelle, M. and Désautels, J. (1991). 'Of course, it's just obvious': Adolescents' ideas of scientific knowledge. *International Journal of Science Education*, 13(4), 373-389.

Leach, J., Lewis, J. and Ryder, J. (1998). Learning about the actual practice of science: three case studies of undergraduate labwork from the UK. Leeds: Centre for Studies in Science and Mathematics Education. Available at http://education.leeds.ac.uk/~edu-jr/references.html

Leach, J., Millar, R., Ryder, J., Séré, M-G., Hammelev, D., Niedderer, H. and Tselfes, V. (1998). Survey 2: Students' images of science as they relate to labwork learning. Working Paper 4, Labwork in Science Education Project. Leeds: Centre for Studies in Science and Mathematics Education. Available at http://education.leeds.ac.uk/~edu-jr/references.html

Lederman, N. G. (1992). Students' and teachers' conceptions of the nature of science: A review of the research. *Journal of Research in Science Teaching*, 29(4), 331-359.

Nadeau, R. and Désautels, J. (1984). *Epistemology and the teaching of science*. Toronto: University of Toronto.

Matthews, M.R. (1994). *Science teaching: The role of History and Philosophy of Science*. New York: Routledge.

Mason, J. L. (1980). Annotated bibliography of field trip research. *School Science & Mathematics*, 80, 155-166.

Millar, R., Lubben, F., Gott, R. and Duggan, S. (1994). Investigating in the school science laboratory: conceptual and procedural knowledge and their influence on performance. *Research Papers in Education*, 9(2), 207-248.

Ryan, A.G. and Aikenhead, G.S. (1992). Students' preconceptions about the epistemology of science. *Science Education*, 76(6), 559-580.

Ryder, J. and Leach, J. (1999). University science students' experiences of investigative project work and their images of science. *International Journal of Science Education*, 21(9), 945-956.

Ryder, J. and Leach, J. (2000). Interpreting experimental data: the views of upper secondary school and university science students. *International Journal of Science Education*, 22(10), 1069-1084.

Ryder, J., Leach, J. and Driver, R. (1999). Undergraduate science students' images of science. *Journal of Research in Science Teaching*, 36(2), 201-219.

Tiberghien, A. and Megalakaki, O. (1995). Characterisation of a modelling activity for a first qualitative approach to the concept of energy. *European Journal of Psychology of Education*, 10(4), 369-383.

Chapter 4

LABWORK AND DATA HANDLING

Introduction

This chapter covers one of the central issues of quantitative labwork in science and in science education: the issue of data handling and deducing conclusions from data. It is related to the problem of measurement errors and students explicit or implicit beliefs about that. Both studies determine the effectiveness of learning outcomes with surveys, getting both results of increased understanding and specific deficiencies.

Leach analyses data handling in a special laboratory setting, where data are not gathered by students but they are supplied with secondary data at the start. This form of lab is called "dry practical". In this setting, chances are very high that the focus will be on issues of data handling. The study was carried out in the field of biochemistry, and more specific in that of enzyme kinetics. The research focus is on the role of theory in data analysis. The method applied to the research question was that of content analysis, while on the other hand the author also gives an assessment of student understanding of data handling and data analysis and addresses epistemological issues, in particular analysing the conclusions that from the students' perspective could be supported by the data.

Evangelinos, Psillos and Valassiades present a new teaching approach to measurement and data treatment that uses "probability" as a fundamental concept. The authors intend to replace students' conceptions relating to "error" by a probabilistic view of "uncertainty". An innovative teaching sequence on single measurements, repeat measurements and correlation of measurements is described. The whole approach is based on a Bayesian and metereological modelling of measurement and data treatment. In the evaluation, results from this innovative approach are compared to results from a more conventional approach. Students' initial views about the exact or approximate nature of theory and experiment were successfully changed to a deeper understanding of the nature of the "true value".

D. Psillos and H. Niedderer (eds.). Teaching and Learning in the Science Laboratory. 163-164.

The Use of secondary Data in Teaching about Data Analysis in a First Year Undergraduate Biochemistry Course

John Leach, University of Leeds, UK

Abstract

This chapter presents a case study of teaching undergraduate students about methods of data handling in biochemistry. The case study aims to identify appropriate curriculum aims about data handling in biochemistry. Methods used include performing a content analysis of the subject matter domain and student surveys and interviews. After teaching, students' understanding of central ideas from the content analysis was limited. Finally, ways of making 'about science' content more prominent in established labwork activities are considered.

Introduction

To date, research on teaching and learning in science education has tended to focus on specific conceptual areas. The (continental) European didactics tradition has worked through established methods to identify and justify conceptual content as appropriate for students at given ages and stages of their science education, and to evaluate teaching approaches which aim at promoting conceptual understanding (see Sjøberg 1996; Tochon 1999; Lijnse, in press). However, as argued by Leach (Chapter 1 *of this volume*) relatively little research has been done to identify what is appropriate curriculum content if our aim is to teach students about the nature and functioning of science itself, or to equip students to conduct laboratory investigations where they have to make decisions about data collection, data analysis and data interpretation. The purpose of this case study is to make a small contribution to the process of identifying and justifying content relevant to student labwork, by focusing on one labwork activity in biochemistry, and considering how that content might be taught more effectively. At the outset, it is worth underlining the fact that although the content to be identified is clearly subject-matter specific - it relates to biochemistry - it goes significantly beyond the conventional approach of listing key concepts and how they relate to one another. Rather, the content to be taught involves teaching students how the biochemical knowledge found in textbooks can be used in the specific situation of the undergraduate teaching laboratory.

Teaching approach

This case study focuses upon students in the first year of an honours degree in Biochemistry and Molecular Biology. It focuses upon one session of 3 hours' duration. The content of the module includes regular lectures and weekly 3 hour sessions of labwork. A small number of labwork sessions are termed 'dry practicals' as their aim is to teach about data analysis using given data sets, rather than using collected data.

D. Psillos and H. Niedderer (eds.). Teaching and Learning in the Science Laboratory. 165-178.

The 'dry practicals' were designed by lecturers in the department of Biochemistry and Molecular Biology to address a perceived difficulty amongst students in data processing during labwork. In particular, laboratory teaching time tends to focus on data collection rather than data analysis, the students taking data sets home with them to work on after the labwork session. By using secondary data the intention was to shift the focus of laboratory teaching time on to data analysis.

The 'dry practical' described in this case study is based upon laboratory data about enzyme kinetics, a standard area of content for students beginning university studies in biochemistry. Students arrive at this labwork session having already attended a lecture on enzyme kinetics. In addition, they are advised to read relevant chapters of a standard biochemical textbook. Students also have handouts for all labwork in advance of the class, though teachers do not assume that these have been read. 'Dry Practicals' are carried out in the same room as all labwork, and students are put into groups of 18 to work with one 'demonstrator' throughout the course. Demonstrators do not generally have a role in designing teaching activities, and their only access to the rationale for the teaching is through printed information supplied in advance. The labwork session starts with the demonstrator giving a short talk about the task to be undertaken. In dry practicals, the students are encouraged to work through the activities in pairs, to encourage collaboration and discussion of difficulties.

The activities that had to be carried out can be summarised as follows:

- Data are presented to students about initial rate of catalysis for an enzyme (V_0), at various initial substrate concentrations ([S]). Two measurements of V_0 are given, with a calculated mean value. Students are asked why V_0 is measured in duplicate.
- Students are asked to define Km and Vm, and relate these constants to a sketch graph of V_0 against [S] for enzymes following Michaelis-Menten kinetics. Students are asked to explain why this graphical method is unsuitable for determining accurate values of the constants.
- Students are told how to draw a Lineweaver-Burke plot (a double-reciprocal plot of $1/V_0$ against 1/[S]), and are asked to draw a plot to determine values for Vm and Km.
- A second data set is presented for the same reaction in the presence of an inhibitor. The students are asked to draw a second Lineweaver-Burke plot and determine values of Vm and Km. Using these values, the students are asked to interpret the nature of inhibition exhibited by the inhibitor, and to explain this in terms of a model of enzyme binding.
- Students asked to determine a value of the inhibition constant KI from instructions, and to use this value to interpret the relative strength of binding of the inhibitor and the substrate to the enzyme according to a given algorithm. All equations are presented for the students, and no information is given about the relationship between the equations and the Michaelis-Menten model of enzyme kinetics.

Research questions

The overarching aim of this case study is to identify appropriate curriculum content about data handling in biochemistry. This aim was addressed by considering the following research questions:

- What are the fundamental ideas used in biochemistry textbooks and by biochemistry lecturers, for presenting enzyme kinetics to first year students for the purpose of measuring and interpreting kinetic data in the laboratory?
- What levels of understanding of content identified as central to the teaching did students show following the 'dry practical'?

Research methods

Content analysis

The data source for this analysis consisted of the standard textbook used on the undergraduate course was analysed for its treatment of enzyme kinetics (Stryer 1995), the lecture notes provided for students, and laboratory notes for the 'dry practical'. These documents were read, and key fundamental ideas were extracted. In particular, the laboratory notes for the 'dry practical' were analysed activity by activity, and a list of knowledge and skills required to complete the 'dry practical', or that might be learnt from the 'dry practical', was proposed. This list was then used for designing diagnostic questions for students, as described below.

Assessment of student understanding

A survey instrument was produced to assess student understanding of the fundamental ideas generated from the content analysis. The design of this survey was informed by ongoing work in the LSE project to design a survey to probe students' understanding of the nature of science (see Chapter 1). The survey was administered to a sample of 48 students, selected randomly from the 500 students who follow the course. Six students, selected from this sample, were interviewed with the aim of validating interpretations of written responses, and providing further insights into the students' reasoning.

Results

Content analysis

The significance of enzyme kinetics

Knowledge about the kinetics of an enzyme's reaction with its substrates is centrally important to biochemists. Information about kinetics is used to interpret the role of particular enzymes in metabolic control. Many drugs and poisons work because of the kinetics of their binding to the active sites of enzymes (or other metabolically important proteins, such as haemoglobin). Students encounter two different models of kinetics during the course (see Table 1). These are the kinetics followed by allosteric proteins, of which haemoglobin is the classic example, and Michaelis-Menten kinetics, which are followed by most enzymes. Students are introduced to the process of obtaining kinetic information about enzymes in the laboratory by considering the case of enzymes which follow Michaelis-Menten kinetics. The maximum rate of reaction is defined as V_m. The Michaelis constant, K_m, is defined for enzymes following Michaelis-Menten kinetics as the substrate concentration at which the initial rate of reaction is $V_m/2$.

Table 1: Knowledge and skills that might be required to complete the 'dry practical', or that might be learnt from the 'dry practical'

Activity presented to students	Knowledge and skills that might be required to complete the 'dry practical', or that might be learnt from the 'dry practical'
Data on V_0 and [S] given	ESSENTIAL: The meaning of V_0.
	POSSIBLE: The experimental determination of V_0[1].
Interpret why two measurements of V_0 at given [S] made, and why mean value used	ESSENTIAL: Measurements have errors associated with them[1].
	POSSIBLE: The sources of errors[1], implications for knowledge claims about values for constants[1], types of inhibition, strength of enzyme-substrate and enzyme-inhibitor binding[2].
Define K_m and V_m, and show significance on a sketch graph of V_0 / [S]	ESSENTIAL: Understand the Michaelis-Menten model to the extent of appreciating the relationship of V_0 and [S][2].
Explain why graphs of V_0 / [S] not useful for experimental determination of Vm and Km.	ESSENTIAL: Understanding of graphical techniques.
Use Lineweaver-Burke plot to determine Vm and Km.	ESSENTIAL: Be able to draw line of best fit[3]. Know the algorithm for determining Vm and Km from Lineweaver-Burke plots[2].
	POSSIBLE: Reasons for using double-reciprocal plot. Relationship of Lineweaver-Burke equation to Michaelis-Menten equation[2].
Use data on V_0 at various substrate concentrations in presence of an inhibitor to determine V_m and K_m, and hence comment on the nature of inhibition in terms of enzyme binding	ESSENTIAL: Algorithm for changes in V_m and K_m in various kinds of inhibition. Understanding of inhibition in terms of enzyme binding.
	POSSIBLE: Significance of K_m as an indicator of the strength of enzyme-substrate binding in terms of the Michaelis-Menten model[2].
Calculate inhibition constant from a given equation, and determine the strength of inhibition according to a given rule	ESSENTIAL: Algebraic substitution into a formula, expressed in standard biochemical terms (which are superficially different from standard mathematical terms)[2].
	POSSIBLE: Relate K_m and K_I to Michaelis-Menten model[3].
All data given for SAME enzyme	POSSIBLE: Not all enzymes follow Michaelis-Menten kinetics.

Footnotes:

1 Students' understanding of this issue was probed in the survey and interviews.

2 Students' confidence about their understanding of this issue was probed in the survey and interviews

3 Probed at interview only.

K_m is significant because it is used to indicate the strength of binding between substrate and enzyme, and changes in K_m in the presence of inhibitors is used to indicate the mechanism of inhibition (i.e. competitive, where an inhibitor competes with the substrate to bind to the enzyme's active site, or non-competitive, where the inhibitor changes the structure of the enzyme so that the kinetics of the active site change).

In order to understand how values for V_m and K_m for a given enzyme and its substrate are determined in the laboratory, students have to appreciate how models of kinetics (such as the Michaelis-Menten model) are used to predict and interpret kinetic behaviour, and appreciate how these models underpin the validity of standard laboratory algorithms for determining the values of constants. For example, they need to recognise that the significance of V_m and K_m derives from the Michaelis-Menten model, and K_m has no meaning for allosteric enzymes. They also need to appreciate the rationale behind algorithms for making and processing repeat measurements, and interpreting data about changes in K_m during inhibition. A further issue relates to the extent to which it is valid to use data about an enzyme's behaviour when buffered at a given pH in a spectrophotometer cuvette to indicate its behaviour in a specific cellular environment. As the 'dry practical' was based on secondary data, the extent to which students appreciated where this data had come from remains open to question.

Fundamental ideas identified

In order to identify fundamental ideas about enzyme kinetics that were central to the teaching, each activity involved in the 'dry practical' was considered in the light of the content analysis presented in the last section. Knowledge and skills that might be required to complete the task, or that might be learnt from the task were then postulated as *essential* or *possible*. This list of knowledge and skills formed the basis of the design of data collection instruments to identify teachers' aims for the 'dry practical', and students' understanding following teaching (see Table 1).

Assessment of student understanding

Survey questions were designed to investigate students' understanding of relationships between kinetic data, and scientific knowledge claims about enzyme kinetics as identified in Table 1.

Two types of data were collected and analysed for this part of the questionnaire. In the first instance, students' written responses to each question were read by one researcher, and recurring statements from students were noted. A similar process was repeated using the transcribed interviews with students. In this way, a coding scheme was generated. The coding scheme was then checked independently by one other researcher, differences of opinion being resolved until a mutually acceptable coding scheme was agreed.

Findings for selected questions are reported in turn, drawing upon both students' responses to the questionnaire and transcriptions of interviews.

Question 1: How do you think a biochemist would go about collecting data from which V_0 could be calculated at a given concentration of the enzyme?

In enzyme-catalysed reactions, values for V_0 are usually calculated by following the disappearance of the substrate, or the appearance of the product spectroscopically. Spectrophotometry is a relatively simple technique that is often used to follow changes in the absorbency of a solution. Samples of substrate of different known concentrations are prepared, and the rate of reaction in the presence of an excess of enzyme is followed. Tangents are then drawn from which initial rates of reaction at the given substrate concentration are calculated.

Students were asked to write an answer to this question in their own words. Responses were coded into 5 categories. Any response indicating a method and data source from which an initial velocity could be calculated (e.g. drawing tangents to a graph of rate against time at a given initial substrate concentration) was deemed to have an acceptable level of understanding, and were coded *Correct.* Of the remaining responses, many made reference to procedures that might well be undertaken, but did not refer specifically to how an *initial* rate would be calculated. These were coded *Ambiguous.* In the interviews, some students who initially gave answers to the question that were coded *Ambiguous*, gave *Correct* responses following probing by the interviewer. It is therefore possible that many students whose written responses were coded *Ambiguous* did have an understanding that would have been coded as *Correct.* Of the remaining responses, some contained a specific error (such as varying the temperature or varying the enzyme concentration). These were coded *Incorrect.* The remaining responses were coded either *No response* or *Don't know*. Table 2 shows the frequency of coding decisions using written surveys only for this question:

Table 2: How Would V_0 Data Be Collected?
[Written Student Responses]

Coding category	No.	%
Correct	5	*11*
Ambiguous	16	*36*
Incorrect	4	*9*
No response	11	*25*
Don't know	8	*18*

These results suggest that many students in the sample did not know how data about V_0 might be collected. Even if many of the students whose responses were coded *Ambiguous* did in fact know how the data might be collected, it appears that their knowledge was such that they were unable to make it explicit in response to the question. The data presented in Table 5 suggest that, at the end of the activity, few students were able to make explicit statements about the likely origin of kinetic data.

In this practical, you determined values of Vm and Km. What do you think are the major sources of error in the values that you calculated?

Students were asked to write an answer to this question in their own words. A similar procedure was used to devise a coding scheme as for the last question. 10 coding categories were generated using students' written responses and transcribed interviews. Students often mentioned more than one possible source of error; a coding decision was recorded for each source of error mentioned.

Many students referred to the process of drawing lines of best fit on graphs, and determining values for constants by extrapolation, as a source of error. Such responses were coded *Graphing*. In most cases, responses did not elaborate reasons for this error. However, a number of written comments and the interview data provide evidence that students may have very different understandings about the nature of the errors involved in graphing. In a number of cases, it seemed that students were referring to human errors in graphing, that could be solved in an unproblematic way, as illustrated by the following response:

> S: *If the points were plotted into a computer you could have much more accurate figures for Vm and Km*[1].

For many students, the process of drawing a line of best fit and calculating values for constants such as V_m and K_m appeared to be viewed as algorithmic. If data did not fit perfectly onto a line, this was seen as being resolvable by taking additional measurements for points that did not fit on the line:

> S: *There's only one point on our two lines that didn't fit wasn't there? There was only 1 point out of the whole of the two lines.*
> I: *OK, but you reckon that if you did enough repeat measures you'd get to a point where you were confident that they would intercept?*
> S: *Yes.*

Making repeat measurements is an appropriate procedure for determining confidence in measured values. For these students, however, it appeared that the implicit assumption was that if enough measurements were made it would be possible to know a 'true' value for the measured quantity. By contrast, other students stated explicitly that drawing lines through points was a process of estimation with associated errors:

> S: *The graphs cannot be entirely relied upon, because the data did not encount [sic] for a true straight line. The straight line hence is only an estimate.*

Such responses were treated as tenuous evidence that students might view values for V_m and K_m as estimates with associated errors.

Some students stated that human errors (e.g. careless pipetting) or errors in instrumentation (e.g. the spectrophotometer, the calibration of pipettes) might be the source of inaccuracies in the calculated values for V_m and K_m. Such responses were coded *Human* and *Instruments* respectively. Other responses referred to experimental errors in general, and were coded *Experimental errors.* Some students suggested that there was error associated with only using 2 measurements of V_0; such responses were coded *Quantity of data.* In the interviews, it appeared that some students assumed that 'true' values for measurements could be determined if enough measurements were made, though in practice no student articulated this position explicitly. The following extract of transcript shows the type of comment that was taken as indicating this point of view:

> S: *... you'd have to take more values to get a more accurate average.*

---

[1] Conventions used for reporting transcript: I denotes talk by interviewer, D denotes talk by a demonstrator, (...) denotes omissions within an utterance; when this symbol is placed in the margin, it denotes the omission of one or more whole comments within an exchange. Comments in square brackets [ ] have been added to ease comprehension by the reader.

I: Supposing we took ten or fifteen values and we still got A coming out at 0.147 and B coming out at 0.149. Would we then know? (...) How many more values do we need to take before we know?

S: I'm not sure (...) Ideally, a lot. (...) If there's more values then (...) the extremes won't be as significant.

Some students stated that rounding errors in calculations might contribute to errors in values for V_m and K_m; such responses were coded *Rounding errors.* One response referred to errors in estimating values of V_0 from measured data, and was coded *Estimation.* Responses which were not understood by the researchers were coded *Other.*

Table 3 shows the frequency of students' responses for this question using the written answers only:

Table 3: What are the major sources of error in the values of V_m and K_m calculated? [written student responses]

Coding category	No.	%
Graphing	28	*64*
Human	4	*9*
Instruments	2	*5*
Quantity of data	4	*9*
Rounding errors	2	*5*
Experimental error	2	*5*
Estimation	1	*2*
Other	2	*5*
Don't know	2	*5*
No response	6	*14*

In general, it appeared that most students saw errors in the values of V_m and K_m as deriving from the experimental procedures used (accuracy of pipetting and calibration of the instruments, the number of measurements taken, the graphical procedures used). Overcoming such errors was seen as a straightforward process involving being careful as an experimenter, obtaining more accurate instruments and drawing graphs more carefully. In effect, it appeared that students thought that 'true' values for V_m and K_m could be determined with appropriate equipment. Of course, such procedures might well increase confidence in estimates of V_m and K_m, though they would not remove uncertainty about values completely. In addition, they would not reduce uncertainties relating to the *in vivo* functioning of enzymes compared to their *in vitro* kinetics. Only a few responses hinted that estimates of V_m and K_m always have errors associated with them.

Reasons why the data collected in science do not always agree exactly with the values that are predicted from theory

Students were presented with 5 possible reasons why values calculated from collected data (such as values for V_m and K_m) do not always agree exactly with the values that are predicted from theory. In each case, students had to tick whether they thought it was a possible reason, not a possible reason, or the they were unsure whether it was a possible reason or not. For each of the 5 reasons, space was given for students to explain their reasoning in their own words. In reading through student responses, it was apparent that the wording used for the 5 possible reasons was understood differently by different students. During interviews, it was also apparent that students were confused by some of the wording. This issue is further discussed later in this section.

Table 4: Reasons for differences between measured values and values predicted from theory [written responses]

Response category	Possible reason		Not a poss. reason		Not sure	
	No.	%	No.	%	No.	%
Human error	43	*98*	3	*7*	0	*0*
Errors associated with the apparatus used to make measurements (pipettes, spectrophotometers etc.)	32	*73*	7	*16*	4	*9*
The process of making a measurement might affect the system and change the results	18	*41*	8	*18*	15	*34*
Theories are created to make predictions about idealised situations. For this reason, real observations will never be the same as theoretical predictions	24	*55*	8	*18*	10	*23*
Other	2	*5*				

The frequencies of students' written responses are shown in Table 4.

- Human error

The majority of students stated that human error during measurement might contribute to differences between theoretical and measured values. Examples of typical mistakes that could be made were often listed, though many students implied that such errors could easily be eliminated.

- Errors associated with the measurement instruments used

Most students stated that errors associated with apparatus might contribute to differences between theoretical and measured values. Students referred to apparatus being 'faulty' or 'going wrong' to justify such responses. It was interesting that no student referred to systematic errors associated with apparatus. A smaller number of students suggested that it should be possible to make perfect measurements if apparatus is sufficiently sophisticated and properly used, as illustrated by the following extract of transcript:

> *S: Well, my tutor always used to say that the spectrophotometer was always right, it can't do anything wrong. (...)*
>
> *I: (...) By actually setting the experiment up in the way that it was set up, and by doing all the things that we do (...) we might actually change the measurements in some way from the measurements that are actually predicted by the theory. Is that possible?*
>
> *S: It could be, but you'd have to set it up so that didn't happen. (...) It's possible to avoid it.*

- Theories are created to make predictions about idealised situations. For this reason, real observations will never be the same as theoretical predictions

This statement was poorly understood by students. However, some responses gave interesting insights into students' reasoning about the relationship between theoretical predictions and actual observations. 18% of students stated that this was not a possible reason why measured values and theoretical predictions may not agree. A number of such responses were justified by statements that seemed to suggest that it ought to be possible for predicted values and measured values to agree. For example, several students stated that the experimental conditions under which measurements were made were not always 'ideal', 'standard' or 'perfect', the implication being that under different conditions theoretical predictions and measured data would coincide. Amongst the 55% of responses saying that this was a possible reason why theoretical predictions and actual measurements differ, a number were justified using very similar reasoning. The frequency counts of student responses cannot therefore be treated as a reliable indication of student understanding.

Some students adopted an even stronger position, as is illustrated in the following quotations:

S: It should be possible if using the exact technique and compensating for any physical errors, under perfect conditions to obtain the theoretical result

S: Real observations can prove theory

S: Have to make [an] experimental observation to get an idea if theoretical answer is correct

Comparison of two data sets for one enzyme

This question was designed to investigate the extent to which students think that confidence in measured values can be judged from a set of measurements, or alternatively whether measurements are thought to be judged by a pre-existing known value. Students were presented with two sets of measurements of V_0 for the same enzyme with mean values, collected by different groups of students. The spread of values in one group was noticeably different from the spread of the other. The means differed by 0.001 μMol NAD min^{-1} g^{-1}. Both the mean values corresponded with a measured value in one or other of the sets of measurements. Students were then asked to tick one of 5 statements, which related to the inferences about the value of V_0 that could be drawn from the data. Space was provided for students to write open justifications of their responses.

The notions of *accuracy* and *precision* are often used in discussing confidence in estimated values. Accuracy refers to the extent to which an estimated value corresponds to an error-free value. An accurate measurement of V_0 would not therefore be subject to systematic errors associated with the assay technique used. Precision refers to the statistical confidence that can be ascribed to the estimate. The precision of a measured value for V_0 would therefore make reference to the spread of the data from which the estimate was made. Although statistical techniques can be used to determine the precision of an estimate, they do not give any insight into its accuracy.

In comparing the two data sets of measurements of V_0 for one enzyme, it might therefore be appropriate to refer to the precision of estimates of the mean values calculated by each group. It might also be appropriate to comment on the number of measurements made, perhaps suggesting that the two groups should pool their measurements to make one larger data set. It might be appropriate not to include certain measured values in calculations of the mean, if a case could be made that those points are anomalous in some way.

Table 5: Frequency of responses about the quality of data collected by groups a and b [written responses]

Response category	Number	%
Group A's results are better, because the range between the largest and the smallest measurement is less	15	*34*
Group B's results are better, because the measurements cover a wider range of values	0	*0*
Both sets of measurements are equally good	9	*20*
From this information it is not possible to tell which group's results are better	5	*2*
The only way to tell which group's results are best is to look up the value of V_0 in a data book	10	*25*
More than 1 box ticked	4	*9*
No response	1	*2*

The 5 statements presented to students were written to reflect reasoning about measured values previously identified amongst students in the literature (e.g. Séré, Journeaux & Larcher 1993; Lubben & Millar 1996). The frequency of students' responses are shown in Table 5.

- Group A's results are better, because the range between the largest and the smallest measurement is less

It was hypothesised that students who think that the quality of an estimated value can be judged from a set of measurements would select this response. In practice, this was the most commonly selected response by students (34%). The justifications offered for this choice by students indicated a variety of understandings, however. Many students referred to the relative errors associated with the measurements of the two groups, assuming that group A's results were therefore 'better', 'more accurate' or 'more consistent' in some way. One student stated that this is the procedure that (s)he had been taught at school. One student commented that group A's measured values seemed to lie on a straighter line than group B's measured values. One student's comment suggested an assumption that the spread of data points relates to the accuracy of measurement rather than precision:

S: If the values are true values they will be closely together (...)

These data do not therefore show that 34% of the sample appreciate the notion of precision, nor that they distinguish precision and accuracy. A more appropriate conclusion is that 34% of the sample looked at the spread of measured values as an algorithm for judging the quality of a data set, and amongst these students a range of epistemological views about the relationship between measured data and estimated values exist.

- Both sets of measurements are equally good

It was hypothesised that some students might think that the two sets of measurements were equally useful in estimating a value for V_0; this statement was therefore written to allow this view to be expressed. 20% of the sample ticked this box. Justifications were very varied. Some suggested that the students were relying upon simple rules and algorithms for judging the data sets. For example, several students said that the groups had made the same number of measurements. Some justifications said that the mean values calculated by the two groups were similar, suggesting that the quality of the data sets were being judged according to the calculated values of the mean rather than the measured values themselves. In general, students did not comment upon the experimental procedures used by the two groups.

The only way to tell which group's results are best is to look up the value of V_0 in a data book

Previous research (Lubben & Millar 1996; Séré 1993) has suggested that some students assume that the only way to judge the quality of an estimated value is to compare the estimate to a value stated in a data book. Such students appear to view values in data books as completely accurate and precise. Furthermore, such students do not appear to have any idea as to the empirical procedures that might be used to arrive at values in data books. This statement was therefore written to allow this view to be expressed. 23% of students ticked this statement. Their justifications seemed to imply that a 'true' value for V_0 was known and could be taken from a data book, and that the two groups' values for V_0 could then be judged against this value. The following response illustrates this position:

S: *You cannot tell which data is better from looking at them if the correct value is not known.*

The following transcript provides further illustration of the point:

S: *I would agree with 'E' I think. (...) Ultimately that would be what the value would be.*

(...)

S: *It might not, ones more (...) precise or whatever but not as, might not be accurate. Even though there is a smaller band you can't tell whether they're right or not. Or look at the individual values and see which one's nearest that correct value.*

Although this student made reference to precision and accuracy in her response, her final comment suggests that individual measured data points can be compared to a known 'true' value, and that this is how the quality of measured data should be judged.

In hindsight, it was not sensible to have contextualised this question in the teaching laboratory. Students might legitimately assume that the people who calculate values for data books use techniques and equipment that is less prone to error than that available to students. However, students' responses to this question provide strong evidence that many assume that accurate values for quantities such as V_0 can be measured in an unproblematic way.

Recommendations

Leach & Scott (1995) present a view of teaching involving the identification of clear learning goals by teachers in terms of the structure of the scientific knowledge to be taught, followed by a comparison with likely student starting points prior to instruction. In this way, the 'learning demands' imposed upon students can be identified and addressed through instruction. Table 6 proposes 5 learning aims for the dry practical, derived from the conceptual analysis presented in Table 1, the learning aims for the activity espoused by teachers, and common student understandings. Suggestions as to how the presentation of the 'dry practical' might be changed in order to improve effectiveness are also included in Table 6.

Table 6: Proposed learning demands and proposed modifications to teaching activities for the dry practical

Learning aim (derived from conceptual analysis)	Common student understandings	Proposed learning demands Proposed modifications to teaching activities
The meaning of *initial rate*, and how it might be measured.	Many students have no idea how V_0 data is collected.	Know how initial rate is determined. *Insert a question to this effect into the existing protocol for students. Prompt demonstrators about likely difficulties in notes.*
Measured values have associated errors.	'True' values are measurable. Errors are due to 'mistakes'. Statistical techniques give insights into accuracy rather than precision. It is only possible to judge an estimate from a known value in a data book, not the quality of the data set from which the estimate was made.	Come to a view of values of V_0 as estimates with associated errors. *Present mean V_o values calculated from 5 measurements. Present opposing statements about what should be taken as the value for V_0. Ask students to form a view and justify it. Prompt demonstrators about likely issues.*
Understand the definition and *significance* of V_m and K_m in terms of the Michaelis-Menten model.	Most students seem to know definitions for K_m and V_m, though few referred to how K_m indicates something about binding.	Come to see K_m as giving information about binding. *Present a short paragraph in students' notes for the practical.*
Relate data about the behaviour of enzymes to the Michaelis-Menten model.	The Michaelis-Menten model is a perfect description of the kinetics of enzymes, rather than a model of the kinetics of some enzymes under defined conditions. The Lineweaver-Burke method is an algorithm, which yields perfect descriptions of the kinetics of an enzyme.	Come to recognise factors that may account for differences between predictions from the Michaelis-Menten model, and data about the behaviour of enzymes. Such factors include *in vivo* and *in vitro* conditions, and whether the enzyme follows Michaelis-Menten kinetics. Come to recognise that the Lineweaver-Burke method has errors associated with it, and that it derives from the Michaelis-Menten model. *Present students with a number of contrasting viewpoints about conclusions that might be drawn from data. Include notes for demonstrators.*
Use kinetic data in the presence and absence of inhibitors to determine the nature of inhibition.	This was well understood, once the presentation of the equation was clarified by demonstrators.	Come to recognise how the equation for Ki relates to other well-understood equations. *Reword the existing text and activity.*

The changes that are being proposed to the teaching involve making epistemological issues involved in handling kinetic data in terms of a model of enzyme kinetics more explicit. For example, it is proposed to ask students about the conclusions that could be supported by the data available on a number of occasions. In addition, the possibility of designing data sets to bring epistemological issues to the fore has been exploited. It is recognised that students' epistemological understanding may well be tacit, and in some cases contrasting arguments are therefore presented to them, to facilitate discussion. As previously mentioned, the only medium available for communicating about teaching with demonstrators is textual.

Acknowledgements

The author thanks staff and students in the Department of Biochemistry and Molecular Biology at The University of Leeds for their contribution to this case study.

References

Leach, J. and Scott, P. (1995). The demands of learning science concepts: issues of theory and practice. *School Science Review*, 76, (277), 47-52.

Leach, J. and Scott, P. (2002). Learning science in the classroom: Drawing on individual and social perspectives. *Studies in Science Education.* In press.

Lijnse, P. (2000). Didactics of science: the forgotten dimension in science education research? In R. Millar, J. Leach and J. Osborne (Editors) *Improving science education: the contribution of research.* Buckingham: Open University Press, 308-326

Lubben, F. and Millar, R. (1996). Children's ideas about the reliability of experimental data. *International Journal of Science Education,* 18 (8), 955-968.

Séré, M-G., Journeaux, R. and Larcher, C. (1993). Learning the statistical analysis of measurement errors. *International Journal of Science Education*, 15 (4), 427-438.

Sjøberg, S. (1996). Science education in Europe: some reflections for the future association. In A. G. Welford, J. F. Osborne and P. H. Scott (Editors) *Science Education in Europe: Current Issues and Themes*. London: Falmer Press.

Stryer, L. (1995). *Biochemistry* (4th Edition). New York: W. H. Freeman.

Tochon, F. V. (1999). Semiotic foundations for building the new didactics: an introduction to the prototype features of the discipline. *Instructional science*, 27 (1-2), 9-32.

An Investigation of Teaching and Learning about Measurement Data and their Treatment in the Introductory Physics Laboratory

Dimitris Evangelinos, Dimitris Psillos and Odysseas Valassiades
Aristotle University of Thessaloniki, Greece

Abstract

This study investigates aspects of first-year Physics undergraduate students' understanding of measurement and data treatment after introductory instruction on measurement and data treatment. In the first part of the study we present key aspects of the design and structure of an innovative sequence, which employs metrological uncertainty and bayesian probability as primary concepts for the modelling of the content instead of the concepts of measurement error and frequentistic probability. In the second part we focus on interpretation of instrument readings by students who were engaged in this sequence. Besides we investigate the interpretation of instrument readings by students who were engaged in a conventional course that is based on a frequentistic scientific framework.

Introduction

Recent research on students' understanding of measurement data and their treatment has revealed that they believe in the existence of a true value, make a second measurement only to confirm the first one, tend to reject the variability of repeat measurements and some times even reject the innate randomness of data (Coelho & Séré 1998). At the university level, where formal statistical analysis of data is usually taught, students do not grasp the necessity of standard deviation and therefore do not use it when expressing a repeat measurement result (Séré, Journeaux & Larcher. 1993). Recent research attempts to interpret why students' observed theoretical, practical and mathematical competence is at odds with their understanding of data treatment by seeking deeply rooted beliefs in this area. Lubben & Millar (1996) suggested a model of progression of ideas concerning experimental data, while Allie, Buffler, Lubben & Campbell (2000) identified a 'point' and a 'set' paradigm, according to whether students consider repeat measurements as isolated points or as whole.

Evangelinos, Psillos & Valassiades (1997, 1999) investigated students' views about the quantitative nature of physical quantities before any instruction about data treatment and claimed that students may use different reasoning schemes in different contexts, evidenced by the use of expressions containing respectively "exactly", "approximately" and "between". According to 'exact' or 'point' reasoning physical quantities are conceived as exact quantities in the sense of real numbers geometrically represented as points on an axis. In 'approximate' reasoning physical quantities are semi-exact but still are considered as representing a unique numerical value, allowing for a small deviation around a central value. To a lesser extent, 'interval' reasoning may also be employed, to describe a quantity that may acquire one of many possible values within an explicitly stated interval. Students seem to shift easily among these reasoning schemes according to the context of the tasks (Evangelinos *et al.* 1997, 1998). Finally, unless prompted to, students do not use the standard terms of 'measurement error' or 'accuracy', but instead they use abundantly the terms "precision" and "approximation". Thus it seems that for students with no

D. Psillos and H. Niedderer (eds.). Teaching and Learning in the Science Laboratory. 179-190.

prior labwork experience "precision" is the only attribute of instruments, which is used as an umbrella term, encompassing both scientific precision and accuracy.

Currently a consensus seems to emerge about the necessity of providing students with an explicit "justification of using statistics" and "an actual purpose to error analysis" (Séré et al. 1993). We maintain that the much sought "true meaning" of the mean and the standard deviation to be taught (Leach 1998), and therefore the epistemic standpoint of introductory data treatment depend on the specific choices made regarding the scientific knowledge to be taught. Currently there are two well-established frameworks regarding data treatment. On the one hand, the so-called 'frequentistic' framework defines probability as the limit of relative frequency of independent trials i.e. the ratio of the favourable over the possible outcomes. We call this the "conventional approach", since it is the one used by most laboratory manuals and is taught in several Physics Department in Greece and elsewhere. On the other hand, the Bayesian framework defines probability as "the degree of belief that an event will occur" and is more prominent in frontier research like high-energy physics and risk assessment (see for example D' Agostini 1999a&b). From the theoretical perspective, the two approaches yield in usual cases numerically similar results, but these are interpreted very dissimilarly within each framework. From a didactical perspective, we argue that the choice of the scientific framework to be used as reference for the introduction of students to data treatment is an important issue, open to investigation and debate.

In our case we are concerned with first-year undergraduate physics students (17 - 18 years) having negligible experience on quantitative experimentation, a situation met in several countries. We developed an innovative teaching sequence for the introduction of the essential concepts and procedures of measurement taking into account both research on students' conceptions in this area and the debates over data treatment within the scientific community. In the present study we focus on aspects of the teaching and learning of measurement data and their treatment in the introductory physics laboratory. In the first part of the study we present key aspects of the design and structure of our innovative sequence as well as on the main differences from conventional teaching that is based on the frequentistic framework; in the second part we explore students' understanding of instrument reading after their engagement in either the innovative or conventional teaching.

Teaching context

The sequence presented here consists of a 12-hour preparative module, taught in the first year Physics introductory laboratory course in the Physics Department of the Aristotle University of Thessaloniki. The development and application of this sequence can be conceived as a case of developmental research (Lijnse 1995). By developmental research we mean the interlacing of research development and application of this teaching sequence in a cycling evolutionary process enlightened by rich research data. Thus the innovative sequence outlined below was the result of several pilot applications led to a successive refinement of the content.

The structure of the sequence

The confusion existing in both the research and the teaching communities about the meaning of fundamental terms such as "precision" and "accuracy" (see for example Thomsen 1997) led us to seek a theoretical frame clarifying the issue. The analysis of the literature, previous research studies and our preliminary research data, as well as the fact that a growing number of scientific and technological bodies currently adopt the metrological and bayesian framework, led us to develop a model-based Bayesian approach to teaching data treatment. Developmental research led us to the didactical reconstruction of the scientific content to be taught in terms of three partial models corresponding to the familiar in the literature phases of "taking data", "treating data" and "using data" (Evangelinos *et al.* 1997, 1998). The modelling of the content that was used deviates from the model implied in most textbooks in the selection of primary concepts, the probabilistic framework used and its field of applicability. Thus the primary concepts we used are measurement *uncertainty* and *bayesian probability* instead of measurement error and frequentistic (i.e. based on frequencies of occurrence) probability.

In the metrological and Bayesian perspective, the objective of measurement is to increase the observer's uncertain knowledge about the true value of the measurand. Therefore reporting a measurement result requires the formulation of a probabilistic statement about the true value, which is based on uncertainty as *"a parameter associated to a measurement result, indicating the range that could be reasonably attributed to the measurand"* (ISO 1993). Thus in this approach one is more concerned with the question *"what can I reasonably claim about the true value?"* instead of *"what is the reliability of my data?"* which is the main focus of the frequentistic framework.

Thus, our approach deviates from conventional teaching like the one taking place in the Physics Department in Thessaloniki, since a) it uses an explicit to the students probabilistic modelling of measurement and data treatment, b) considers the necessity of using probabilistic reasoning in science as fundamental and not as a mere technique for treating data; c) uses as primary concepts uncertainty and bayesian probability, while undermining the concept of "error" and d) puts special emphasis on clarifying the quantitative nature and the field of applicability of the primary concepts, taking into account student's initial views.

According to these guidelines, we structured our innovative experimental sequence after several applications and changes. The main structure of the present version is shown in Table 1. A main difference of our sequence from conventional teaching is the special emphasis put in the case of single measurements (Units 1&2). Usual curricula undermine this case, because on the one hand frequentistic probability cannot be applied to single events, which are not repeatable, and thus the core statistical concepts can only be introduced in the repeat measurement phase. On the other hand, it seems that the implicit hypothesis in usual curricula is that in scientific and educational laboratories always more than one measurement are taken, which is often not the case in for example frontier research. In contrast, our approach is based on the assumption that deeply anchored students' initial views about the "precision" of single measurements may condition the understanding of both repeat measurements and regression line. In the work presented in this study we focus only on Units 1 to 4 and not the case of correlation.

In the first two units the primary concepts are introduced first qualitatively and later quantitatively. Students' initial views about the exact or approximate nature of theory and experiment are made explicit and challenged by for example discussion on published values of physical constants over the decades. The concept of uncertainty is built on the differentiation between discrete measurement numbers having limited number of digits and continuous arithmetic numbers that are real numbers. Thus concrete features of measurement and the cognitive aspects of uncertainty are linked by the absence of knowledge regarding the "missing figures" representing a measurand. Students apply the new knowledge by performing simple measurements using for example rulers and stopwatches and by discussing demonstrations of electrical measurements of varying precision and accuracy.

Table 1: The structure of the innovative sequence

UNITS	CONTENT / OBJECTIVES
	Case of Single Measurement
UNIT 1 (2 hrs)	Familiarisation and introduction of uncertainty and probability in measurement. Modelling of the measurement process. Concept differentiation: accuracy vs. precision, data vs. conclusions. Universal use of probabilistic intervals.
UNIT 2 (2 hrs)	Data processing using formal uncertainty and probability. Formal probabilistic confidence intervals. Data assessment and inference making. Linking between measurements, data processing and drawing conclusions. Applications.
	Case of Repeat Measurements
UNIT 3 (2 hrs)	Familiarisation with the non-repeatability and non-reproducibility of measurement. Introduction and elaboration of the statistical modelling. Standard deviation. Differentiation: data modelling vs. conclusions modelling. Linking between the single and repeated measurement cases.
UNIT 4 (2 hrs)	Probability distributions and confidence interval representations of uncertainty. Uncertainty Types A and B. Linking between measurement, data processing and drawing conclusions. Theory and experiment concordance.
	Case of Correlation of Measurements
UNIT 5 (2 hrs)	Correlation and independence of physical quantities. Extension of parameter uncertainty into model uncertainty. Measurement, data processing and conclusions drawing in the case of correlation.
UNIT 6 (2 hrs)	Elaboration of the regression line and the uncertainty of its parameters. Non-linear correlation using linear regression. Linking between measurements, data processing and drawing conclusions. Summary: relation of Theory to Experiment.

We allocate two units on single measurement in order to: a) provide the same probabilistic meaning to the uncertainty of the classical "of the least scale reading" rule and the standard deviation; b) to introduce probabilistic reasoning as early as possible in the sequence in order to address students' views; and, c) to fully elaborate from the beginning of the teaching the three fundamental aspects of scientific reasoning in the laboratory: measuring, data processing and conclusion drawing.

Units 3&4 deal with extending the field of applicability of the primary concepts and models taught in the previous Units in the case of repeat measurements. It should be noted here that conventional approaches, like the one taught in our Physics Department, strive to emphasise the variability of data by treating measurement as a random process, model it statistically and then apply probability theory. Therefore these approaches have to start by demonstrating variability, modelling it statistically and finally introduce probability as a necessary tool to describe repeat measurements. Our approach uses an inverse course: probability, which has already been taught as a primary concept, is directly applied to interpret repeat measurement sets as a whole. In particular, we address students' tendency to interpret varying results as a set of "approximately equal" measurements needing no probabilistic representation by a demonstration of a 6-digit instrumentation voltmeter giving fluctuating readings of the voltage of one and the same battery. This type of experiment helps students disassociate the modelling of data (reasoning about the readings as evidence) from the modelling of the conclusions drawn from data (reasoning about the probable true value). Thus two Gaussian distributions are introduced: the frequency distribution of the data and the probability distribution of the true value. The respective graphs are constructed, juxtaposed and bridged by the fundamental Bayes formula. In this way Units 3&4: a) promote the need for evaluating and using standard deviation in relation to the taught necessity of using uncertainty in Units 1&2; and, b) promote the idea that all measurements should be considered and treated as a set.

Finally, Units 5&6 extend the applicability field of the concepts of uncertainty and probability in the case of correlation of physical variables, in a way similar to Units 3&4. It should be noted here that as in the ending phase of each pair of units, a special recapitulating and linking phase is included, involving whole class discussion, aiming at emphasising the role and global applicability in science of the concepts and models treated so far.

Research design

Our study is part of a broad research program focusing on the teaching and learning of introductory data treatment. Our research is based on a series of investigations aiming at the diagnosis of students' views and reasoning about the nature of measurement and data treatment and on monitoring their development before, during and after conventional and innovative teaching in this area and. (Evangelinos *et al.* 1998). In this study we focus on the following two research questions:

1. "What concepts and types of reasoning do students use to interpret instrument readings after their engagement in an introductory innovative teaching sequence based on a Bayesian and metrological modelling of measurement and data treatment?"
2. "What concepts and types of reasoning do students use to interpret instrument readings after a conventional introductory teaching based on the frequentistic modelling of measurement and data treatment?"

The innovative sequence was applied to 16 students randomly selected from a population of 220 students from the first year of the Physics Department following two previous pilot applications. The present population comprises a typical entry to the introductory laboratory in the Physics Department like the ones in which our previous studies took place (see above). The students have entered the Department after national examinations that are strongly competitive. Normally they have a good theoretical knowledge of Physics but negligible prior laboratory experiences a situation that is met in several countries. Data in the innovative group were collected by several techniques including, interviews, video recordings and specially developed written questionnaires administered anonymously before, during and after instruction. In the present study we discuss aspects of questionnaire results focusing on both students' answers and their justifications.

Regarding the second research question, a group of 41 students was also randomly selected. These students participated in several classes taught by experienced lecturers, who run labwork in the Physics Department. The introductory conventional course is based on a frequentistic framework as mentioned above. It focuses on the concepts of measurement error, the Gaussian distribution of errors and the classification of errors into random and systematic. Regarding correlation, students are taught about graphing techniques, linear regression and its applications. In addition, these students had completed laboratory sessions about mechanics, electricity and heat involving the use of a variety of analogue and digital instruments experiments. For each session they had to compile a detailed written report involving data treatment.

The specially developed post-tests comprise semi-open written tasks, all asking students to justify their responses. The tests aims at elucidating students' understanding of taught core concepts and procedures concerning measurement and data treatment. The students who have attended either the innovative or the conventional teaching may provide correct responses to several common tasks. These tasks focus on the differentiation between accuracy and precision, the interpretation of a measurement results, the understanding and use of the formula <mean>±<standard deviation>. Most tasks comprise a number of items depicting simple measurement situations using analogue and digital instruments.

Due to space restrictions, in this study we present selected post-test results from one task, which is presented below, concerning specifically how students conceptualise the process of interpreting instrument readings. One main objective of this task is to investigate which forms of expressing a measurement result the students consider as scientifically valid, and specifically whether students consider confidence intervals as a plausible form.of measurement result. Besides it investigates whether a measurement result has a deterministic or a probabilistic status for the students. As part of the broad research an equivalent task was feasible to be addressed as a pre test too to the innovative group.

Results and Discussion

In terms of our modelling of the content we discriminate between three physical quantities involved in every measurement process: a) the true value of the measurand, which is an exact but unknown quantity; b) the instrument reading, which consists the perceptible data coming from measurement and is a known quantity; and c) the conclusion drawn from the reading, which is a probabilistic statement describing the information that is deduced from the data. Accordingly, the interpretation of readings requires a transformation of physical quantities from the level of readings to the level of conclusions drawn from them, thus involving a change of type of numerical representation from exact and certain quantities to probabilistically interpreted intervals. Therefore, in a number of pre- and post tasks we compare the quantitative properties students attribute to the physical quantities represented by instruments readings to the respective quantitative properties of the conclusions that can be drawn based on these readings. This type of comparison allows us to identify the types of reasoning students use in order to draw conclusion based on measurements.

Results from the innovative group

Pre-test task 2 (see fig. 1) asks students for a detailed interpretation of a digital instrument reading. It should be noted here that the design of this task takes into account the already diagnosed reasoning patterns mentioned above. Thus alternative (a) corresponds to 'exact' or 'point' reasoning, (b) corresponds to 'approximate' reasoning while (c) and (d) correspond to 'interval' reasoning. The difference between (c) and (d) lies in the probabilistic or not nature of the interval.

Task 2. An automated measurement set-up using a high precision chronometer gave a reading of 1.55834 sec. Which conclusion can a researcher draw from this reading and which not? Justify in each case.

(a) The true value of time is exactly 1.55834 sec.

(b) The true value is approximately 1.55834 sec.

(c) The true value definitely lies between 1.558335 and 1.558345 sec

(d) The true value probably lies between 1.558335 and 1.558345 sec

In the post test the chronometer was substituted by a manometer and in addition the following sentence (e) was added:

(e) If we repeat this measurement, we can determine the true value of pressure with as many decimal figures as we please.

Figure 1: The digital instrument task (Pre- and post Task 2)

From the perspective of the scientific model, either the conventional or the innovative one, the only correct alternative is (d), which describes a probabilistic conclusion in terms of a confidence interval. However, all four alternatives were included in the task and a justification is asked for each, since it was observed from previous investigations that students are not always aware that these cases are mutually exclusive. Thus the proportions of students agreeing to each alternative may not add up to 100%.

The majority of the innovative group (13/16) chooses an approximate representation, agreeing with (b): "Although very modern equipment was used for the measurement of time, we cannot ignore the slight possibility of an error or a deviation from the actual value. It's a fact that the value of time is very close to 1.55834". At the same time these students disagreed with (a) because *"several factors beyond the chronometer can affect measurement"* or because *"no instrument has infinite precision"*. Only 2/16 students agreed with the "exact" alternative (a), on the grounds that *"this conclusion is compatible with the capabilities of the instrument"*.[1]

Regarding the plausibility of intervals as a conclusion, only half (8/16) of the students provided a clear response to (c) and (d). These students do not seem to realise that the two cases are mutually exclusive and consider them as *"the most safe conclusion"* because *"we include a safety margin and thus increase the probability of reporting a correct result"*. Such arguments imply the existence of intuitive rules regarding the validity of results. Moreover, it seems that interval notation bring forward the intuitive notion of confidence in experimental results. As several students agreeing to (d) put it, *"we trust the precision of the instrument"* and *"modern technology offers high precision and increases the possibility of a successful experiment"*.

Responses from these and other pre-test items support the following tentative hypotheses regarding how students approach and interpret measurements:

(1) According to a naïve approach, a scientifically made measurement in a research environment using a high precision instrument can in principle provide

[1] These are verbal citations of students' statements.

the true value. This point of view represents 'exact' reasoning as applied to instrument reading. In an analogous way it is applicable to physical quantities calculated theoretically (Evangelinos 1998).

(2) According to a pragmatist view, "in practice" such an exact determination is not feasible for practical reasons: instruments have not infinite "precision", often are subject to "bad experimental conditions" and also "the human factor" may affect measurement. Therefore "approximately x" is considered as the only possible and scientifically valid measurement result. This "approximate" quantity is conceived as a small region around the true value, but it is not an interval, in the sense it is a unique but somewhat vague value.

(3) According to a criteria-based view, one is allowed to report the result as an interval, in cases where one is interested in reporting a "successful experiment" or when we want to ensure "trust in the result of an experiment". This type of reasoning may involve intervals, but it does not encompass probabilistic notions since the interval as a whole is conceived as a single "experimental result" which may be either "correct" or "wrong".

As is evidenced in students' responses to this and other tasks from the pre-test questionnaire, students may use simultaneously two or all three types of reasoning, because these serve different purposes and represent different aspects of experimental work. Thus students on the one hand use exact reasoning to accommodate experimental results with their views of physics as an exact science. On the other, students feel the need to allow for practical limitations during measurement thus "inventing" and using approximate reasoning to describe real-life measurement situations. Finally, although in their previous schooling they were never explicitly taught about criteria of agreement between theory and experiment, students intuitively are aware that some kind of criteria must exist and express them either in the form of a "close approximation" or in terms of "intervals" having a meaning close to concept of scientific confidence intervals.

After instruction according our innovative sequence, students were administered post test Task 2, which is equivalent to the respective pre-test, using a manometer for taking pressure measurements and including an additional question about the role of repeat measurements (see fig.1). Thus the task comprises the same four alternatives presented in the pre-test version including similar numerical values plus sentence (e) which represents a well-known misconception in the context of teaching about data treatment, namely that repeating measurements ad infinitum may lead to zero "error" or zero standard deviation.

All 16 students of the innovative group correctly disagreed with alternatives (a) and (b), indicating that they have successfully rejected 'exact' and 'approximate' representations for the conclusions about the true value. This was further evidenced from the level of argumentation in their justifications, which did not involve pragmatic arguments but made use of the taught knowledge about the nature of experimental results:

He cannot deduce this conclusion because all he can do from one measurement is to state an interval for the true value, since the true value has infinite decimal figures.(a) [1]

[1] Here and on the following pages, all text in *Italics* are verbal citations of students' statements.

"The only correct conclusion he can draw is that the true value lies within an interval e.g. from a to b with a probability of such percent". (b)

Regarding (c) and (d), 15/16 students also reply correctly, rejecting the deterministic character of (c) and agreeing with (d). This indicates that students have noticed that these alternatives are mutually exclusive which was not the case in the respective pre-test task. Students here also tend to use the knowledge taught:

In this case the experimenter uses an uncertainty of ±0.00005, therefore he states an interval which is plausible. However he should also use the concept of probability, since even for the interval he states one cannot be 100% certain. (d)

It should be noted that most students in their justifications explicitly name the interval as 'uncertainty' and use the concept of probability, although these concepts were purposely not included in the wording of the task. Thus, even the one student that agreed with both (c) and (d) stated that *"He is entitled to provide this result because the uncertainty he states has reasonable magnitude"*, indicating that he found case (c) plausible in terms of the data provided in the task.

Finally, in (e) all students correctly recognised that the repetition of measurements cannot improve the precision of a measurement ad infinitum, using either the taught difference between arithmetic and measurement numbers or the features of the measuring instrument:

No, because the manometer provides values of pressure having a specific number of digits, while the true value of pressure can have an infinite amount of digits.

No, this is not possible. Repeating a measurement cannot improve the precision of a single measurement.

Summarising the post-test results from the innovative group, almost all of the students replied correctly to this task and provided arguments using the knowledge taught. There was little evidence of persistence of approximate representations of the measurement results. Interval representations were successfully used by students instead of approximate reasoning, which was one of the main objectives of the sequence. In addition, students applied the new concepts of uncertainty and probability in a way that indicated that their field of applicability is the level of conclusions drawn from data and not the data themselves, which is the essence of our innovative sequence.

Results from the conventional group

Post task 2 was also administered to the 41 conventional students described in 4. In this task, 38/41 of the students disagreed with (a) and at the same time agreed with (b), providing numerous justifications having a strong pragmatic character:

No, because there always are errors, even when using digital or fully automated instruments. (a)

No, because the error is not due exclusively to the instrument but can be attributed to other factors too, like e.g. the observer. (a)

Yes, because we did use a high precision instrument but there are also some errors in the measurement (b)

Most of these 38 students (35/38) also correctly disagreed with (c) and agreed with (d). Thus 35 students disagreed with (c) on the grounds that the measurement error is not known:

"We cannot assert this because we don't know what errors exist in the instrument and in our readings" (c)

No, because the true value probably lies between these limits, but could as well be a bit higher or lower. (c)

The rest 3/38 students found the interval reasonable agreeing with both (c) and (d):

Yes, since the instrument has high precision, its error will be very small. However there are other factors that can influence a measurement, so these can lead to a larger error. (d)

Finally, the rest 3/41 of the students think that (b) cannot be justified unless more measurements are taken in order to calculate a standard deviation or imply that the specific measurement cannot be evaluated unless it is related to more measurements. These students possibly are aware of the necessity to interpret results in a probabilistic way, but are not able to evaluate a confidence interval without access to repeat measurements.

Regarding question (e), only 2/41 state correctly that the precision will not improve:

No, the instrument gave 5 decimals, so whatever measurements we take, it will give us 5 decimals.

About one fourth of the students (10/41) still think that it is possible to determine the true value through repetition:

Yes, because the more measurements, the more precision.

The more we repeat a measurement, the more we minimise error. Therefore, after a sufficient number of measurements, we can determine the true value.

These students conceive data treatment as a means to find the true value, thus indicating the still strong influence of exact reasoning on how students conceptualise the formalism of data treatment: the true value is an exact quantity, thus the process of measurement must somehow allow its determination.

Finally, more than half of the students (25/41) suggests that repeating measurements will increase the precision of this measurement, although not ad infinitum:

Yes, because the more measurements, the higher precision is obtained

No. The only thing that we will manage is to calculate the mean, thus approaching infinitesimally the true value.

Such views are compatible to the approximate reasoning scheme: repetition does reduce errors but cannot totally eliminate them, because for these students a measured quantity is by definition an approximate quantity.

Summarising the results of post-test task 2, the students of the conventional sample did show a high percentage of correct responses, but the justifications provided revealed an abundance of arguments using approximate reasoning. It should be emphasised here that the majority (35/41) of the students did not realise that the four cases (a) - (d) are mutually exclusive, but agreed both with both (b) and (d), providing similar arguments. We consider this as evidence of the fact that for these students, after conventional instruction, approximate, interval and probabilistic reasoning share common properties and use common arguments. Results from this and other post-test tasks indicate that for these students a measurement result is considered simultaneously as an approximate one, because it contains unavoidable errors, as an interval because errors result in an enlargement of the instrument reading, and as probabilistic because the sources and the magnitude of errors are unknown.

Conclusions

Our findings from the exploratory study of students' learning after conventional instruction presented in this study support the hypothesis that students were able to assimilate the concept of measurement error as well as the taught data treatment procedures within their initial approximate reasoning scheme. These students seem to employ an 'enriched' approximate reasoning scheme, which manages to use the concept of error for a deterministic (i.e. non-probabilistic) explanation of the discrepancies between experiment and theory. Thus, although the conventional approach attempts to model measurement errors probabilistically, the students prefer to view "error" as a quantity that merely expresses how well the measured value approximates the true value, ignoring its probabilistic nature. Thus students' probabilistic reasoning is limited and serves to describe the presence or non-presence of small or large errors. Accordingly, data treatment is viewed as a formal means to achieve either a "better approximation to the true value" (within approximate reasoning) or the true value itself (within 'exact' reasoning).

The innovative approach presented here attempted to attack the core of students' intuitive reasoning patterns mainly by making two choices driven by the analysis of the content to be taught. The first one was to abandon in the teaching the use of measurement "error", and replace it with the metrological concept of uncertainty. The second feature was to avoid presenting and introducing probability in the context of treating repeat measurement results, but introduce it as a primary concept, as a valid feature of scientific reasoning, and as an interpretative tool for quantifying uncertainty. Results from the specific innovative group presented in this study, as well as results from its previous applications suggest that the sequence helped students conceptualise the difference between exact and uncertain quantities and use successfully the taught concepts of uncertainty and probability in all cases of measurement, as was manifested in students' justifications.

Recommendations

According to the experience gained and our results, recommendations may be formulated for the selection of concepts and models to be taught in preparative labwork, especially in the case of students having negligible laboratory experience.

Preparatory labwork instruction has the crucial role of contributing to the establishment of firm links between theoretical and experimental approaches to physical phenomena. However, learners' intuitive concepts and reasoning patterns, dominated by 'exact' and 'approximate' reasoning, are very resistant to teaching. We suggest that the acceptance of measurement uncertainty and probability as valid features of scientific thought by the students requires more attention than is usually paid for in certain conventional approaches to introductory undergraduate labwork especially for students having negligible previous laboratory experience. We suggest that more attention may be facilitated if the university teachers gain insight on the characteristics and the potential of *the possible scientific frameworks to be used as reference for the introduction of students to measurement, data of treatment and evaluation of measurement results as well as the related educational reconstruction of scientific knowledge.*

Our findings seem to suggest that adopting an alternative to the widespread frequentistic theoretical framework for approaching laboratory data, which makes explicit the nature of scientific inferences based on them may help undergraduate physics students to understand data treatment. More specifically we suggest that an introductory sequence about data treatment should explicitly address students' initial views about instruments and measurement data. Elaborating more than usual the treatment of single measurements may offer opportunities to challenge students' 'exact' and 'approximate' reasoning schemes that represent deterministic images of scientific enquiry. Introducing uncertainty and probability as valid features of scientific reasoning from the beginning of an introductory course on measurement and data treatment may address views resistant to teaching and promote probabilistic reasoning.

References

Allie, S., Buffler, A., Lubben, F & Campbell B. (2000). Point and set paradigms in students' handling of experimental measurements. In: R. Duit (ed.) *Science Education: Past, present and future*. Kluwer academic publishers (in press).

Allie, S., Buffler, A., Kaunda, L, Campbell, B. & Lubben F. (1998). First-year physics students' perceptions of the quality of experimental measurements. *Int. J. Sci. Educ.* 20 (4) pp. 447-459

d'Agostini, G. (1999a). Bayesian reasoning in high-energy physics: principles and applications. CERN Yellow report CERNREP 99-03. Available in electronic form at http://library.cern.ch/cern_publications/yellow_reports.html

d'Agostini, G. (1999b). Teaching Statistics in the Physics Curriculum: unifying and clarifying role of subjective probability. Contribution to the Theme Issue of the American *Journal of Physics on statistical physics*, 1999.

Coelho, S. & Séré, M.-G. (1998). Pupils' reasoning and practice during hands-on activities in the measurement phase. Research in Science & Technological Education 16 (1), pp. 79-96

Evangelinos D., Psillos D. & Valassiades O. (1997). Measurement uncertainty in the introductory university laboratory: students' views and teaching objectives. Paper presented at the 1st International ESERA Conference "Science Education Research in Europe" Rome, 2-6 September 1997.

Evangelinos D., Psillos D. & Valassiades O. (1998). Students' introduction to measurement concepts: a metrological approach. In: D. Psillos & H. Niedderer (Eds.) Case Studies on Labwork in five European Countries (pp. 563-587). Working paper 7, Labwork in Science Education Project PL 95-2005.

Evangelinos, D. & Valassiades, O. (1999). Undergraduate students' views about the approximate nature of measurement results. In: M. Komorek, H. Behrendt, H. Dahncke, R. Duit, W. Gräber & A. Cross (Eds.) *Proceedings of the Second International conference of ESERA* (pp. 208-210). Kiel: IPN.

ISO (1993). Guide to the expression of uncertainties in measurement. International Organisation for Standardisation (ISO), 1 rue de Varembé, CH-1211 Genève 20, Switzerland.

Leach (1998). The use of secondary data in teaching about data analysis in a first year undergraduate Biochemistry course. In : D. Psillos & H. Niedderer (Eds.), Case studies on Labwork in five European Countries. Working paper 7, Project PL 95-2005 " Labwork in Science Education ". European Commission.

Lijnse P. L. (1995). "Developmental Research" As a Way to an Empirically Based "Didactical Structure" of Science, *Science Education* 79(2), 189-199.

Lubben F. & Millar, R. (1996). Children's ideas about the reliability of experimental data. *Int. J. Sci. Educ.* 18 (8), pp.791-806.

Séré, M.G., Journeaux, R. & Larcher, C. (1993), Learning the statistical analysis of measurement errors. *Int. J. Sci. Educ.* 15 (4), pp.427-438.

Thomsen, V. (1977). Precision and the Terminology of Measurement. *The Physics Teacher*, Vol. 35, pp. 15-17

Chapter 5

LABWORK BASED ON INTEGRATED USE OF NEW INFORMATION TECHNOLOGY

The new possibilities of using computer information technologies in science teaching are one of the most promising developments in the last decades. Yet, little research is known about the effectiveness of using computer technology in the context of labwork. To use computers integrated into labwork is one of the special possibilities, which is given through this technology. Computers can serve as tools for measuring, graphing, simulation, and modelling, to list only few of the rich possibilities. In this chapter all five studies report about new teaching approaches to integrate computer information technology into labwork and all of them report research results about their evaluation of effectiveness. Most of the studies determine both kinds of effectiveness. For effectiveness 1, mainly the CBAV method is used. For determining effectiveness 2, different instruments like tests, concept maps, or categorisation of levels of understanding are applied. In four of the studies, this has been part of a doctoral dissertation.

In the five studies of this chapter, the computer is used as a tool for various purposes: for data collection (MBL), for analysis and graphical representation of data, for model building with model building software (MBS), for simulation of a physical model and for demonstration of an interactive microscopic model. In all five studies, the contribution of integrated use of computers in labwork is analysed with respect to the aim "to link theory to practice". That means the computer is always seen as a tool to foster conceptual understanding of experiments. All five studies deal with different contents, but all from physics. Contents such as heat and temperature, mechanical oscillations, Newton's second law, optics, and electric polarization are studied.

Different empirical methods are used for evaluation of effectiveness in these five studies. Three of the studies use the category based analysis of videotapes (CBAV) described as method in Chapter 1 already. Other methods used are concept maps,

D. Psillos and H. Niedderer (eds.). Teaching and Learning in the Science Laboratory. 191-192.

semi-structured interviews, transcripts and their interpretive analysis, questionnaires, and qualitative video analysis.

In the study of Bisdikian, the computer is used for working with graphical representations. These graphical representations are seen as a separate world between the world of theories and models and the world of objects and events. Graphical representations are used in three different forms, coming from prediction of students, from simulations of a model, and from real time measurements. The learning process of students is analysed with semi-structured interviews, focussing on a deeper understanding of the concepts of heat and temperature.

The two following studies by Hucke and Fischer and by Sander, Schecker and Niedderer focussed on the integrated use of model building in labwork. One important result out of both studies is that model building in close relation to labwork helps for more use of conceptual knowledge in relation to the experiments. In the study of Hucke and Fischer, one result is, that knowledge acquisition in labs with integrated use of computer is not higher than in traditional labs. Results like this show that we must be careful to not exaggerate the expectations for better learning from modern information technology. In the study of Sander, Schecker and Niedderer, a detailed example shows how the computer model building can add to a theoretical view of the experiment, which is carried out.

In the study of Buty, model building is used with a special software Cabri-Géomètre used for geometrical optics. Again, the main research question is related to the link between theory and practice. Students modelling constructions are seen as a level between the world of theories and models and the world of objects and events. This intermediate level is called the materialised model. In his results, Buty sees an improvement of relations to physics concepts, whereas the link to the objects in the experiment needs further development.

In the study of Barbas and Psillos, the computer is used for a microscopic model of polarization in insulators in electrostatics. The learning effects of the use of this microscopic computer model is analysed. The microscopic models are studied in interrelation with real experiments with charged and uncharged bodies. Main results are about conceptual changes from a macroscopic view to different kinds of microscopic views to explain the results. The small-scale interactive simulations seem to support the intended mental representation.

Enhancing the Linking of theoretical Knowledge to Physical Phenomena by Real-Time Graphing

Garabet Bisdikian and Dimitris Psillos
School of Education, Aristotle University of Thessaloniki, Greece

Abstract

In the first part of the study we present key aspects of an innovative laboratory-based teaching sequence in the area of heat, which is characterised by interactive and real-time graphing features. The sequence aims at enhancing primary education student teachers' ability to construct links between heat theory and phenomena with the help of graphical representations. In the second part we present results concerning the development of student teachers' graphing skills, from a partial towards an integrated interpretation of graphs, combining both theoretical and phenomenological aspects of the domain.

Introduction

Science teaching and learning deals not only with the acquisition of knowledge about models and theories but also with the development of procedures which enable students to carry out experimental investigations and apply scientific knowledge in the description and interpretation of physical phenomena. Quite recently, science educators have investigated ways to improve the linking of theoretical models to practical activities by engaging students in labwork activities which develop both conceptual and procedural knowledge (Seré 1999). One approach towards such a linking, worth investigating, is graphing. Graphical representation can be considered as a bridge facilitating the linking between physical phenomena and the related content theory during any data handling process connected to school science laboratory work.

Attaching physical meaning to graph characteristics by linking physical phenomena with the relevant content theory is important both for comprehending graphical symbolic representations and understanding the models which a graph represents (Bisdikian & Psillos 1998a). Several studies, however, show that students, at different levels, can easily learn to read graphs but fail to link graph characteristics with conceptual models, often seeing graphing as a procedure disconnected from practical activities (McDermott 1987, Thornton 1995). It appears that content-related graph construction and interpretation are high cognitive processes which students sometimes find difficult to perform. The limited access to phenomena, the inability to isolate variables under investigation and the cognitive load of processing quantitative data points as abstract representations of quantities in a mechanical way are considered to be obstacles that deter students from attaching physical meaning to graphs (Leinhardt et al. 1990).

By introducing new technologies in labwork, mainly by real-time data acquisition from real or simulated phenomena, the relation of content theory with physical phenomena seems to be placed on a new potential basis (Rogers 1995). The capacity of computers to construct real-time graphs in parallel with the evolution of phenomena can help graphing skills development (Rogers 1995) and content knowledge acquisition (Linn et al. 1991). Graphical representations become a dynamic instrument, especially by introducing parametric interactive computer

D. Psillos and H. Niedderer (eds.). Teaching and Learning in the Science Laboratory. 193-204.

simulations and on-line data acquisition techniques (Thornton 1995). In spite of all these potential advantages, not much research has as yet been done on teaching approaches that help students to comprehend graphs and to develop skills in attaching physical meaning to specific content graphs (Roth et al. 1997).

In this context, the purpose of the present study is twofold. In the first part, we present key aspects of an innovative laboratory-based teaching sequence in the area of heat, characterised by interactive and real-time graphing features. In the second part, we present selected results concerning the development of student teachers' skills towards an integrated interpretation of graphs.

The teaching context

The case study was carried out at the School of Education of Aristotle University of Thessaloniki as a part of a wider research program about graph comprehension. In the frame of such research, a four session laboratory-based teaching sequence (total 12 hours) had been developed and taught several times to 3rd semester primary education student teachers, as part of a six-month compulsory course on "Experimental Science Teaching".

Modeling the acquisition of physical meaning of graphs

Research suggests that interpreting and constructing graphs are the essential skills involved in understanding and using graphical representation (Leinhardt, Zaslavsky & Stein 1990). We have argued elsewhere that in the case of Science, both skills need to be employed by students in order to link theory to physical phenomena. Such linking can be represented by a three-level model which relates physical phenomena, graphs and theories (Bisdikian 2000). We call this model the "GraPhys" model, after the words *Graphs* and *Physics* (Figure 1). The linking of the three different levels is represented by a set of arrows corresponding to specific interpretation and construction skills which a student should employ in order to bridge content theory with physical phenomena by means of graphing (Bisdikian & Psillos 1998b).

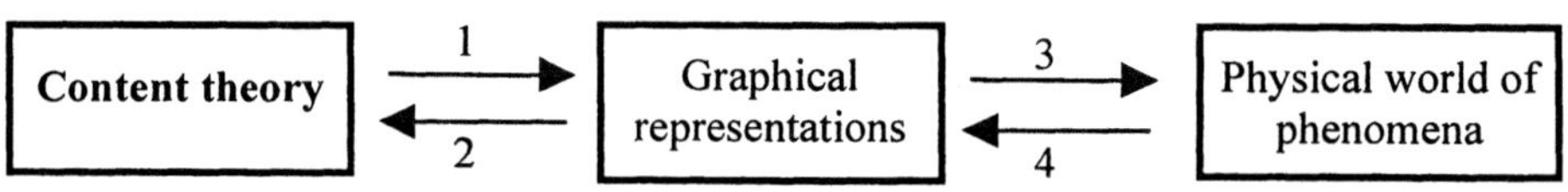

Figure 1. The "GraPhys" model

More specifically, according to the "GraPhys" model, students should be able to:

1) construct graphical representations to support conclusions or predictions concerning relations between variables (the relation is symbolised by arrow1);
2) read, interpret and compare graphs in order to extract relations between physical variables (arrow 2);
3) after interpreting graphs, plan experimental activities, predict or reconstruct events (arrow 3);
4) construct graphical representations to describe the evolution of physical phenomena or experiments (arrow 4).

Two main categories of skills which are related to graph interpretation and construction appear in the model. Interpretation at the *level of theory* is carried out in order to represent theoretical knowledge and relations between variables. At the *level of phenomena*, interpretation is necessary to represent the evolution of physical phenomena or practical activities. Similar skills refer to graph construction.

If a student argues at the level of theory and at the same time at the level of phenomena, then he/she comprehends scientific knowledge and attributes to graphs the role of a bridging link between the world of theory and the world of physical phenomena. Such a student is able to employ both categories of skills and attach physical meaning to graphs.

Concerning pre-instruction skills, our data suggest that primary education student teachers employ partial approaches to graph interpretation. This situation can be revealed by the variety of specific expressions, which these students use in interpreting several types of graphs (Bisdikian 2000). Broadly speaking, a number of students mainly refer to graphs in theoretical terms and as relations between variables. For example, in a task involving the interpretation of a temperature graph representing a water heating situation, a typical response is:

... here, the heat supply is doubled, so temperature increases more rapidly..,

In this case, graphs mean theory to this student. In terms of the "GraPhys" model, this approach can be represented by arrow 2.

Other students' responses mainly describe the actions of a person or events, the evolution of phenomena or practical activities, e.g. in the same task, another typical response is:

... we open the water tap more and the amount of water in the vessel increases...

In this case the student prefers to employ actions and events, which indicates that in his/her perception graphs are more connected to the physical world of phenomena. In terms of the "GraPhys" model, this approach can be represented by arrow 3.

Interlacing heat content and graphs

Our teaching aims at elaborating students' conceptions both at the level of content and at the level of graphing procedures and at establishing links between the world of phenomena and the world of theoretical models. In such a perspective, it is reasonable that the specific content to be taught should refer to such changes in the physical variables, that the introduction of graphs must be important both for content understanding and graphing skills development. Heat is considered to be an appropriate domain, a choice that does not exclude other areas of science. The fundamental physical variables (Q, T) are time dependent and interrelated, at an introductory level, by simple relations. Temperature and heat amount changes can be represented by time graphs, which are considered to be more comprehensible by students than non-time graphs (Leinhardt et al. 1990).

Research shows that several students hold alternative ideas prior to instruction and face considerable difficulties in understanding heat theory and phenomena at an introductory level (Kesidou, Duit & Glynn 1995). In particular, the differentiation of heat and temperature constitutes a major conceptual obstacle for students at several levels. In this respect, we consider that constructing and interpreting heat graphs is important, since by means of specific graphs, the students may be introduced to reasoning that can facilitate their understanding of the differentiation between heat and temperature.

In this frame, using a simplified "*heat flow*" model (Linn & Songer 1991), we developed four laboratory sessions, each lasting three hours, covering heat theory, the related physical phenomena and the graphing issues to be taught. The teaching sequence is expected to help students comprehend and differentiate the concepts of heat and temperature. Specific aspects of heat are covered, namely the concepts of heat and temperature, the relationship between heat energy applied, the mass and material of the heated objects and temperature change, changes in state and heat conduction. The relevant experimental field includes heating processes for different amounts of water or other materials, ice melting, water boiling and thermal exchanges between various objects in different temperatures.

At the same time, the sequence aims at promoting the appropriate skills which will help students carry out the fundamental graph interpretation procedures according to the "GraPhys" model, as described before.

Laboratory units involving real-time graphing

The slow evolution of heat phenomena which does not allow repeated heating experiments, the difficulty of parametric investigations and the absence of real-time multiple representations of changes are considered to be limitations of a conventional science laboratory. Such a weakness can be alleviated by employing real-time computer graphing (Rogers 1995, Linn & Songer 1991). In the present case study, the above limitations set the criteria for developing small scale parametric simulations of heat phenomena and for using temperature and heat flow sensors, connected via an analog/digital interface to a computer.

In this context, we developed a series of structured labwork episodes. The episodes constitute the core of the teaching and are carried out by pairs of students following worksheets or teacher instructions. Each episode aims at an integrated study of a concept or a phenomenon and its graphical representation and can include the following steps:

1) Step 1 (Graphical prediction): Students explicitly predict, by means of qualitative graphs, the evolution of familiar heat phenomena. This step may provide information about their initial ideas regarding graphs and content knowledge.
2) Step 2 (Hands-on experimenting): Students carry out hands-on experiments concerning the concept under study and construct paper-and-pencil graphs, to familiarise themselves with the procedures of experimentation.
3) Step 3 (On-line measurements): Students observe and reflect on the teacher performing the same experiments, using sensors in a computer-based environment. The on-line techniques help in relating real world variations with representations and enable a smooth transition to the simulated phenomena.
4) Step 4 (Parametric simulations): Students run a set of parametric simulations in order to investigate the relations concerning the variables of the domain. They witness the consequences of manipulating the parameters as relevant graph responses, allowing thus a qualitative study of the concepts.
5) Step 5 (Graph comparison): Students compare the graphs, which were obtained during the above steps and work out graphing drills. The comparison of graphs representing student predictions with graphs representing results from real or

simulated experiments is a procedure which may lead students to strengthen or reconsider their ideas (Linn & Songer 1991).

Such a structure of the laboratory episodes, by the use of graphs as a common method for describing variations, allows the didactical interlacing of heat content theory and phenomena and their graphical representation.

Research methodology

The present study explores the learning evolution concerning the development of graphing skills and content knowledge. The question which is investigated is:

What patterns of development regarding the ability to link content theory to physical phenomena by using graphical representations appear in students who participate in a teaching sequence characterised by interactive and real-time graphing features

In the study we refer to 20 4th-semester students (age 19 - 22), who were selected at random from a population of 80 students. Data for eliciting initial and final students' reasoning were obtained by interviewing each student before and after applying the teaching sequence.

Semi-structured interview tasks were developed, tested and gradually improved during pilot applications. During the pre and post interviews, the interviewer asked the students to describe and interpret temperature over time graphs connected to water-heating experiments and then to consider what possible causes could have resulted in such variations in temperature. Before instruction, the students were shown a graph with horizontal and inclined straight line parts, like the one shown in Figure 2, representing simplified temperature variations of a given amount of water in a vessel. After instruction the interviewer used a graph which represented the temperature variations of an ice-water system (Figure 3). Students had to take into account the temperature of the environment and that at some time interval, which they were asked to locate, melting of ice had taken place.

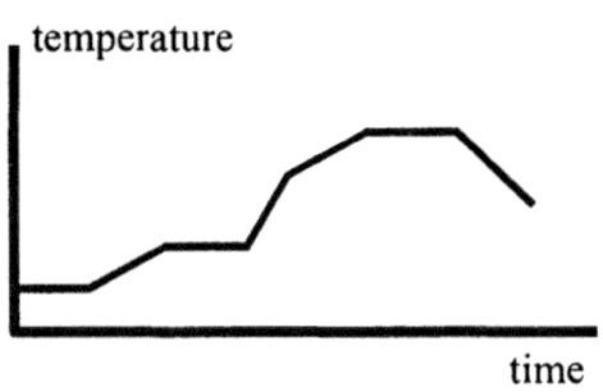

Figure 2. Initial temperature graph

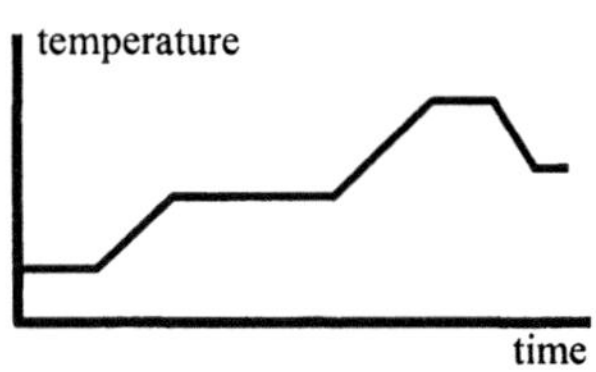

Figure 3. Final temperature graph

In order to foster unbiased reasoning concerning the changes, the interviewer posed open questions like:

... explain how you would describe the graph,.. why do you think this happens..

All pre and post interviews were audio recorded, transcribed and decoded to individual protocols. The analysis of the protocols is based on students' initial and modified responses on the specific tasks, according to vocabulary used. Responses are further analysed to reveal which mode of expression is followed, i.e. whether a student connects the changes to theoretical relations or to phenomena evolution and practical activities.

A "GraPhys" pattern was formed after characterising arrows 2 and 3 of the model according to the approach each student followed (Figure 4). Comparing each student's initial and final pattern facilitates the visualisation of his learning evolution. Intra-group comparison of the students' protocols and their "GraPhys" patterns revealed a series of different trends within the group.

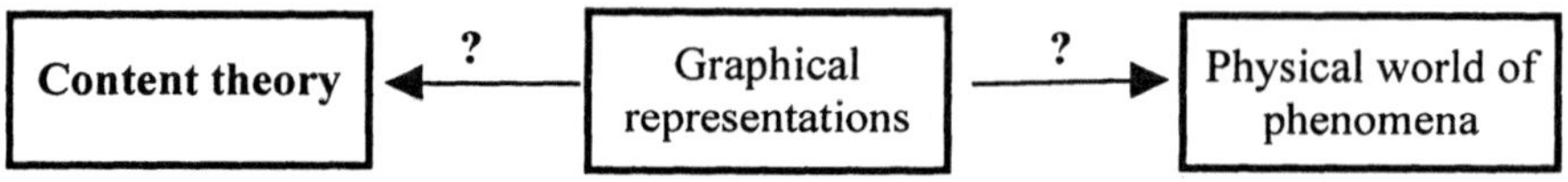

Figure 4. Frame for "GraPhys" model-pattern for each student

Research results

Three students, showing characteristic developmental patterns, were selected as representative cases. The first two students, Bill and Niki, had shown a low initial background regarding graphing knowledge, while Maria was a student with a more formal background. We present the first case in more detail so as to demonstrate the method we followed. The two other cases are discussed in less detail, focusing mainly on the differences they present between pre and post interviews.

The case of Bill

Bill is a typical student with little previous experience in graphing and science, but with good communication skills which provided a rich protocol. We present extracts from the initial and final interviews which show that this student underwent a remarkable development towards the intended linking process.

A. Pre-interview

Responding to the interviewer's question about graph interpretation, Bill stated that (numbers in brackets refer to parts of the graph in Figure 2):

Bill: ...suppose that water starts at this temperature [1] which is not zero, otherwise it would be ice, and has a temperature which we try to keep through a small hot surface which gives little heat...
...from [2] we start supplying the water with heat... mm... temperature from the heater,
...at [3] we raise the supply a little and we give greater heat at [4] and the temperature is rapidly up ..with a higher frequency
...at [5] we give 10° more heat,
...at [6] we keep it steady at 30°,
...and at [7] it goes down slowly..

At first sight, it is obvious that Bill was able to recognise the shape of the graph and to describe what kind of changes it represents. His expressions show that he connects parts 1, 3 and 6 to stable temperature and the other parts to a temperature rising or falling to greater or less degree.

We further analysed the student's response in order to get an insight into how he understood the meaning of the graph parts and which mode of reasoning he followed to interpret the changes. We noticed that Bill preferred to use terms implying the manipulation of variables as if these were physical entities. For describing changes in variables he adopted theoretical expressions containing both qualitative (*..we give greater heat at [4] and the temperature rises rapidly..*) and quantitative descriptions

(*..at [5] we give 10° more heat, ..at [6] we keep it stable at 30°*). A few expressions about practical activities were also met (*[1].. we try to maintain it through a small hot surface*).

The above initial explanations were in most cases wrong, probably due to a limited knowledge of heat theory. Bill not only confused the concepts of heat and temperature, but he also presented misconceptions regarding their proper use and meaning (*..which is not zero otherwise it should be ice, ..we give 10° more heat, ..the temperature rises rapidly ..with a higher frequency*).

B. Post-interview

The extracts from Bill's responses in the post interview presented below follow the same notation as in the pre interview:

Bill: ... there is melting, ... mm... where is zero? ...the graph should be stable till the last part of ice melts and then it goes up... (it may be) here [1]. ..at the end of [1] melting is over and from now on [2], as we provide heat .the temperature rises.

...I suppose that part 3, as horizontal, may also correspond to melting, so that point is 0 °C (extending part [3] to the left), but what happens below [3] worries me. Mmm, the water is ice, the temperature of water is below zero, ..so parts 1 and 2 are ice. At part 1, the temperature is steady, so there is no heat supply... here at [2] we apply heat to take it to zero and keep it there during melting [3]

... After melting, temperature rises [4]

...here [5] we stop the heater and temperature remains constant,

...then at [6] the supply is interrupted. But if there were only the heat losses (to the environment), part [6] should be less steeply inclined. That means that here we have heat absorption [6]

...at [7] we maintain a steady supply which keeps temperature at a steady value (because of the environment)

...mm... also at [5] we keep a little supply to balance the losses, since the environment is colder.

Interviewer: Are there no heat losses to the environment at [1]?

Bill: No, the environment gives heat to the ice

Interviewer: So, to preserve the temperature of the ice steady and lower than that of the environment, what must be done?

Bill: Put it in a fridge. It will absorb heat...

During the final interview, we noticed that Bill presented a different kind of expression in relation to his initial interview and described the graph correctly. We noticed that he did not simply refer to parts of the graphs as changes in variables but he used combined expressions referring both to physical phenomena (*we stop the heater and temperature remains constant, ..put it in a fridge..*) and also theoretical relations between heat and temperature (*as we apply heat the temperature rises, the graph should be steady till the last part of the ice melts, we keep a little supply to balance the losses*). These expressions were consistent with heat theory and the explanations were correct both at the level of content and at the level of graphing. Bill assigned successfully an accepted theoretical meaning or phenomena evolution to all parts of the graph. It is evident that while interpreting the graph, Bill showed that he had a sound knowledge about heat. Bill's initial and final patterns of reasoning are visualised in the "GraPhys" model in Figure 5.

Comparing the initial and final patterns, we may note that the student shifted from a poor theoretically oriented initial situation to a final integrated approach to graph interpretation. We consider that in the case of Bill, the description of physical variables using graphical representation helped him to link successfully theoretical models with practical activities, a situation which was also achieved by 5 more out of the 20 students.

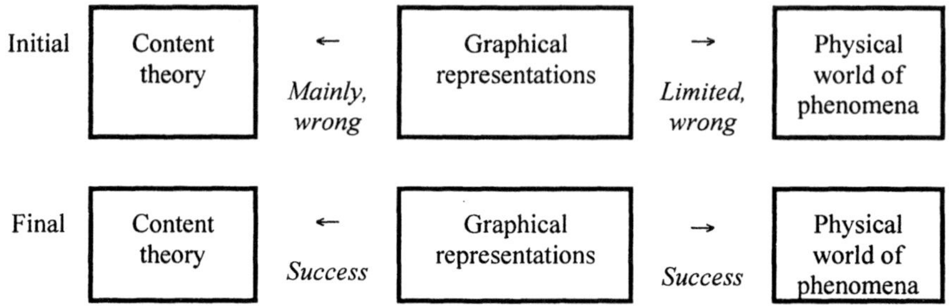

Figure 5. Initial and final "GraPhys" patterns from student Bill

The case of Niki

It seems though that linking heat theory to practical activities and phenomena by the mediation of graphical representations, although feasible as cases like Bill proved, is a demanding cognitive task for several students. Niki, for example, the second student examined in this study, presented a considerably improved interpretation in her final interview. However her reasoning was not quite acceptable on both sides of the "GraPhys" model, as was the case with Bill.

Niki initially referred only to practical activities, probably because of her inability to interpret the graph in terms of heat theory. Furthermore, her description was not consistent with the kind of temperature changes that were represented, since she related a steady temperature to a constantly operating heater:

> *...it is as if we put a gas-burner under the vessel. For part [1] temperature is constant so the burner is constant. From the end of [1] to the beginning of [3] we increase the burner, and during [3] we keep the burner constant again...*

Unlike in her initial approach, in the final interview Niki managed to employ expressions concerning the evolution of phenomena or practical activities. Moreover she included theoretical references concerning the kind of changes that the final graph represented, an approach she had not followed at all initially:

> *...below 0 °C we have negative heat, ..in part [2] the temperature is increasing, that means we turn on the burner and we have an increase in supply, .. at 0 °C, part [3], we keep the burner on, for the ice to melt, ..in [4] the heat supply is increased, so the temperature rises. In [5] the temperature is stable so the heat supply is also stable.. .*

We notice that for the same parts of the graph which she had initially referred to as activities, she later used theoretical descriptions. For example, she connected the horizontal parts of the graph which represent a constant temperature to a theoretical relation between the physical variables (*in [5] the temperature is stable so the heat supply is also stable*), while initially the same situation was connected to an activity (*...we keep the burner constant..*). Nevertheless, both cases were incorrect since however expressed, either as a constantly operated burner or as a constant heat supply, the result should be an increase in temperature. It is worth noting that during the final interview the student was able to successfully correspond parts of the graph to practical activities (*..at [2]..we turn on the burner, ..at [3] we keep the burner on, for the ice to melt*).

The "GraPhys" pattern for Niki, as resulting from her initial and final interviews, is given in Figure 6. Analysing her pattern, we notice that her initial wrong approach concerning the phenomena side of the model turned into a final successful situation. Concerning the theory side of the model, while she initially did not refer to at all, she finally managed to include theoretical expressions, which were, however, only partially acceptable. Niki showed, to a certain extent, an improper use of heat concepts. We accept that this progress indicates a shift towards an integrated approach of interpreting graphs, yet linking is only partially achieved.

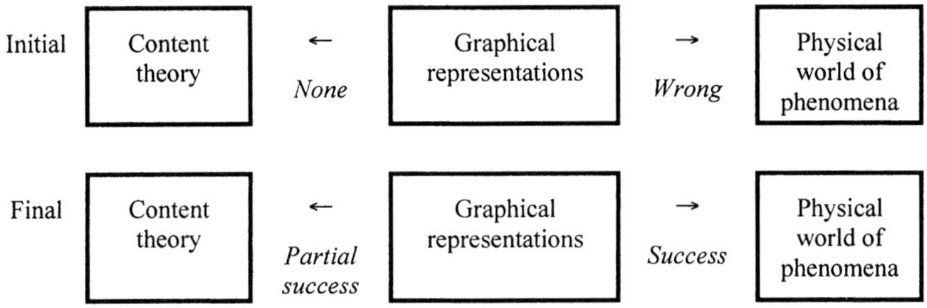

Figure 6. Initial and final model-patterns of student Niki

Twelve out of the 20 students were included in this category. They all showed a "*none*" - "*wrong*" initial "GraPhys" pattern which, after instruction, turned into a "*successful*" - "*partially successful*" pattern concerning the two sides of the model.

Data analysis did not reveal a unique initial to final developmental pattern. Instead, there were several individual patterns emerged, all implying a tendency to integrated interpretation. In another case, for example, the "GraPhys" pattern of a student showed that an initial "*wrong*" approach regarding theory and "*none*" approach to phenomena, finally turned into a situation according to which the student referred "*successfully*" to theory and "*partially successfully*" to phenomena. These individual developmental patterns, from an initially narrow to a final partially successful interpretation indicate a shift towards an integrated approach to interpreting graphs by all the students belonging to this category.

The case of Maria

The desired linking of theoretical aspects to practical activities in students' understanding cannot always be achieved by graphically representing the changes, as shown in the following characteristic case of student Maria. Describing the initial graph of Figure 2, this student stated that:

> *In part [1] of the graph, the temperature remains unchanged, so no heat is being applied, ..in [2], the temperature is increasing so we have an increase in heat supply, ..in [3] the temperature is stable so the supply is not increased.. If we have no supply, so there is no temperature change...*

With continuous reference to heat content theory, Maria interpreted the parts of the initial graph in terms of variables and expressed her ideas as a relationship between temperature and heat supply, formulating quite an acceptable hypothesis, which probably derived from her previous formal background.

In spite of her initial moderately successful interpretation, in the final interview Maria argued that:

> *I suppose ice melts a little above zero degrees. Melting starts at the end of [3] and is completed somewhere near the end of [4],... interrupting heat supply causes temperature fall, like if I put a piece of ice in the water, it absorbs heat..*

These are ideas which are neither considered as correct at the level of heat theory nor at the level of describing physical phenomena. Maria's "GraPhys" pattern, resulting from the above extracts and the entire interview, is shown in Figure 7.

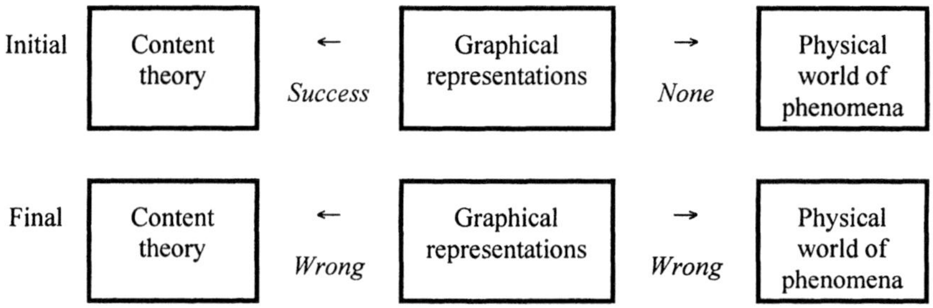

Figure 7. Initial and final model-patterns of student Maria

In this case we consider that linking theoretical and practical aspects of the domain was not accomplished by graphically representing variations. The inability to relate heat knowledge with graph characteristics was a possible reason which caused deviation from successfully interpreting changes in physical variables. Patterns similar to Maria's response were met in 3 out of the 20 students.

Conclusions

The present study explored the results of a laboratory-based teaching sequence in enhancing students' ability to establish links between heat theory and phenomena by representing variations with graphical means. Graphs and theoretical or experimental approaches to phenomena were not used independently, rather they were interrelated. Graphs had constituted a symbolic means of expressing ideas and were used as a conceptual bridge, linking the phenomena investigated and the related theory. This function seem that has been enhanced by the real-time graphing feature which characterises the teaching sequence that we developed and applied to our students. It seems that the students, having approached phenomena investigation using interactive real-time graphing activities, learned a symbolic language to communicate in physics. After the sequence, as shown by the representative cases, they described changes in the physical variables using expressions regarding graph characteristics and in the same time they described the parts of the graphs referring to content theory, at different levels.

The focus of the present research was not to investigate whether students develop partial graphing skills or content knowledge, but to describe possible patterns of development, according to which the two kinds of knowledge intertwine.

Using the "GraPhys" model, which we developed for describing graphing procedures, as a teaching tool, we determined the skills to be taught connected to a specific content theory and set of phenomena. According to the "GraPhys" model, we considered students' approaches as being effective in linking theory with phenomena when the expressions they used during graph interpretation referred consistently both to theoretical and to phenomenological descriptions of the changes in the represented physical variables. This focus of consideration turned the "GraPhys" into an operational research tool for modelling linking, upon which students' developmental patterns were drawn.

The results show that a teaching sequence in the area of heat which involved extended computer-based, real-time graphing units, allowed students to connect phenomena or theoretical relations between variables with what the graphs represent. Broadly speaking, the students who participated in the computer-based activities improved their graphing skills and their knowledge about the domain of heat. However, they presented different patterns of linking heat content theory to physical phenomena. Analysing in depth the pre and post interviews from three representative cases as well as from the rest of the students, we located three major trends.

The advanced category concerns cases whereby an initial partial graph comprehension led to an integrated approach to interpretation. The majority of the students reached an intermediate state concerning the formulation of accepted expressions which combine both theoretical and phenomenological aspects of the domain, showing a tendency to shift from a partial towards an integrated approach of graphing. These students showed minor interpretation problems either at the level of heat content knowledge or at the level of graphing skills. It is possible that a more extended intervention than the one followed in the present study might help these students to overcome the problems. Finally, there were a few cases in which the interweave between content knowledge and graphical representations deviated from linking. It is possible that the short duration or other aspects of the teaching, were rather not adequate for broadening certain students' interpretation skills, a situation which needs further investigation.

Reflecting on the above results we can assume that with respect to the students who developed the skill of linking theoretical models to physical phenomena, the knowledge corresponding to the side of the "GraPhys" model which is first revealed as being consistent to the scientific theory may later form a possible background basis for developing the complementary side of the model. Extending this assumption we may say that lack of ability to combine the two kinds of knowledge may result in deviation from an integrated and successful graph interpretation and may lead the students either to inability to figure out the correct meaning of the graph characteristics or to an inadequate content comprehension.

Recommendations

Graphing can be enrolled dynamically in learning situations. In this frame, graphical description should be conceived not only as a technique for data treatment but also as a teaching tool for facilitating the construction of links between the theory taught and the practice experienced. Graphing skills and procedures should be taught using changes in physical variables in a specific content, approaching both theoretical relations and experimental activities. Students can be encouraged to use expressions concerning graph characteristics to described changes in the physical variables, otherwise may not be able to link theoretical models with physical phenomena.

It is reasonable that science teachers should take into account students' alternative ideas, concerning graph characteristics besides scientific content, since this could influence both their graphing performance and content understanding. Students' development towards an integrated approach to graph interpretation may follow different patterns of linking content theory to physical phenomena which teachers should take into account in laboratory work. Laboratory-based teaching incorporating considerable teaching time and taking into account the different patterns that students develop may be necessary, instead of short interventions to enhance the process of linking theory to phenomena by a wide range of students.

The teaching sequence used in the present case study is characterised by a continuous and gradually more complex transition between experiments and graphical representations. The mediating function of graphing between theoretical and experimental approaches to scientific topics can be enhanced when computer-based on-line data acquisition techniques for real-time graphing are appropriately introduced, in combination with short-scale simulations for parametric investigation of phenomena. For this reason, computer-based graphing literacy could be adopted in science education and widely used as a common symbolic language for communicating science.

References

Bisdikian, G. & Psillos, D. (1998-a). A computer-based approach to relating graphs and physics: The case of heat and temperature. In Niedderer, H. & Psillos, D. (Eds.), *Case studies on labwork in 5 European countries, Working paper No 7*. LSE Project, No PL95-2005, (European Commission DG XII).

Bisdikian, G. & Psillos, D. (1998-b). Developing content knowledge end acquiring graphing skills: The case of Heat and Temperature. *Proceedings of 1st Greek Conference on Science Education*. May 1998, Aristotle University of Thessaloniki.

Bisdikian, G. (2000). A study of multimedia applications on the teaching of graphs and physical concepts. *Doctoral dissertation*. Aristotle University of Thessaloniki.

Kesidou, S., Duit, R. & Glynn, S. (1995). Conceptual Development in Physics: Students' Understanding of Heat. In Glynn, S. & Duit, R. (Eds.), *Learning Science in the Schools: Research Reforming Practice*. LEA Pub, N. Jersey.

Leinhardt, G., Zaslavsky, O. & Stein, M. (1990). Functions, Graphs and Graphing: Tasks, Learning and Teaching. *Review of Educational Research,* Vol 60, (1), p 1-64.

Linn, M. & Songer, N. (1991). Teaching thermodynamics to middle school students: What are appropriate cognitive demands? *Journal of Research in Science Teaching*, 28, p 885-918.

McDermott, L. (1987). Students difficulties in connecting graphs and physics: Examples from kinematics, *American Journal of Physics*, 55 (6), June 1987, p 503-513.

Rogers, L. (1995). The computer as an aid for exploring graphs. *School Science Review*, 76, 276, p 31-39.

Roth, W. & McGinn, M. (1997). Graphing: Cognitive ability or practice? *Science Education*, 81, p 91-106.

Seré, M.G. (1999). Learning science in the laboratory: Issues raised by the European Project "Labwork in Science Education". In Bandiera, M., Caravita, S., Torracca, E. & Vicentini, M. (Eds.), *Research in Science Education in Europe*. p 165-175, Kluwer Academic, Nederlands.

Thornton, R. (1995). Conceptual Dynamics: Changing Student Views of Force and Motion. In Bernardini, C. et al (Eds.), *Thinking Physics for Teaching*, Plenum Press, New York.

The Link of Theory and Practice in Traditional and in Computer-Based University Laboratory Experiments

Lorenz Hucke and Hans E. Fischer, University of Dortmund, Germany

Abstract

This work deals with learning in the beginners' physics laboratory at the university level in Germany. It examines whether the students acquire knowledge of physics and of scientific experimentation by performing a laboratory experiment, and if a relationship exists between the students' knowledge acquisition and their actions during laboratory work. Thus, the learning expectations are compared with the actual results of beginners' physics laboratories. Furthermore, it is examined to what extent the use of the computer for data capture and model building can affect students' action regulation and learning outcomes. Constructivist theories of learning constitute the theoretical basis for the investigation. They are used in combination with concepts of action theory, with aspects of physics education taken into account. Video recordings of laboratory work are analysed by a category-based analysis, which was specifically developed for this research. Concept maps are used for investigating students' knowledge and are analysed on the basis of a reference map. The results show that the traditional beginners' physics laboratory at the university level is not a learning environment particularly well suited for applying and acquiring knowledge of physics actively, and that in this case the use of the computer hardly effects either the students' action regulation or knowledge acquisition. Based on the results of this research, ways to improve the effectiveness of physics laboratory work are discussed.

Introduction

The goals of scientific laboratory work are often discussed in detail (i.e. Boud, Dunn, Kennedy & Thorley 1980; Hegarty-Hazel 1990; Wellington 1998; Welzel et al. 1998). The main goals of labwork seem beyond doubt and have repeatedly been published over the last 35 years. According to Welzel, Haller, Bandiera, Hammelev, Koumaras, Niedderer, Paulsen, Robinault & Aufschnaiter (1998) and AAPT (1998), the link of theory and practice is seen by experts as one of the most important goals of laboratory work in physics education. Nevertheless, among researchers, there seems to be a consensus that the existing laboratory work achieves the postulated goals only in an imperfect way (i.e. Toothacker 1983; Lazarowitz & Tamir 1994; White 1996; Wellington 1998). Lunetta (1998, 250) claims: "To many students, a lab means manipulating equipment and not manipulating ideas".

Theoretical background

The theoretical background of the study refers to constructivist theories of learning. Whereas radical constructivism (Glasersfeld 1995) is discussed controversially, a pragmatic interim position is now widely accepted. It is seen as an integration of constructivism and cognitive theory, as it accepts learning as a process of individual cognitive construction and claims the dependence of this process on adequate learning environments (Weidenmann 1993; Derry 1996).

D. Psillos and H. Niedderer (eds.). Teaching and Learning in the Science Laboratory. 205-218.

In this investigation, learning is seen as an individual cognitive construction of action schemes and networks of concepts which are to become viable through interaction and which depend as well on motivational and knowledge related cognitive states as on environmental conditions. In order to describe a learner's cognitive constructions related to the action regulation in a certain situation we distinguish between a descriptive and an abstract cognitive level (see Fischer 1994; Horstendahl 1999; Hucke 2000). Regarding physics, abstract cognitive performance is characterised by combining different physics concepts (see Table 1). In this case, the action[1] is guided by knowledge of physical theory. Descriptive cognitive performance is related only to objects and their properties. In the following sections this parameter is described as "action regulation on the descriptive or abstract cognitive level".

Table 1[2]: Simplified model of complexity levels of cognition and the related action regulation (cognitions) of a learner (Fischer 1994)

	Description	Action	Cognition	Example
High level of complexity	Shared features of the object matter are categorised and used as principles and systems	Concept related	Abstract cognitive performance (manipulating ideas; Lunetta 1998)	"In the case of a harmonic oscillation the back-driving force is proportional to the elongation."
Low level of complexity	Objects are perceived, described, manipulated and related to other objects	Object related	Descriptive cognitive performance (manipulating objects)	"Hanging a weight on a spring, it is drawn up."

A physics laboratory experiment can be structured related to three content areas: (1) the theoretical background (physics), (2) methodological aspects (measurements, data analysis) and (3) material objects and observable events (equipment, experimental set-up) (see contributions of Bécu-Robinault or Buty *in this volume*). Thus, for the investigation, a threestage science educational perspective is combined with the theoretical assumptions described above. For a more detailed description of the theoretical background of the study see Fischer (1993), Fischer & Aufschnaiter (1993), Fischer & Horstendahl (1997), Horstendahl (1999) and Hucke (2000).

As described in the introduction, experts expect students to link theory and practice during laboratory work by manipulating ideas (physics concepts) rather than objects (Welzel et al. 1998; Lunetta 1998). Therefore, an investigation of student laboratory work related to the conceptual goals of labwork deals with the question:

[1] In this szudy, the term action does not only mean actions on material objects but every observable activity of a person (including communication, loud thinking,...).

[2] The examples may help to illustrate the categories used for video analysis (see later sections of this article). As the lower example refers neither to physics concepts nor to aspects of measurement (which would be coded by category M), it would be coded by category O which is similar to the category KT as described in Niedderer et al. (see page 36). If this statement contained a physics concept, it would be coded by category P (descriptive cognitive level; compare figure 2). Contrarily, in the upper example different physics concepts are combined. Therefore it would be coded by category PP (physics theory, abstract cognitive level; compare figure 2).

Do students use their knowledge of physical theory to regulate their actions in the laboratory?

Teaching approach

Conditions of laboratory work

The investigation was carried out under usual conditions of the beginners' physics laboratory at the University of Dortmund. This laboratory work is compulsory and it lasts two semesters after two semesters of an intensive physics lecture and seminar. Each student has to conduct one experiment per week. The students work in groups of two and they are supervised by a post graduate (doctoral) student. The tutors, who usually have no teaching experience and are not trained to teach, supervise two to four experiments and are assisted by a technical expert who is responsible for the overall organisation of the laboratory.

The students have to prepare each experiment using a detailed written instruction (labguide) which includes the physical theory, the methodology, a description of the apparatus and sometimes the expected results of the experiment (Finke 1992). It takes the students one afternoon per week to do their practical work. At the beginning of each session the tutor tests the students' knowledge of the experiment and its theoretical background. During the session he/she assists in case of technical and content related problems. The students finish an experiment by writing a report at home. A certificate for the laboratory work is handed out for 24 successfully passed examinations and reports.

Pieces of labwork investigated in this study

For this investigation, two experiments of 42 were chosen. They are typical experiments of the traditional beginners' physics laboratory in Germany regarding the organisation of the laboratory work and the arrangement of the learning environment. Both experiments were implemented in a traditional setting (without computer) and in a computer-based setting.

Experiment A: Relaxation behaviour of an RC-circuit

This experiment illustrates a typical relaxation problem. The time constant of an RC circuit is measured in different ways. The circuits are assembled by the students using the detailed description in the labguide. In the traditional setting an X-Y-plotter is used for measuring and plotting. The data must be read and processed "by hand". In the computer-based setting the plotter is substituted by an interface (CASSY, Leybold 1994) and a PC. The data are captured automatically and are processed by means of a specific software (ORIGIN, Microcal 1995). In addition to interface and PC, a model building system (STELLA, High Performance Systems 1994) is used in the third setting for modelling, simulating the experiment and comparing the results of the simulation with the experiment. To learn more about model building systems see, for example, Doerr (1996), Schecker (1998), and the contribution of Sander et al. (Chapter 5 *of this volume*).

Experiment B: Non-linear oscillations

In this experiment, the characteristic behaviour of a forced non-linear oscillation is analysed by means of a Pohl wheel driven by a motor. A small mass can be attached to cause non-linearity. Students measure the amplitude of the oscillation and the phase shift between the frequency of the oscillation and the driving force. In the traditional setting the amplitude is measured "by eye" on an angle scale and the frequency by using a photoelectric barrier and a digital watch. The students compare the experimental curve to a theoretical curve calculated on base of the parameters of the oscillating system.

In the computer-based setting a rotary potentiometer is fixed at the pendulum's axle to measure the amplitude. The data are captured using an interface and a PC (see above) and can be processed immediately. By help of the computer, the beginning and the phase shift of the oscillation as well as the characteristic abrupt change of the amplitude and the phase can be visualised. In addition, some students construct a STELLA model which simulates the experiment to predict the frequency at which the amplitude and the phase jump can be expected, and compare their predictions with the experimental result.

Research questions, design and hypotheses

This study aims to answer the following questions:

- Does performing a physics experiment in the beginners' laboratory help students to acquire knowledge of physics and of experimentation?
- Do students use knowledge of physics to regulate their actions during laboratory work and, if so, under which conditions?
- Are there differences regarding both knowledge acquisition and action regulation between the traditional setting and a computer-based setting of the same laboratory experiment?

Research design

The following three settings of the two laboratory experiments described above were implemented in order to control influences on students' acting and learning which are not related to the use of the computer:

1. The students of the "traditional" group (TRAD) conducted the experiments as it was done over the past 20 years.
2. The students of the second group used the computer for data capture and processing (MBL; Microcomputer Based Laboratory) as described above.
3. In addition to MBL, the students of the third group used a model building system (MBS) to construct a model of the experiment, to simulate it and to compare the simulation results with the experimental results.

The differences between the three learning environments confine to the use of the computer. There are no considerable differences concerning the other conditions such as the labguide, the tutor or the schedule. Each group consisted of 6 students (3rd semester, physics). That is, each experiment in each setting was carried out by three pairs of students.

Hypotheses

As described above, students' action related cognitive performance can be characterised as being on a concept related (manipulating ideas) or on an object related level (see Table 1). Many studies about learning during laboratory work lead to the assumption that in the traditional laboratory students regulate their actions on a low level of cognitive complexity, and that only little knowledge of physics and of experimentation is acquired (e.g. Tamir & Lunetta 1981; Okebukola 1985; Lazarowitz & Tamir 1994; White 1996; Lunetta 1998). On the other hand, the use of MBL and MBS seems to be helpful for raising laboratory work on a more conceptual level (e.g. Thornton & Sokoloff 1990; Doerr 1996; Redish, Saul & Steinberg 1997; Linn 1998; Schecker 1998). This leads to the following hypotheses:

(1) The cognitive performance of students during laboratory work depends on the learning environment.
   a. In the TRAD setting the students regulate their actions mainly on the descriptive cognitive level regarding physics.
   b. In the MBL and MBS setting the students regulate their actions more often on the abstract cognitive level regarding physics as compared to the traditional setting.
   c. If in addition to the use of the computer for data capture and processing a MBS is used, the students regulate their actions more often on the abstract cognitive level as compared to MBL.

(2) Performing a laboratory experiment students acquire knowledge of physics and of experimentation. The increase in knowledge depends on the learning environment.
   a. In the MBL and MBS setting the increase in the students' knowledge is higher as compared to the TRAD setting.
   b. In the MBS case, the increase in the students' knowledge is higher as compared to MBL.

(3) Students who regulate their actions more often on the abstract cognitive level (guided by knowledge of physical theory) have a higher increase in knowledge of physical theory than other students.

Hypotheses (1) will be tested by means of video analysis, hypotheses (2) will be tested by means of concept mapping and hypotheses (3) will be tested by correlating the results of video analysis and concept mapping. This methodology fits the model of effectiveness of laboratory work described by Psillos et al. (Chapter 1 *of this volume*).

Research methods

Video analysis

The video data are subjected to a category based analysis. The basic idea of the method and the relations to science education are described in detail by Niedderer et al. (Chapter 1 *of this volume*). The categories described by Niedderer et al. were modified in order to fit the theoretical background of this study. The reliability of the categories between different raters was tested. Two sorts of categories are applied:

- Students' activities are coded using nine categories (e.g. writing, manipulating, model building, etc.). This system of categories is derived inductively from the data (Mayring 1997). Its application provides results concerning the time devoted to different laboratory activities.
- Students' verbal actions are coded using six categories which refer to the content area and, regarding physics, to the complexity of a verbal expression (object or concept related, see above). This system of categories is derived from the theoretical assumptions described above. Its application provides results concerning the content area (measurement or physics) which students focus on, and the level of cognitive complexity on which students regulate their actions during laboratory work.

Thus, the stream of actions (Asendorpf & Wallbott 1979) of each student is coded by 16 categories. The categories are applied to sequences of thirty seconds. The time related correlation of both systems of categories provides results about which activities might foster the cognitive development related to a certain content area.

The rules for applying the categories were developed iteratively by two raters. The agreement was calculated by means of the statistic kappa (Cohen 1960). The actual analysis of the video data began after values for kappa higher than 0.7 were achieved.

Altogether 110 hours of laboratory work of 18 students were analysed. An analysis of variance (ANOVA) method was applied to the results of the video analysis to check the statistical significance of differences between the three learning environments and to test hypothesis (1).

For more details concerning this method of video analysis see Hucke (2000, 53).

Concept maps

Concept maps can be used to detect changes in the learners' knowledge caused by instruction (e.g. Markham, Mintzes & Jones 1994; Fischler & Peuckert 2000). They consist of single concepts (nodes) and relations between these concepts (propositions). Concept maps allow insights into the conceptual knowledge of learners which are not accessible by other methods (Hasemann & Mansfield 1995). In particular, the quality and the structure of the knowledge of a topic can be examined in detail.

In this research, not only knowledge of physics but also knowledge of experimental set-ups and processes is concerned. Therefore the knowledge represented in the concept maps can be evaluated as belonging to different content areas (equipment, measurement or physics) and to different levels of complexity regarding physics: object related or concept related (physical theory). Thus, it is assumed that the levels of cognitive performance postulated above (Table 1) can be identified by analysing the concept maps.

One concept map was made by each student before and after conducting one of the laboratory experiments investigated. The maps are analysed using the same category system as used for the video analysis. Only propositions representing correct statements from an expert's view are considered.

An analysis of variance (MANOVA) method was applied to the 36 maps relating to experiment B ("Non-linear oscillations") to check the statistical significance of the differences between the three learning environments and to test hypothesis (2).

Both, the entire increase in knowledge and the increase in knowledge within the different categories is regarded. The results of the concept maps are correlated with the results of the video analysis.

Additionally, the concept maps were analysed qualitatively by comparing the students' concept maps regarding certain propositions or certain aspects of the laboratory experiment. Thus, differences and mutualities in the increase in knowledge between individual students, as well as the knowledge of physics and of experimentation typically acquired by performing a laboratory experiment, can be described.

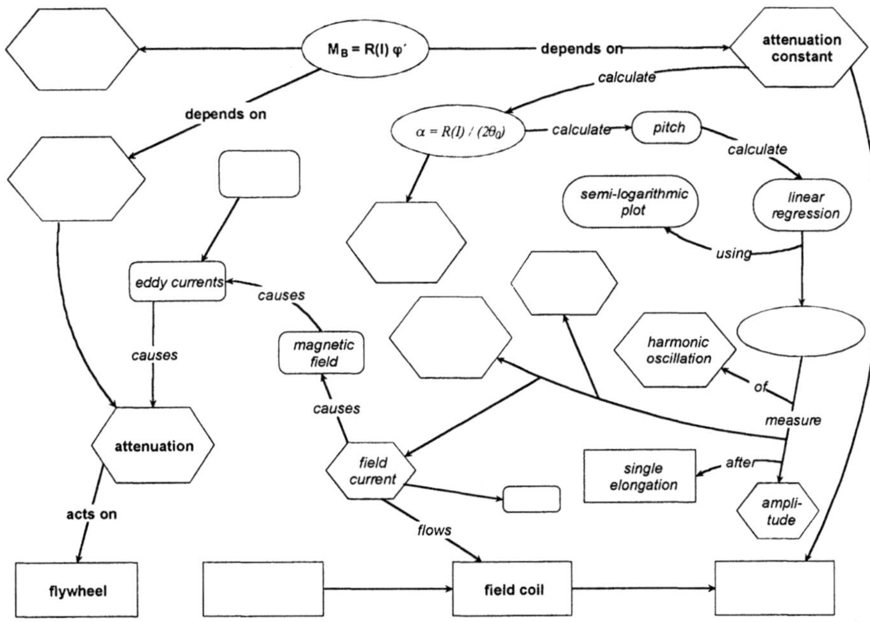

Figure 1: Reconstructed student map based on the reference map

Figure 1 represents a small extract of the reference map. The different shapes describe different types of concepts (physical, methodological, objects, etc.). Empty shapes and propositions without description mean that this individual student did not mention them in his concept map, but other students did. Therefore they appear in the reference map. For a more detailed description of the concept mapping method applied see Hucke (2000) or Hucke & Fischer (2000).

Research results and conclusions

Results concerning the students' action regulation during laboratory work

These results refer to the category based analysis of video data of student laboratory work (experiments A and B) (see above and the studies of Niedderer et al., Sander et al. and Buty (all *in this volume*). The students have also been videotaped during the data analysis and interpretation which usually occurs outside the laboratory, one or two days after the experiment.

Traditional laboratory work

In the TRAD setting the students' actions are to a considerable degree determined by the labguide. Due to the detailed instructions in the labguide and the temporal and social basic conditions of the laboratory students have a very small scope of action. Therefore, the distribution of different activities within the labtime (time spent while conducting an experiment in the laboratory, typically 3 to 5 hours) is nearly the same for all students.

Between 40 and 60% of the labtime are spent on measurements. It is the main activity of the laboratory work and within approximately 40% of the labtime the action-leading cognitions refer to measuring. Furthermore, within another 40% of the labtime the students regulate their actions without thinking about physical or experimental aspects of the experiment. This shows that the students do many routine activities. The time needed for these activities is influenced neither by the topic of the experiment nor by the use of the computer. Therefore, we assume that it is caused by the basic conditions of traditional laboratory work.

In the TRAD setting, only within about 20% of the labtime the action-leading cognitions refer to physics (sum of categories P and PP, see Figure 2). Thus, the physics related aspects of the experiment are far less important than the aspects related to measurement. Moreover, action regulation on the abstract cognitive level (using physics theory, manipulating ideas) occurs only in 5% of the labtime (category PP, Figure 2). It was found that this is even less than the time referring to subjects not related to the experiment and physics at all. That is, in the TRAD setting, the students regulate their actions on the descriptive cognitive level regarding physics. Thereby, hypothesis (1a) is confirmed.

Another important result is that the activities 'manipulating' and 'measuring', which take most of the labtime, do not promote the use of physics concepts. In the TRAD setting, the only activity which promotes the use of knowledge of physical theory is interaction and discussion with the supervisor.

In traditional laboratory work, the analysis of experimental data is usually separated from the measurement procedure and done outside the laboratory (see above). However, the results do not confirm the assumption often made by experts, that students use their knowledge of physical theory more often during the period of interpretation of data than during measurement. Like during the actual laboratory work, only a negligible part (< 5%) of the verbal actions refers to physical theory. Obviously, students focus on calculating the final results required by the labguide even during the analysis of experimental data.

Computer-based data capture and processing (MBL)

Hypothesis (1b) has to be rejected. The results do not support the assumption that the use of the computer for data-capture and -processing promotes the examination of physics concepts. In the observed learning environment the cognitions refer to physics concepts in only about 20% of the labtime (Figure 2). That is, that even during MBL action is regulated on a descriptive cognitive level regarding physics.

According to the results, the main advantage of MBL is the possibility to discuss experimental results immediately after measurement. Thereby, the students get a feedback, which they cannot receive in the traditional laboratory. In the case of experiment A this leads to the autonomous discovery of measurement errors by students and to physics-related discussions.

5% of labtime was spent on computer-specific activities (saving data, formatting graphs, etc.). Nevertheless, the verbal actions during MBL refer less often to the experiment as compared to students in the TRAD group because these students spent a lot of time waiting for data during which they talked about private topics. In contrast, in MBL, the students' attention is permanently directed to the computer. This may have negative consequences: In experiment B, all students missed the characteristic jump of the pendulum's amplitude and phase at the critical frequency, because they were occupied with computer-related activities.

Model building and simulation (MBS)

Hypothesis (1c) is confirmed. The results show that the development and application of a physics model by MBS promotes the examination of physical theory. The action-leading cognitions of the MBS-group refer significantly more frequently to physical theory as compared to the students of the groups TRAD and MBL (category PP, Figure 2).

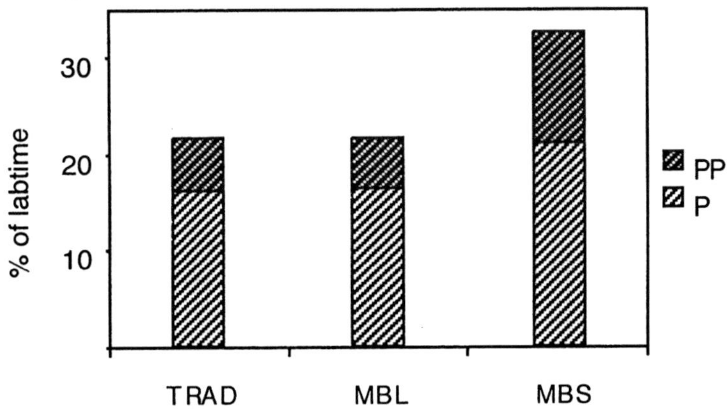

Figure 2: Frequency of physics verbalisations (P and PP) in three different types of labs for experiment B[3]

That is, the students using a model building system regulate their actions significantly more frequently on the abstract cognitive level. Figure 2 shows that, in the groups TRAD and MBL, the students' action leading cognitions refer to physics in only about 20% of the labtime (P+PP), and to physical theory (PP, manipulating ideas) in only about 5% of the labtime. Using a model building system (MBS) leads to a significantly higher frequency of physical theory (PP).

The interrelation between the students' activities and their verbal actions allows to explain this result. A lot of physics-related actions are caused by model building, which is the only activity in the laboratory which clearly requires the use of physical theory.

[3] Categories P: descriptive cognitive level (object related); category PP: abstract cognitive level (concept related). These categories can be understood as subcategories of the category KP, knowledge of physics, in the paper of Niedderer et al. *in this volume*; compare also table 1 and footnote 2 on page *206*.

However, this does not apply to simulating the experiment. During simulation the students only modify the parameters systematically until they receive an adequate result. Nevertheless, the simulation of the experiment before the beginning of the measurement helps students to direct their attention towards the critical range of frequency of oscillation. In experiment B, contrary to the students of the groups TRAD and MBL, all students of the MBS-group pay attention to the characteristic jump of the pendulums' amplitude at the critical frequency.

Results concerning the students' knowledge acquisition

These results refer to the analysis of 36 pre- and post-concept maps related to experiment B. We find an increase in the students' knowledge in the subject area of the experiment (Figure 3). Figure 3 shows the number of propositions (correct statements) in the students' pre- and post-concept maps.

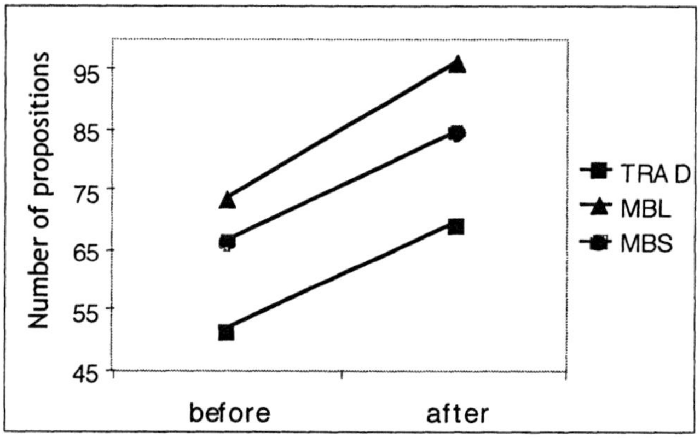

Figure 3: Increase in knowledge for the groups TRAD, MBL and MBS (Data for experiment B)

The hypotheses (2a) and (2b), however, have to be rejected: Students of the MBL- and the MBS-group have no higher increase in knowledge than students of the TRAD-groups, although the students of the MBS-group had used their knowledge of physical theory more frequently during the labwork than the students of the other groups (as showed by the video analysis). In conclusion, the offered learning environments seem to have no influence on the students' knowledge acquisition. Moreover, the correlation of the frequency of different verbal actions during labwork with the increase in knowledge in the related categories gives no significant result. That is, there seems to be no relationship between the students' action regulation in the laboratory and their learning outcomes. Therefore, hypothesis (3) has to be rejected, and it must be assumed that there are other (hidden) variables influencing the students' knowledge acquisition.

The qualitative analysis of the concept maps provides more detailed results about the students' knowledge acquisition. In particular, it is striking that some of the students seem to acquire only very little or no knowledge, while for other students a high increase can be stated. Since these differences do not depend on the learning environment (see above), it must be assumed that they are caused by the students' individual prerequisites. The more detailed analysis leads to the conclusion that affective variables such as attitude, interest or motivation play an important role.

Additionally, the results reveal that the labguide plays a crucial role for knowledge acquisition in the TRAD setting. Due to the labguide, the students are able to give a detailed description of the physical theory, the experimental set-up, the measurement methods and the expected results even before conducting the laboratory experiment, which is mostly sufficient to pass the colloquium. But the students' average increase in knowledge is low, and predominantly those knowledge elements are acquired which are not illustrated by the labguide. Thus, the labguide anticipates an active acquisition of new knowledge by students. This helps to explain the result that students obviously do not link theory and practice during laboratory work and that students using a model building system do not perform better in pre- and post-tests than the others.

Conclusions

Matching the goals of laboratory work

One of the most important goals of physics laboratories is the application of physics concepts and the active acquisition of knowledge by experimentation. The results of this study reveal that the goal to promote a deeper understanding of physics by the active use of knowledge of physical theory (to link theory and practice) is not achieved with the traditional organisation of the physics laboratory at German universities. In conclusion, two alternatives can be deduced:

The first is to give up this goal. In this case, laboratory work should be focused on the students' experimental skills and knowledge about experimentation. As students at university should be prepared for a vocational activity as physicists, it is important that they get used to manipulating technical devices and to applying experimental methods. However, organising all laboratory experiments in the same way means accepting that, despite the considerable technical, financial and organisational expenditure, the educational potential of a science laboratory seems not to be exploited.

The second possibility is holding to the goal of linking theory and practice. In this case, the organisation of laboratory work should be changed. The results of this study underline the need to orientate laboratory work explicitely towards different goals (cf Séré, epilogue *of this volume*). That is, not every experiment has to be conducted following the same organisational and behavioural patterns, but, embedded into an overall conception, different parts of laboratory work (data handling, modelling, interpreting results, applying physical theory, manipulating technical equipment, etc.) have to be targeted in different settings.

Using the computer in the laboratory

The results of this investigation reveal that the use of the computer in the laboratory does not automatically improve students' action regulation and knowledge acquisition. It is not sufficient to simply add single computer related tasks to existing laboratory experiments. Hence, the computer should be used in the context of an overall educational framework, which specifically focuses on the potential of the computer to improve science learning (e.g. Linn 1998). Then, the advantages, which the computer offers as compared to a traditional laboratory, can be effective.

Different experiments should be co-ordinated regarding a framework of all different goals of the laboratory work, and the use of the computer has to fit in the framework. Thus, the introduction of the computer into the physics laboratory can help to clarify and to explicit the specific goals of laboratory experiments.

Recommendations

The following suggestions are made to support students to link theory and practice:

1. Reduce activities which do not lead to the use of physics concepts.

Traditional laboratory consists mainly of measuring data and manipulating equipment. Reducing those activities in favour of, for example, model building or activities such as planning an investigation, avoids students being involved merely in the adjusting of devices, technical problems and "sitting out" lengthy series of measurements.

2. Use a model building system.

This explicitly leads students to use physical theory. It is important that the software used must allow to model not only the parameters but the physics of the experiment.

3. Improve the supervision by training the tutors.

Appropriate questions and targeted discussions can lead students to use their knowledge of physics. As the results show, this is the only way to improve the use of physical theory during laboratory work if the current traditional laboratory is maintained.

4. Use MBL in order to analyse experimental data immediately after measurement.

MBL enables students to receive feedback immediately. This never happens in the traditional laboratory, since measurement and analyses are separated. However, to make MBL effective it is necessary to arrange the laboratory in a way that students can calculate results, check for errors, and discuss and repeat measurements without time pressure. Furthermore, as learning processes are triggered by cognitive conflicts between expected and occurring events (e.g. Fischer 1993), students should be allowed to use their non-resilient concepts. This requires a more open organisation of the laboratory (see below) and the opportunity for students to discuss and ask questions fearless of being examined.

5. Simulate the laboratory experiment before measuring.

This can help students to focus on and to specify the goals of their actions. Any software for simulation, or even a targeted discussion before the beginning of the measurements, can be used. A model building system, however, appears particularly well suited, since students are actively involved in modelling. However, no complicated and lengthy series of measurements should be conducted in the same experiment; otherwise other goals of action become too important (see above). Furthermore, allowing students to predict the results of their actions is only meaningful if they receive feedback whether the expected results are achieved. Therefore, the use of MBS should always be combined with MBL.

6. Raise the degree of openness of the learning environment.

This might be the most fundamental suggestion. Most of the recommendations presented so far can only be enforced if students have a chance to act autonomously. As a consequence, labguides should not be as detailed and the experimental settings should not in all cases be as complete and perfect as they usually are. Moreover, planning and designing experiments should be taken into consideration as one goal of laboratory work.

7. Consider the influence of affective variables (motivation, attitude and interest).

There are only very few studies which take the role of affective variables in science laboratories into consideration. Preparative sessions some days before the actual laboratory work begins, co-operative learning in groups of several students, or alternative procedures for the evaluation of laboratory work are some suggestions research puts forward (cf Johnstone, Watt & Zaman 1996; Nicol, Kane & Wainwright 1994; Tobin 1990). Furthermore, raising the degree of openness of the laboratory and offering students complete tasks including planning and designing experiments as well as getting feedback from other "researchers" (peers) may also improve students' attitude and motivation (e.g. Schmidt & Kleinbeck 1990).

References

AAPT (American Association of Physics Teachers) (1998). Goals of the introductory physics laboratory. *American Journal of Physics*, 66(6), 483-485.

Asendorpf, J. & Wallbott, H.G. (1979). Maße der Beobachterübereinstimmung: Ein systematischer Vergleich. *Zeitschrift für Sozialpsychologie*, 10, 243-252.

Boud, D.J., Dunn, J., Kennedy, T. & Thorley, R. (1980). The aims of science laboratory courses: A survey of students, Graduates and practising scientists. *European Journal of Science Education*, 2(4), 415-428.

Cohen, J. (1960). A coefficient of agreement for nominal scales. *Educational and Psychological Measurement*, XX(1), 37-46.

Derry, S.J. (1996). Cognitive schema theory in the constructivist debate. *Educational Psychologist*, 31(3/4), 163-174.

Doerr, H.M. (1996). STELLA ten years later: A review of the literature. *International Journal of Computers in Mathematical Learning*, 1, 201-224.

Finke, S. (1992). *Versuchsanleitungen zum Anfänger-Praktikum in Physik*. Dortmund: Wulff.

Fischer, H.E. (1993). Framework for conducting empirical investigations of learning processes. *Science Education*, 77 (2), 131- 152.

Fischer, H.E. & Aufschnaiter, St. v. (1993). The development of meaning during physics instruction. *Science Education*, 77 (2), 153- 168.

Fischer, H.E. & Horstendahl, M. (1997). Motivation and Learning Physics. *Research in Science Education*, 27 (3), 411-424.

Fischer, H.E. (1994). Physiklernen: Eine Herausforderung für Unterrichtsforschung. In Nachtigall, D. (Ed.): *Didaktik und Naturwissenschaften*, Band 3. Frankfurt: Lang.

Fischler, H. & Peuckert, J. (Eds.) (2000). *Concept Mapping in fachdidaktischen Forschungsprojekten der Physik und Chemie*. Berlin: Logos.

Glasersfeld, E. v. (1995). *Radical constructivism - A way of knowing and learning*, Washington: Farmer Press.

Hasemann, K. & Mansfield, H. (1995). Concept mapping in research on mathematical knowledge development: Background, methods, findings and conclusions. *Educational Studies in Mathematics*, 29(1), 45-72.

Hegarty-Hazel, E. (1990) (Editor). The student laboratory and the science curriculum. London: Routledge.

High Performance Systems (1994). STELLA II, Version 3.0.5 for Windows. Hannover, NH: High Performance Systems (http://www.hps-inc.com).

Horstendahl, M. (1999). *Motivationale Orientierungen im Physikunterricht*. Berlin: Logos.

Hucke, L. (2000). *Handlungsregulation und Wissenserwerb in traditionellen und computergestützten Experimenten des physikalischen Praktikums.* Berlin: Logos. http://eldorado.uni-dortmund.de:8080/FB2/ls11/forschung/1999/Hucke

Hucke, L. & Fischer, H.E. (2000). Wissenserwerb und Handlungsregulation im physikalischen Praktikum. In: Fischler, H. & Peuckert, J. (Eds.): *Concept Mapping in fachdidaktischen Forschungsprojekten der Physik und Chemie.* Berlin: Logos.

Johnstone, A.H., Watt, A. & Zaman, T.U. (1998). The students' attitude and cognition change to a physics laboratory. Physics Education, 22-28.

Lazarowitz, R. & Tamir, P. (1994). Research on using laboratory instruction in science. In Gabel, D.L. (Editor): *Handbook of Research on Science Teaching and Learning.* New York: Macmillan.

Leybold (1994). CASSY, Version 2.13, Universelle Messwerterfassung. Hürth: Leybold Didactic GmbH (http://www.leybold-didactic.de).

Linn, M.C. (1998). The impact of technology on science instruction: Historical trends and current opportunities. In Fraser, B.J. & Tobin, K.G. (1998) (Eds.): *International Handbook of Science Education.* Dordrecht: Kluwer, 265-294.

Lunetta, V.N. (1998). The school science laboratory: Historical perspectives and contexts for contemporary teaching. In Fraser, B.J. & Tobin, K.G. (Eds.): *International Handbook of Science Education,* Dordrecht: Kluwer, S. 249-264.

Markham, K.M., Mintzes, J.J. & Jones, M.G. (1994). The concept map as a research and evaluation tool: Further evidence of validity. *Journal of Research in Science Teaching,* 31(1), 91-101.

Nicol, D.J., Kane, K.A. & Wainwright, C.L. (1994). Improving laboratory learning through group working and structured reflection and discussion. *Educational and Training Technology International,* Vol. 31(4), 302-310.

Mayring, P. (1997). Qualitative Inhaltsanalyse. Weinheim: Deutscher Studien Verlag.

Redish, E.F., Saul, J.M. & Steinberg, R.N. (1997). On the effectiveness of active engagement microcomputer-based laboratories. *American Journal of Physics,* 65(1), 45-54.

Schecker, H. (1998). Integration of experimenting and modelling by advanced educational technology: Examples from nuclear physics. In Tobin, K. & Fraser, B. (Eds.): *International Handbook of Science Education.* Dordrecht: Kluwer, 383-398.

Schmidt, K.H. & Kleinbeck, U. (1990). The role of goal setting and feedback in job design. In Kleinbeck, U., Quast, H.-H., Thierry, H. & Häcker, H. (Eds.): *Work Motivation.* Hillsdale, N.J.: Lawrence Erlbaum.

Tamir, P. & Lunetta, V.N. (1981). Inquiry related tasks in high school science laboratory handbooks. *Science Education,* 65, 477-484.

Thornton, R.K. & Sokoloff, D.R. (1990). Learning motion concepts using real-time microcomputer-based laboratory tools. *American Journal of Physics,* 58(9), 858-867.

Tobin, K.G. (1990). Research on science laboratory activities: In pursuit of better questions and answers to improve learning. *School Science and Mathematics,* 90, 403-418.

Toothacker, W.S. (1983). A critical look at introductory labwork instruction. *American Journal of Physics,* 51, 516-520.

Weidenmann, B. (1993). Multicodierung und Multimodalität im Lernprozeß. In: Issing, L.J. & Klimsa, P. (Eds.), *Information und Lernen mit Multimedia.* Weinheim: Psychologie Verlags Union, 65-84.

Wellington, J. (1998). Practical work in Science: Time for a reappraisal. In Wellington, J. (Editor): *Practical work in school science - Which way know?* London: Routledge.

Welzel, M., Haller, K., Bandiera, M., Hammelev, D., Koumaras, P., Niedderer, H., Paulsen, A., Robinault, K. & Aufschnaiter, S. v. (1998). Ziele, die Lehrende mit experimentellem Arbeiten in der naturwissenschaftlichen Ausbildung verbinden - Ergebnisse einer europäischen Umfrage. *Zeitschrift für Didaktik der Naturwissenschaften,* 4(1), 29-44.

White, R.T. (1996). The link between laboratory and learning. *International Journal of Science Education,* 18(7), 761-774.

Computer Tools in the Lab – Effects Linking Theory and Experiment

Florian Sander, Horst Schecker, Hans Niedderer, University of Bremen, Germany

Abstract

This study investigates the effectiveness of labwork with the integrated use of computer tools for data collection and for model building. The approach was trialed in a first year university physics course in Newtonian mechanics. Student understanding of basic physics concepts increased at least as much as in a comparative traditional course. During model building phases the students talked a lot more about lab-related physics than in most other lab activities. This supports the assumption that the use of model building software (STELLA) in the lab contributes to the objective "to link theory to practice". In order to fully exploit this learning potential, labguides and tutors have to demand the comparison between experiment and model explicitly.

Introduction

Many researchers have reported results that students fail to relate their labwork activities to the theoretical background (Toothacker 1983). "To many students, a 'lab' means manipulating equipment but not manipulating ideas" (Lunetta 1998). On the other hand many teachers and students consider "linking theory to practice" to be one of the most important objectives of labwork (Welzel, Haller, Bandiera, Bécu-Robinault, Hammelev, Koumaras, Niedderer, Paulsen, & v. Aufschnaiter 1998; Haller 1999). According to Tiberghien (Tiberghien 1994, Bécu-Robinault & Tiberghien 1998) understanding physics means to relate the "world of objects and events" to the "world of models and theories". Our approach towards strengthening the relationship between practical lab activities and theorising is to combine computer tools for experimenting and modelling in an integrative lab curriculum (see Table 1; Schecker 1998a).

Table 1: Computer tools for theory and experiment

	Computer tools
Experiment	MBL Microcomputer Based Labs for data collection and processing e.g. ULI (Vernier), CASSY (Leybold)
Theory	MBS Dynamic Model Building Systems software for modelling and simulating e.g. STELLA (High Performance Inc.)

Microcomputer based labs (MBL) help to collect data and display the results in graphs on the computer screen online or with only a small time gap (Thornton 1987). Positive learning effects are well proven (Thornton & Sokoloff 1990). As cited above, it seems to be even more important to foster students' work on models and theories in the context of lab activities by implementing interactive tools for *modelling (model building systems, MBS)*. Modelling results are then to be

D. Psillos and H. Niedderer (eds.). Teaching and Learning in the Science Laboratory. 219-230.

compared with results from equivalent measurements (Schecker 1998a, b). Producing graphs with MBL is seen as an experimental activity, whereas their interpretation would belong to the theoretical domain.

In this study, we introduced both kinds of tools (MBL, MBS) to the students in an innovative introductory physics course at university. The tools were integrated into both lecture and labwork. The focus of this study is our overall research question: Does the integrated use of computer tools for modelling and experimenting help students to develop a deeper understanding of physics and, particularly, to establish a stronger link between theory and practice? [1]

Teaching approach

Innovative first year introductory physics course in Newtonian mechanics

This study analyses laboratory work during a first semester introductory physics course at the University of Bremen (winter term 1996/97). The content was Newtonian mechanics for prospective high school physics teachers. The course lasted for 13 weeks and included lectures, tutorials and a beginners' laboratory with one lab session per week (about two to three hours). The content of the lab was closely related to the lectures. Students worked in pairs. A graduate student guided them as a tutor. In most of the experiments the students were given a prepared set of apparatus to observe and determine physical quantities and to explore their relations. Writing reports was done at home. An overview of lecture, lab, and computer tools is given in Table 2.

Table 2: Lectures and labs in the course with two types of computer tools

Lecture content	Lab	Lab content	Tools Exp.	Tools Theory
Kinematics: v=ds/dt, a=dv/dt	1	Linear motion on air track: s(t), v(t), a(t)	Traditional	
Constant force (F=m∗a)	2	Measuring accelerations on air track	MBL	trad.
Non-const. force F(x) or F(t)	3	Modelling accelerations on air track	none	MBS
Oscillations	4	Linear oscillation with spring	MBL	MBS
Forced oscillations	5	Forced linear oscillation	MBL	MBS
Motion in two dim.	6	Modelling movement in the plane	none	MBS
Rigid bodies 1	7	Torque and circular motion	traditional	
Rigid bodies 2	8,9	Torsion pendulum (Pohl's wheel)	MBL	MBS
Momentum	10	Collision in 1 dimension	trad.	MBS
Work and energy	11	Collision in 2 dimensions	trad.	MBS
Presentations	12	Mini-projects (2 lab sessions)	Free choice	

For each session the participants had to write a lab report containing a description of the theory, a description of the experimental procedures and the results. The reports were checked and feedback was given. Sometimes improvements were demanded, but the reports were not graded. Lectures contained interactive elements and the use

[1] A more comprehensive description can be found in Sander (2000).

of computers as well. New, more open-ended labguides were written for this innovative course with the integrated use of computer tools.

The computer served as a tool both for data collection (MBL) *and* model building (MBS). Seven labs included MBS (with the software STELLA), in four labs students also worked with MBL. The experiments were designed with a varying degree of openness: The core question, the apparatus, and sometimes a brief description of the procedure and the methods to handle data were given. Furthermore the students were asked to formulate and investigate additional own questions. At the end of the term, students worked on mini-projects, setting their own questions and designing their own experiments in the domain of mechanics.

In parallel there was a second first year course in Newtonian mechanics at our university for physics majors. It also consisted of lectures, tutorials and weekly labs. This course, with a different lecturer and more time per week, had more traditional features, such as no use of computers in lecture and lab, and more traditional labguides. Both courses were calculus-based and covered the same domain.

We used the second course for contrasting our test findings from the innovative course. However, the study does not have a strict control group design. The two courses differed in too many aspects, e.g. the amount of teaching time.

Modelling with model building software tools (MBS)

We used the software STELLA for model building. The tool is based on the system dynamics approach. Incremental increases or decreases in the values of variables are described by the relationship between a variable and its rate of change ($x_{new}=x_{old}$+rate_of_change_of_x*Δt). The rate of change $\Delta x/\Delta t$ relates mathematically to the differential quotient dx/dt.

The special powerful feature of STELLA lies in iconic representations of the model variables and their relations. Stella allows students to start working on a model on its conceptual layer. Similar to a concept map the model quantities are placed as objects on the screen. Their relationships (functions, incremental changes) have to be quantified in a second step of model construction (physics equations layer). STELLA generates difference equations automatically. The model is then transformed by STELLA into a simulation program, so that it can be used for quantitative simulation runs.

Thus STELLA contributes to *thinking physics* while a model is built and to overcome mathematical difficulties. The multiple representations - qualitative graphical structures as well as mathematical equations - are expected to provide new perspectives for understanding theoretical structures. STELLA guides students towards concentrating on the "power tools" of physics, i.e. the most general definitions (like a=dv/dt) and laws (like F=m*a). By reducing mathematical boundaries computer-based modelling tools also open up more complex topics for teaching. Quantitative investigations of real world problems, like the motion of a parachutist, that are otherwise restrained by the students' insufficient mathematical competence, can thus be included. The students can concentrate on the *physical* aspects of the model (i.e. conceptualisation and applying principles) while the computer numerically solves the differential equations.

A guide how to work with STELLA in physics can be found in Schecker (1998 a, b) and Niedderer & Schecker (1996).

Research questions

Our general research question is:

Does the integrated use of computer tools for modelling and measuring in lecture and lab lead to a better understanding of physics and to a stronger link between theory and experiment?

This can be broken down to more detailed questions:

- What is the *overall learning outcome* with respect to conceptual understanding?
- What is the effect of computers on the *types of activities* in the lab?
- Do the students have problems in handling the model building software?
- Does working with model building systems (MBS) contribute to *link theory to experiment*?

To what extend do students *interrelate theory (MBS) and experiment (MBL)*?

The study is embedded in research about the use of computer tools for data-gathering (Thornton & Sokoloff 1990, Lazarowitz & Tamir 1993, Thornton 1995), and for modelling (Doerr 1997; Hucke 1999) and research about the combination of both aspects (Niedderer & Schecker 1996, Schecker 1998a, b).

Research methods

Overview

Data were collected in both courses from October 1996 till February 1997 over the whole winter term. Thirteen students took part in the experimental course and about 30 students in the physics majors course. We gained quantitative data from pre and post tests in both groups. In the innovative course all the lab activities of four students were videotaped. We analysed protocols of all the 13 students in this course.

Data sources

Tests: Two standard tests, referring to the understanding of basic Newtonian concepts were applied pre and post to all the students in the study. The Force Concept Inventory (FCI; Hestenes, Wells, Swackhamer 1992a) and the Mechanics Baseline Test (MBT; Hestenes & Wells 1992b) were developed as two complementary tests. The FCI items are more qualitative; they can be solved without formal mathematics knowledge, whereas the MBT requires knowledge about physical formulas and mathematical skills. The pre and post-test mean scores and their differences (gains) were calculated to compare the overall learning effectiveness of the two courses.

Video-recordings (VT): We videotaped two pairs of students through the whole term during their labwork sessions, documenting the process of modelling and data analysis. These data were used for a Category-Based Analysis of Videotapes from labwork (CBAV). In this study we present two cases of model development from these data.

Lab reports: We collected all the lab reports in the innovative course. They were used to analyse the quality of students modelling abilities.

Category Based Analysis of Video Tapes (CBAV)

Videotapes from the first seven lab sessions were analysed with the CBAV method. Every 30 seconds the students' actions were classified using two category systems: "labwork context" and "verbalised knowledge". "Labwork context" refers to the resources that the students draw upon for their work (see Table 3). "Verbalised knowledge" categorises the type of knowledge that the students apply (see Table 4). For a detailed description of the CBAV method see the contribution of Niedderer et al. (Chapter 1 *of this volume*). The context categories however have been simplified in this study to show some effects more clearly.

Table 3: Selected CBAV categories of labwork context

Category	Description
Tutor	Interaction with 3rd person, in this study only the tutor
Experiment	All types of activities related to the process of measurement with and without computer, including manipulation of apparatus
Modelling	All types of activities with the computer model building system

Table 4: Selected CBAV categories of verbalised knowledge

Category		Description
Physics knowledge	KP	Apply physical terms and/or notions
Technical knowledge	KT	Use technical terms concerning apparatus/software
Technical and physics k.	KTP	Use physical ideas and technical arguments
Mathematical knowledge	KM	Use mathematical arguments

To compare the amount of time used for knowledge verbalisation in different contexts, we define "verbalisation density in %" as the percentage of time used for verbalising a specific category of knowledge while working in a certain labwork context. For example, a high density of KP in the context tutor would mean that students talk a lot about physics while having contact with the tutor.

Categories for the analysis of lab reports

Lab reports were analysed to evaluate the relationship between theory (MBS) and experiment (MBL). We defined three levels for quality of interrelations (Table 5):

Table 5: Categories for interrelation between theory and experiment

Category	Description
A	False models or missing interrelations
B	Graphs from measurement and simulation are compared; no consequences are discussed.
C	Active comparison leading to one of the following consequences: a) check physics laws; b) describe the system's behaviour with new parameters; c) develop a broader theoretical frame for the model.

According to our research question, only labs with the use of MBL *and* MBS were selected. We wanted to find out to what extend students *interrelate theory (MBS) and experiment (MBL)*. The effects of different contents were not analysed in this explorative study.

Research results

Overall learning outcome

Physics educators tend to expect motivational effects from computers in physics instruction. They are much more sceptical about contributions to the conceptual understanding of physics. They are afraid that students "play around" with the machine and do not engage in sound physical reflection. Our data shed some light on the issue of learning effects, seen as "effectiveness 2" (Psillos & Niedderer, Chapter 1 *of this volume*). The FCI (force concept inventory) and the MBT (mechanics baseline test) were used to examine and compare learning outcomes in the domain of Newtonian mechanics. The results of the pre and post-tests for both courses are given in Table 6.

Table 6: Pre and post test mean scores in two different courses

		N	Pre	Post	Gain	g
FCI	Innovative course	12	46%	62%	16%	.30
	Traditional course	19	53%	64%	11%	.23
MBT	Innovative course	10	39%	50%	11%	.18
	Traditional course	14	55%	62%	7%	.16

g ("Hake factor") is a weighted gain index that relates the attained gain (Post – Pre) to the maximum possible gain (100-Pre). These results show that (in terms of the FCI and the MBT) teacher students in the innovative course profited at least as much from their course as the physics majors in the conventional course. Their gains in the FCI were significantly higher (Mann-Whitney U test; $p=0,05$). In absolute values the majors were still better in the post-test, particularly in the MBT. A plausible reason is that the traditional course had more teaching time and the students started at a higher level.

The FCI gains of the traditional course and of the trialed interactive one are similar to results that Hake (1998) calculated from tests in many other "traditional courses". Our gains are higher than the ones that Schecker & Gerdes (1999) found in courses that focused on computer-based modelling only, and they are in the same order of magnitude as "interactive courses" that Heller & Huffman (1995) report about.

We thus see an overall positive learning effect in our new course approach. The results indicate a certain potential of increasing effectiveness by the integrated use of computer tools for measuring and modelling. Yet, the improvement was not as high as we had hoped for. The results of the post-tests suggest that students in both courses did not develop a deeper understanding of basic Newtonian concepts and basic skills for solving mechanics problems.

What is the effect of computers on the types of activities in the lab?

If we inspect the CBAV results about lab contexts, we can see that students draw on a greater variety of resources when the computer is part of the labwork setting. The time is spread over more types of contexts, so labwork becomes more complex. Therefore it was a good decision to start with labs that do not involve computers (lab

1); then to continue with MBL only (lab 2), with MBS only (lab 3), and end up in the integrating use (labs 4 and 5).

Do the students have problems in handling the model building software?

One condition for the model building system to promote the link of theory and experiment is that students quickly learn to handle the software. Figure 1 compares the verbalisation densities of mere technical related knowledge KT and theory-related forms of knowledge (KP+KTP+KM) during the construction of the first seven models (in four successive labs). In this special context, the meaning of KT is talking about software problems. The graph shows that talking about software problems takes quite some time at the beginning, but becomes less important with more applications. In parallel the amount of time talking about theory increases. It takes about two model constructions to learn the software. This is in line with our experiences from other studies (cf. Schecker 1998b). It must not be underestimated that teaching and learning time has to be invested for introducing the MBL- and MBS-tools to the students. Even after several weeks, students sometimes returned to inappropriate strategies in dealing with the modelling software.

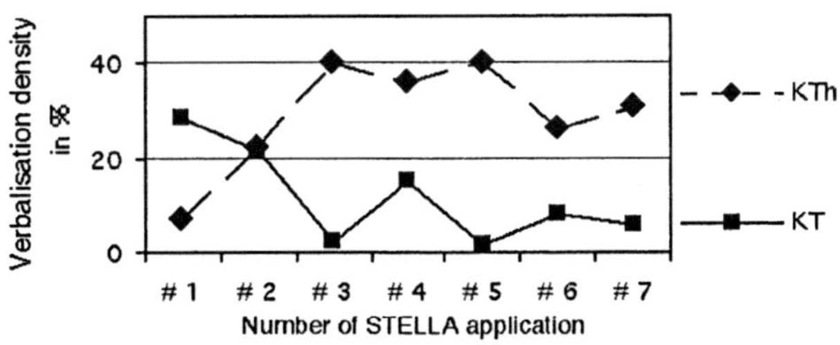

Figure 1: Talking about software problems (KT) versus talking about theory (KTh) during the first seven model building phases

Does computer-based modelling lead to stronger cognitive links between theory and experiment?

The main purpose of producing CBAV data was to investigate the relationship between labwork contexts and the verbal use of knowledge. We consider talking about physics in a lab context to be a good surface indicator for the intended cognitive activity "linking theory to practice", seen as "effectiveness 1" (Psillos & Niedderer, Chapter 1 *of this volume*).

We found that different lab contexts contributed differently to talking about physics. Some results are shown in Figure 2.

From this figure we can conclude:

Working in experimental contexts contributes little to talking about physics and "to link theory to practice". On the other side, these activities use up a lot of lab time.

During model building phases we can see a rather high density of physics knowledge (about 20%). This is the second highest density behind talking with the tutor. This strongly supports the assumption that the use of model building software

(STELLA) in the lab contributes to the objective "to link theory to practice". Hucke & Fischer (Chapter 5 *of this volume*) found KP densities in model building contexts up to 75%.

The high verbalisation density in the context 'tutor' (55%) shows the important role of the tutor to link theory and experiment. Note that only *students'* contributions are counted in the knowledge verbalisation densities, not the contributions of the tutor talking himself.

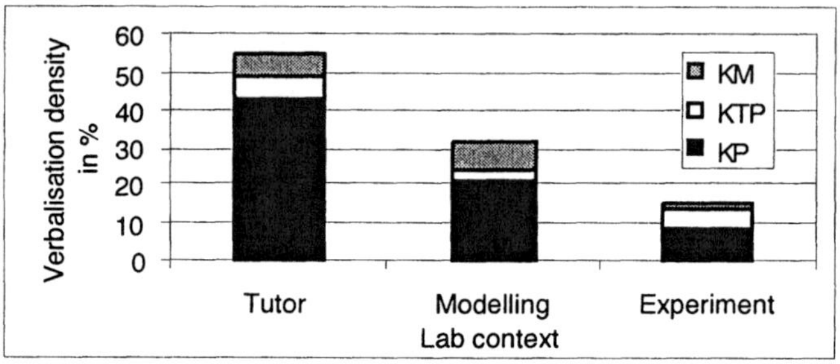

Figure 2: Density of verbalising knowledge in three types of lab contexts

To what extend do students interrelate theory and experiment?

The goal of integrating modelling and measuring tools is to enable students to change freely from theory to experiment and vice versa so that the two activities mutually profit from each other (Schecker 1998a, b). Our analyses of lab reports and videotapes show that these intended active interrelations are not found very often. Nevertheless, these are important activities. Below we describe five types of those interrelations found in the data.

(1) Identify relevant quantities: in the course of modelling students reflect upon which quantities have to be measured in the experiment for the model to work properly.

(2) Correct model parameters: students change experimental parameters that are covered in the model, sometimes affording new measurements.

(3) Explore the system: the model is run in a simulation or the experiment is carried out in order to observe the behaviour of the system; the findings are compared.

(4) Change basic model structures: the model (conceptual layer and equations layer) is reconsidered and adapted, e.g. in order to include new experimental influences.

(5) Develop new experimental ideas: students extend the experimental setting or apply new measurement techniques that were not given in the labguide; this also results in model structure changes.

Some of these interrelations could only be found in the lab reports that the students wrote at home after the lab sessions. Quantitative results of the lab report analysis are given in Table 7.

The table shows the quality of comparisons between model and measurement found in the text. In about 77% of the analysed cases, lab reports show some relevant relations (categories B and C). In about 36% equivalent graphs from measurement and from model simulation are shown, but without drawing further consequences (category B). In another 41%, the comparisons between measurement and theory show a deeper understanding of the experiment (category C). Category A stands for false models or missing interrelations. The aim would be to have more reports with active comparisons between measurement and computer models (category C).

Table 7: Quality of comparisons between theory (model) and experiment in the lab reports (categories A, B, C: see text above)

Lab	Category: Lab content	A	B	C
2&3	Measuring and modelling accelerations	1	3	2
4	Spring oscillations	1	2	3
5	Forced linear oscillation	2	2	2
9	Torsion pendulum (Pohl's wheel)	1	1	2
	Percentage	23%	36%	41%

A case study: Movement with non-constant acceleration - spring oscillations

We illustrate the integrated use of computer tools for measuring and modelling with a case from a lab about spring oscillations. Lab 4 started with the task to measure the free movement of a bob hanging on a spring (see Figure 3). Afterwards, two students develop their own computer model in many small steps, starting from an empty STELLA desktop. Let us keep in mind that the aim is to foster a qualitative understanding of physics phenomena. Students are expected to learn how to use basic physics concepts (like force or momentum) in relation to a specific experiment.

The three most important steps in this development are to:

- develop the acceleration -> velocity -> distance sequence of variables and rates of change (kinematic part of the model),
- apply the force-mass-acceleration relation (Newton 2) as a power tool for all force and motion problems, and
- introduce the specific forces for the spring problem.

The students performed the following lab activities:

a. With computer measurement (MBL) the students got a graph, which due to friction had decreasing amplitude (see Figure 3).
b. Afterwards they built a model with STELLA (MBS). Their discussion centred on the acting forces. They held two different points of view: A "Newtonian" point of view, due to which forces reduce velocity, and a more "Aristotelian" point of view due to which force is proportional to velocity. They especially discussed the lower point of the movement.
c. An intermediate model already showed the total force as the sum of spring force and weight force. The spring force related to the distance. No friction force was considered. This resulted in an oscillation with no damping.
d. They compared the model's prognosis with the graph from measurement (see Figure 3). Although not explicitly demanded in the labguide, the students engaged in a longer session of modelling activities in order to get a decreasing graph like the one measured. They started to discuss friction forces, especially air friction. They needed some time until they got at the principle that the friction force is always in the opposite direction of velocity.

e. Finally, they got a good agreement between measurement results and the simulation graph from their model (see Figure 3).

Altogether, this model building phase took about 30% of the time of the whole lab. During this discussion, the density of talking about physics was about 25%.

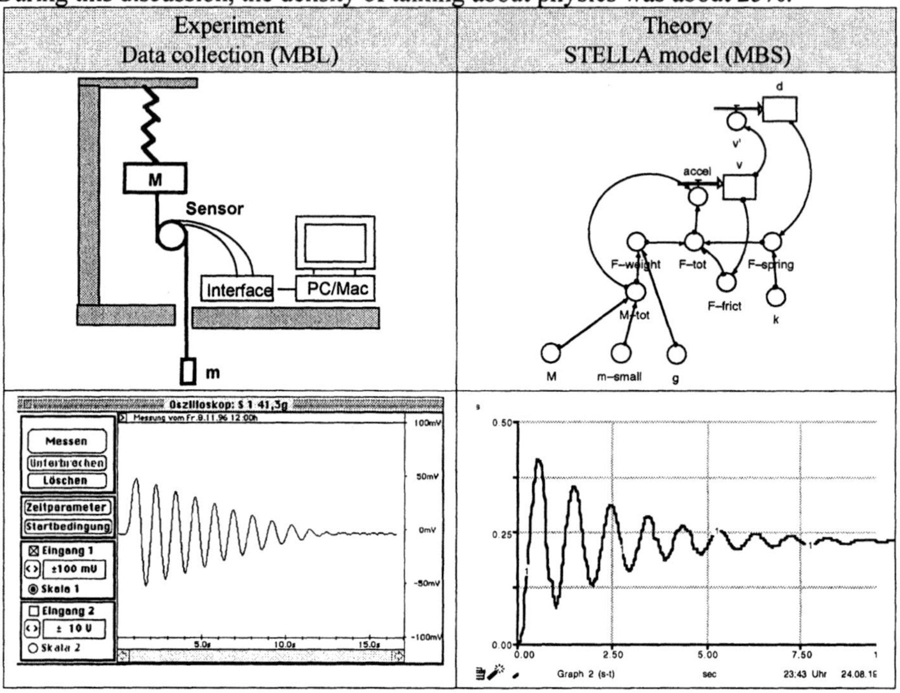

Figure 3: Experiment, computer model and results from both in lab 4

During this discussion the students talked about the following concepts of physics related to their specific experiment:

- weight force and spring force
- acceleration at different points of the oscillation
- balance between different forces
- relation between position of the moving object and the force of the spring
- directions of forces and signs of their values in the equations
- friction as an extra force in addition to the other acting forces; its direction with respect to movement
- magnitude of the friction force: Is it constant or air friction related to velocity.

The example shows that students are able to actively build their own computer model. During this process, a lot of theory is used and developed, especially in situations, where measurement graph and simulation graph are essentially different.

Conclusions

We investigated the effects of integrating computer-based modelling into a labwork course on Newtonian mechanics. The results can be comprised in three main points:

- Under the innovative approach student understanding of basic physics concepts increased at least as much as in a traditional course. Still, conceptual tests revealed remaining deficits in understanding force and motion. This holds for both approaches.

- The integrated use of computer-based modelling has a potential for improving the link between theory and practice during labwork. This was shown by a category-based analysis of students' actions in the lab and by the qualitative analysis of single cases.
- In three quarters of analysed lab reports, students' lab reports discussed some relevant relations between the experiment and the corresponding model. In 41% of the reports these comparisons led to a deeper understanding of the experiment.

The general research question "Does the integrated use of computer tools for modelling and measuring in lecture and lab lead to a better understanding of physics and to a stronger link between theory and experiment?" can be answered in the affirmative although there are some efforts left to fully exhaust its learning potential.

Recommendations

We recommend the integration of computers into labwork courses as tools for collecting and processing experimental data in close connection with computer-based modelling. Based on our empirical results, we see the following advantages:

- With micro-based labs, measurements are done faster and results can be seen immediately. This makes it easier to vary experimental parameters and observe their consequences.
- Computer-based modelling triggers theoretical reflection about the related experiment. Students talk a lot about physics during the process of modelling, so that the link between theory and experiment can be improved.
- By combining computer-based measuring and modelling, students have the chance to compare their own theoretical approach to their own experimental results. This can lead to a more profound reasoning and to the mutual adaptation of model and experiment. In order to fully exploit this learning potential, labguides and tutors have to demand the comparison explicitly.
- Integrating computer tools into labwork raises the cognitive load. The learning environment becomes more complex. It is therefore necessary to go through stages: starting with labs without computer, then introducing measuring and modelling tools separately, before they are used in an integrated approach.
- We do not recommended to employ computer tools in all the lab sessions. Of course students should also learn to use conventional techniques.

References

Bécu-Robinault, K., Tiberghien, A. (1998). Integrating experiments into the teaching of energy. *International Journal of Science Education*, Vol 20, n° 1, pp. 99-114.

Doerr, H. M. (1996). STELLA ten years later: A review of the literature. *International Journal of Computers in Mathematical Learning*, 1, 201-224.

Haller, K. (1999). Über den Zusammenhang von Handlungen und Zielen. Eine empirische Untersuchung zu Lernprozessen im physikalischen Praktikum. Dissertation Universität Bremen. In H. Niedderer, H. Fischler (Eds.): *Studien zum Physiklernen*, Bd. 5. Berlin: Logos.

Hake, R.R. (1998) Interactive-engagement versus traditional methods: A six-thousand-student survey of mechanics test data for introductory physics courses. *American Journal of Physics*, 66, 64-74

Heller, P., Huffman, D. (1995). Interpreting the force concept inventory. A reply to Hestenes and Halloun. *The Physics Teacher*, 33(8), 507-511.

Hestenes, D., Wells, M., Swackhamer, G. (1992a). Force Concept Inventory. *Phys. Teach.* 30, 141

Hestenes, D., Wells, M. (1992b). A Mechanics Baseline Test. *Phys. Teach.* 30, 159-166

Lazarowitz, R. & Tamir, P. (1993). Research on using laboratory instruction in science. InGabel, Dorothy L,: *Handbook of Research on Science Teaching and Learning*, Vol. 33 (7), pp. 693-707

Lunetta, V. N. (1998). The School Science Laboratory: Historical Perspectives and Contexts for Contemporary Teaching. In: Tobin, K. & Fraser, B. (Eds.), *International Handbook of Science Education*. Dordrecht: Kluwer.

Niedderer, H., Schecker, H. (1996). Laboratory Tasks with MBL and MBS in Introductory Physics Classes for Prospective High School Teachers, in Proceeding of the International Conference on Undergraduate Physics Education (ICUPE), 461-474

Niedderer, H., Tiberghien, A., Buty, C., Haller, K., Hucke, L., Sander, F., Fischer, H.E., Schecker, H., v. Aufschnaiter, S., Welzel, M. (1998). Category Based Analysis of Videotapes from labwork (CBAV) - the method and results from four case studies. Working Paper 9 from the European project Labwork in Science Education (Targeted Socio-Economic Research Programme, Project PL 95-2005), 41 pages
http://didaktik.physik.uni-bremen.de/niedderer/projects/labwork/papers.html

Sander, F. (2000). Verbindung von Theorie und Experiment im physikalischen Praktikum - Eine empirische Untersuchung zum handlungsbezogenen Vorverständnis und dem Einsatz grafikorientierter Modellbildung im Praktikum. Dissertation University of Bremen. In H. Niedderer, H. Fischler (Eds.): *Studien zum Physiklernen,* Bd. 13. Berlin: Logos

Schecker, H. (1998a). Integration of Experimenting and Modeling by Educational Technology: Examples from Nuclear Physics. In: Fraser, B.J. & Tobin, K. (Eds.): *International Handbook of Science Education*. Dordrecht: Kluwer

Schecker, H. (1998b). *Physik Modellieren.* Stuttgart: Klett (See also Schecker's homepage at http://didaktik.physik.uni-bremen.de/niedderer/projects/)

Schecker, H. & Gerdes, J. (1999). Messung von Konzeptualisierungsfähigkeit in der Mechanik. Zur Aussagekraft des Force Concept Inventory. In*: Zeitschrift für Didaktik der Naturwissenschaften* 5, H. 1, 75-89.

STELLA [software]. Richmond, B. (1985) Lyme, NH: High Performance Systems.(http://www.hps-inc.com/)

Toothacker, W.S. (1983). A critical look at introductory laboratory instruction. *American Journal of Physics,* 51 (1983), 6, 516-520

Thornton, R.K. (1987). Tools for scientific thinking -- Microcomputer-based laboratories for teaching physics. *Physics Education* 22, 230-238.

Thornton, R. K. & Sokoloff, D. R. (1990). Learning motion concepts using real-time microcomputer-based laboratory tools. *American Journal of Physics*, Vol. 58 (9)

Thornton, R. (1995). Conceptual dynamics. In C. Bernardini, Tarsitani, C. , Vicentini, M. (Ed.), *Thinking physics for teaching* (pp. 157-183). New York: Plenum Press.

Tiberghien, A. (1994). Modeling as a basis for analyzing teaching-learning situations. *Learning and Instruction*, 4, 71-87. // g1

Welzel, M., Haller, K., Bandiera, M., Bécu-Robinault, K, Hammelev, D., Koumaras, P., Niedderer, H., Paulsen, A.C. & v. Aufschnaiter, S. (1998). Teachers' Objectives for Labwork. Research Tool and Cross Country Results. Working paper 6 from the European project LABWORK IN SCIENCE EDUCATION (TSER Programme, Project PL 95-2005). Electronically published:
http://didaktik.physik.uni-bremen.de/niedderer/projects/labwork/papers.html

Modelling in Geometrical Optics Using a Microcomputer

Christian Buty, CNRS - University of Lyon II, France

Abstract

This analysis concerns the help given to students by a computer-based model, in the field of geometrical optics. We consider that physics learning is necessarily the establishment of links between two worlds, the world of objects and events, and the world of theory and models. Our study aims to understand whether the computer-based model has enriched these links in the verbal productions of students. We relate the resources given to students during a rather long teaching sequence, and the kinds of their verbal productions. The results show that the computer-based model has favoured the use of physics theory, with more efficiency than the performing of experiments. But the link with the world of objects and events needs accurate instructions from the teacher.
Keywords Physics labwork, modelling, geometrical optics, learning, use of computers in science education.

Introduction

Many previous studies have claimed that one of the main problems raised by "classical" labwork in science education is that students can successively follow the steps described in the labwork sheet and perform experiments without understanding the physical model which justifies their activity (see for example Pernot 1993 p. 102; Saltiel 1994; Lunetta 1998). We aim to study in which extent and under which conditions there could be a link between what they do and what they understand and learn.

Previous research suggests also that "new technologies can complement and support student collaboration and engagement in school laboratory experiences" (Lunetta 1998). In a particular manner, computer-aided-modelling is expected to help students to "actively construct meaning" during labwork (Schecker 1990; Niedderer, Schecker, Bethge 1991) or to foster the link between theory and practice (Sander et al., Hucke & Fischer, Chapter 5 *of this volume*). In physics classes, a computer can play many roles; most of the time, especially in France, it is mainly used in a labwork session to take measurements or to process data (Durey & Beaufils 1998). In this work we propose another way to use computers during labwork activities, more innovative.

Theoretical frame of the study

Modelling is an essential process in Physics; basically, it consists in establishing relations between two worlds, that we shall call the world of objects and events (or real world), and the world of theories and models, or theoretical world. These relations are established under the control of a coherent theory, accepted by the community of physicists.

When facing a material disposal, an individual, especially a student in classroom activities, constructs his/her own model of the situation too, under the control of his/her own previous theory (Tiberghien 1994). This personal theory may be quite

D. Psillos and H. Niedderer (eds.). Teaching and Learning in the Science Laboratory. 231-242.

different from what could be accepted by a physicist, of course; it is the trace of individual understanding and experiences, in everyday life and in school time as well. It involves initial conceptions, spontaneous ways of reasoning, analogies with similar phenomena, and so on. In this perspective, the aim of physics teaching must be to allow students to construct physics-conform meanings to concepts, by establishing links between the two worlds.

We have grounded our work on the hypothesis that a dynamic computer-based representation could help students to establish these links, because it constitutes a "materialised model", as it was defined and used in another context by Quintana-Robles (1997, p. 27); the materialised model is a set of correspondences between an element of the theoretical world, here the model of geometrical optics, and a real object perceptible to students' eyes, on the computer screen; it is a pathway between the world of objects and events and the world of models and theories (as symbolised on Figure 1). The materialised model has two aspects:

- first it gives students a material interface, associating a way to act on it (by the mouse) and a perception on the computer screen, showing the results of the action;
- secondly its behaviour, coming from the implemented rules used for its construction, is supposed to be coherent with the physics laws students have to learn; in this sense it is a learning tool.

The support of the materialised model is expected to be especially efficient in two classical ways of linking the theoretical world and the real world: interpretation and prediction. In both cases, the possibility to see a representation of theoretical elements (such as rays, in our case) is a strong help for thinking.

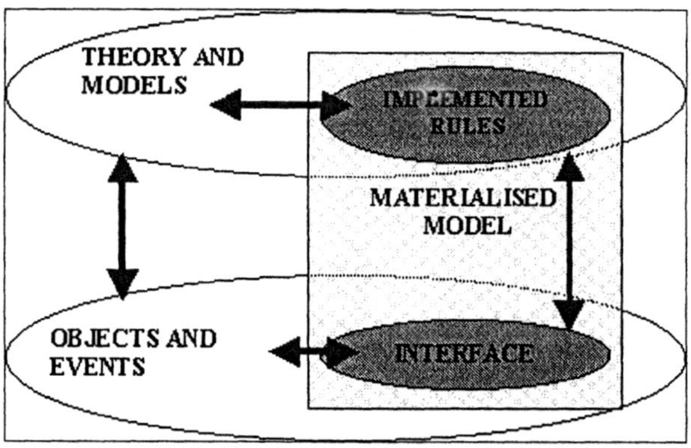

Figure 1: The potential help of a materialised model

Coming back to our initial purpose to study in which extent and under which conditions students could establish a link between what they do and what they understand and learn during labwork activities, our theoretical frame leads us to formulate our research question in the following form: "Did the use of a materialised model help students to make relations between the world of objects and events and the world of theories and models?"

Teaching approach

The choice was made to construct an appropriate teaching sequence in order to answer our research question, in the field of geometrical optics. The software used to realise the materialised model has been Cabri-Géomètre II (©Texas Instruments).

General information

This case study concerns geometrical optics in the last class of the French upper secondary school (grade 12). The main centre of interest of the official curriculum is the formation of images through an optical instrument.

The teaching sequence is expected to last for seven or eight weeks, a session of two hours a week. The lecture is done in the same time as the experiments (in France it is called a "cours-TP"). It means that the teacher is present all the time, giving either instructions for experiments or some theoretical insights, or some commentaries on the results and measurements, and so on.

The study was done in a real class-context, so that the extensibility of results could be better founded. There were fourteen students in the class, working by pairs. Each pair of students had some experimental device and a computer; files made with Cabri-Géomètre (let us say "cabri-files") were implemented in the computer, each file corresponding to a particular experiment.

Teaching objectives and the use of the computer

The global objective of the teaching sequence is that the students construct the meaning of the concept of optical image. This concept can be specified in three main statements: an object is a set of points; the image of an elementary point is the point where all the emergent rays pass after going through the optical system; the image of an object is the set of the images of the various points of the object.

The theoretical model of geometrical optics thus appears as an analytic one (Viennot 1996, p. 31): it splits up the reality of a light flux in separate rays, and an object in discrete elementary sources; it allows to have a global view of phenomena only by a cognitive operation - to pass from one ray to all the rays, from one elementary source to the whole object - which is usually not represented on the schemas drawn by students in a pen-and-pencil environment.

On the contrary, the Cabri-environment of the materialised model it is possible to represent the three aspects of the concept of image, and to develop procedures to symbolise the cognitive operation re-constructing the continuous reality from the discrete theoretical elements:

- A light object can be represented by a segment, and an elementary point source can be represented by a point eventually moving in this segment.
- Some geometrical constructions (called Snell's constructions) allow to draw a generic light ray, coming from a single point source, which obeys the laws of refraction; the emergent part of the generic ray passes, for every position of it, through a single point which represents the image of the elementary source.
- When the point source is moved on the object-segment, the image point covers the image of the segment.

Organisation of the whole teaching sequence

The whole sequence has been organised in 15 "situations". Each situation is defined by a specific teaching content. Table 1 gives a brief description and some characteristics of these situations.

We shall here examine the essential part of the second situation, in order to document the kind of articulation between the knowledge content of this situation and the expected way for students to use the computer.

The knowledge aim in this second situation is the concept of main image focus of a converging lens, that is, the point where all the emergent rays gather after the lens, when the incident beam is parallel to the axis. This concept of main image focus is defined only under one condition: the rays must be near enough of the axis. If not, what we know from the refraction law, applied to the two faces of the lens, allows to predict that all the emergent rays will not converge in a small point, but in a rather large area (as it can be observed in Figure 3). The consequence in the world of objects and events is that the lens must have a diaphragm.

Table 1: Teaching content of the different situations [1]

Sit. #	Teaching content	N of E	N of F	Order
1	Introduction	none	none	unfounded
2	First Gauss condition, main image focus	1	1	interpretation
3	Optical centre	none	1	unfounded
4	Second Gauss cond., secondary image focuses	1	1	interpretation
5	Main and secondary object focus	2	2	prediction
6	Image of a point source as intersection of rays	none	1	unfounded
7	Image of an object as set of image points	none	1	unfounded
8	Descartes' law for the position of the image	1	1	prediction
9	M. of focal length by autocollimation method	1	1	prediction
10	Measurement of focal length by Bessel's meth	1	1	prediction
11	M. of focal length by Silbermann's meth	1	1	prediction
12	Properties of a magnifying glass	1	1	prediction
13	Properties of a diverging lens	1	1	prediction
14	Optical properties and defects of the eye	1	3	prediction
15	Study of an astronomical refracting telescope	1	1	prediction

For the corresponding experiment the students are given (see Figure 2):

a. a lamp which produces a parallel beam of light;
b. a semicylindrical lens, which is too wide, so which gives a bad convergence of the emergent beam;
c. a sheet of white paper, to observe the curved shape of the emergent beam.

[1] The situations' numbers are those used later in the data analysis; the column "N of E" indicates the Number of Experiments achieved during the situation; the column "N of F" indicates the Number of cabri-Files used during the situation; the column "Order" indicates whether the experiment is performed before the use of the dynamic representation ("explanation") or after ("prediction"); in this column is indicated "unfounded" when the lack of experiment does not allow to use such categories. One example of the articulation between teaching content and the use of the computer

This experiment takes place before any teaching about the concept of main image focus; the question asked to students is "how can we manage to have a punctual converging area after the lens?"

Figure 2: The experimental device in situation 2

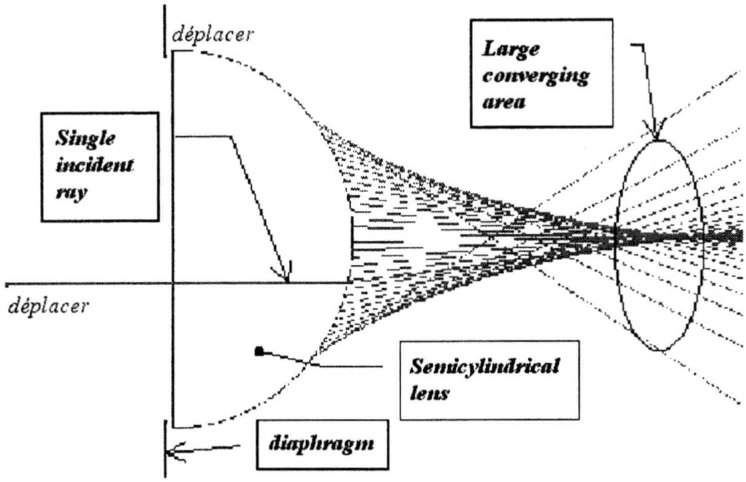

Figure 3: Materialised model associated to the experiment in situation 2

To help them to answer this question, they may use a cabri-file modelling the experimental situation (Figure 3).

When they open the file, they see on the screen:

1. a representation of the semicylindrical lens, as a half-circle;
2. a single incident ray, parallel to the axis of the lens, which goes straightforward through the plane face of the lens, but is refracted on the curved face; this ray may be changed by moving its origin (word "déplacer");
3. the corresponding emergent ray, which is drawn according to physics laws from the incident ray; students may see the hidden geometrical construction by using some command of the software, so they can be convinced that this emergent ray obeys the laws of Physics.
4. A representation of a diaphragm was also drawn, which can be more or less opened by moving the points at its edges; if a ray arrives on the diaphragm, it stops;

This drawing on the computer screen is perceived by students as a *materialised model* of geometrical optics:

- because it shows (materialised) theoretical objects of optics, such as rays (model);
- because it obeys refraction laws (model), that students are expected to know from previous teaching, and this relation of drawings to refraction law is visible (materialised);
- because they can modify it, moving the incident ray (materialised), but the result is always conformable to physics theory (model).

Furthermore, students may make the emergent beam appear, by another command of the software, so they can see how the model accounts for the result of the experiment, especially the curved shape of the emergent beam and the non-punctual converging area, as can be seen in Figure 3.

The materialised model allows also to answer this question. It makes visible that the more external rays are responsible for the non-convergence of the emergent beam. So by diaphragming the lens these rays vanish, and the convergence area is punctual, the image focus exists.

The interplay between the world of objects and events and the world of theories and models, as students were expected to experience it, was the following: students were supposed not to find the answer of their task only if staying in the experimental field; they were supposed to find the answer by using the materialised model, when producing an explanation in terms of the model. As far as cognitive operations are concerned, the first step should have been an *interpretation* of what happened in the experiment (the shape of the emergent beam) by the model (in terms of rays); then the problem should have been *solved inside* the materialised model; at last a *prediction* should have taken place, leading to the action in the real world (to put a diaphragm in order to have a punctual converging area).

Research methods

The research methods are directly determined by the aim of the research, namely investigating the impact of a materialised model on modelling activity and learning among students.

Collecting data

We judged it necessary to follow the activity of the same students all along the different sessions. It was probable indeed that an evolution could be expected in the way for students to use the materialised model, because some habit is certainly useful to handle this rather sophisticated software, and because the increasing understanding by students of the involved topics (image formation) was supposed to influence their ability to apply the materialised model to experimental situations.

Consequently, one pair of students has been videotaped during all the labwork sessions, recording thus their actions on the experimental device, their verbal productions, and the events and actions on the computer screen. The analysis in this study concerns the activity and verbal productions, all along the fifteen situations, of *one particular student* among the two in the observed pair: a young man named Emmanuel. Thus we were able to follow step by step the joint evolution of his modelling activities and of his understanding of involved physics concepts. This study deals mainly with the modelling activity; the evolution of student's conceptions about image formation can be found elsewhere (Buty 2000).

Methodology for the analysis

For analysing these data, an appropriate methodology was established in collaboration with two German groups (in the Universities of Bremen and Dortmund) participating to the LSE project. The shared aim was to emphasise the relations between the kind of resources the student has recourse to during his activity, and his use of physics theory: this was the path to investigate the efficiency of practical activities (with all the resources they involve) upon physics understanding. A more detailed description of this method was given by Niedderer et al. (Chapter 1 *of this volume*).

Nevertheless, some adaptations were necessary, to fit the particular context of a "cours-TP" and, the theoretical background of our study, which has been explained before.

The first step is naturally to describe and categorise the various resources offered to students, and to define indicators for the use of physics theory.

Inventory and coding of resources

Several resources were accessible to the observed student. For a detailed description see Niedderer et al. (Chapter 1 *of this volume*, p. 36). In our particular case, we have modified the meanings of some categories:

- The teacher has a special role in our study, comparing to classical labwork sessions : he is often speaking to the whole class, he gives written questions or summaries from time to time;
- The computer-based model, in our case, is mainly used without constructions made by students; students' actions are supposed to be almost exclusively moves in positions of points or objects; consequently, this resource is coded MM (for materialised model), without distinguishing "computer model building" and "computer model use".

Categories for verbalised knowledge

Deeper changes have occurred in the categories for knowledge verbalisation, if compared to "standard" description as it can be found in the contribution of Niedderer et al. (Chapter 1 *of this volume*, p. 36). The difference comes from our special theoretical framework, but the general idea is the same: elaborating tools to examine the link between theory and practice during labwork. Table 2 below gives our categories for verbalised knowledge; examples are taken from transcripts. To understand these examples, it is necessary to remember that students are working in pairs, discussing what they are doing, often looking at the computer screen.

The first three categories of verbalisation are produced by the student when staying in a given modelling level; the other ones describe the establishment of links between different levels. Such a methodology, which fits to our theoretical frame, is particularly adapted to verify whether the activity of the observed student corresponds to what could be expected in an environment constructed also according to this frame.

Table 2: categories for verbalised knowledge

Category		Description	Examples
World of objects and events	WOE	Student refers to real objects and events	because look that's easy you put the rule and you see what you get that is nothing but the glass *(using a rule as an obstacle to take off a parasite part of a light beam)*
World of physics theory	WPT	Student expresses a physics concept	we must put it at the image focus (when drawing a pen-and-paper schema)
Characteristics of the Materialised Model	CMM	Student expresses something he is seeing on the screen of the computer, or manages the software functionnalities	OK go on the half-circle there where is written "move", no there on the screen *(he is guiding the action of his fellow student)*
Relations between theory and objects	RTO	Student establishes a relation between objects/events and the physics knowledge	the image given by the beam of the lamp upon a well upon something on a table
Relations between the materialised model and physics	RMMP	Student interprets what he sees on the computer screen in terms of physics theory	[it corresponds] to the refraction; yes this point go go go go go here have you seen ?) Ha that's too bad they have not done [the second part of the reflected ray]
Relations between the materialised model and the world of objects	RMMO	Student establishes a relation between the aspect of the materialised model and the objects or events he has observed on the real experiment	So that is the bea(: :) how do you call it ? How did we call it? The ray which was around? His fellow student: oh yeah the parasite light here.

As Figure 4 shows, the choice and definition of these categories are related to our theoretical frame.

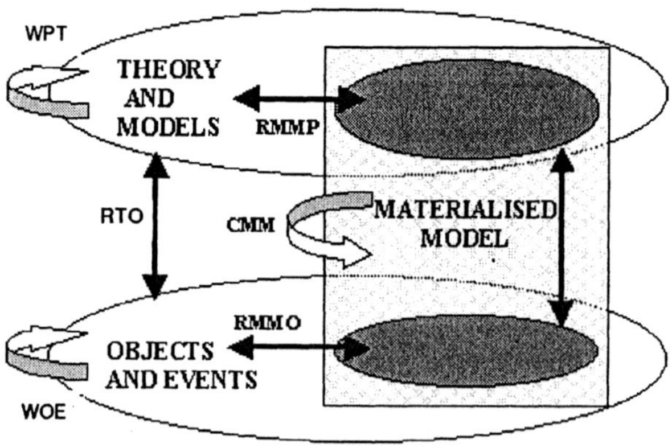

Figure 4: The categories of analysis for expressed knowledge

The activity of the observed student is thus ticked in a grid (see an example in Niedderer et al. (Chapter 1 *of this volume*, p. 37). The columns of the grid correspond to the different categories described above. The rows of this grid are used as the time progresses. This grid facilitates recording information while watching the videotapes in real time.

Results: correlations between resources and verbalised knowledge

After having documented the grid of analysis for each situation, we can derive, as result, a correlation between the used resources and the verbalised knowledge, in order to see whether such and such kind of resource facilitates or not the verbalisation of such and such kind of knowledge.

Our approach

Our aim in this study is to verify the pertinence of a materialised model for facilitating the relations between the experimental situations and the use of physics theory. In our research context, it means asking whether students have or have not a greater tendency to verbalise physics knowledge when using the experimental apparatus or when using the computer-based model.

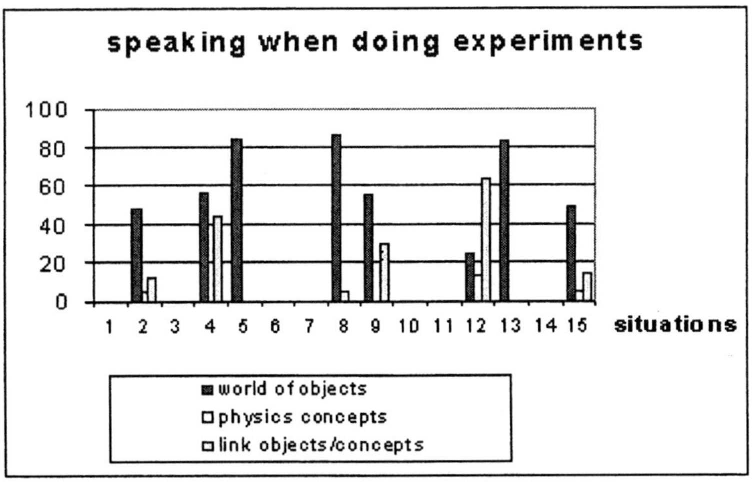

Figure 5: Speaking during experiments

To visualise this comparison between the two kinds of resources, we draw two graphs, each one corresponding to one resource. The first graph (speaking when doing experiments, Figure 5) deals with experimental activity: for each situation, we represent the relative frequency with which each of the three pertinent kinds of knowledge is verbalised by the observed student when he is handling the experimental apparatus. These relative frequency are calculated as if all the situations had the same duration. The pertinent kinds of verbalised knowledge are the ones referring to the world of objects, to the world of physics concepts, to the links between the two worlds.

The indications of this graph must be understood as follows: in situation 5, for instance, during 80% of the duration of the experiment, the student produces verbalisations related to the world of objects and events.

From this graph, we can notice that:

- The main category of verbalisation is obviously related to the world of objects.
- The category of physics-theory related verbalisation is very poor. It means that when manipulating, the student rarely speaks physics, except in relations with the objects or events.

The second graph (speaking when using a computer, Figure 6) deals with computer-based activity: for each situation, we represent the relative frequency for verbalising the three pertinent kinds of knowledge when the observed student is handling the materialised model on the computer. These pertinent kinds of verbalised knowledge are the ones referring to the pure description of the objects on the screen, to the links between the objects on the screen and the world of physics concepts, to the links between the objects on the screen and the world of objects and events.

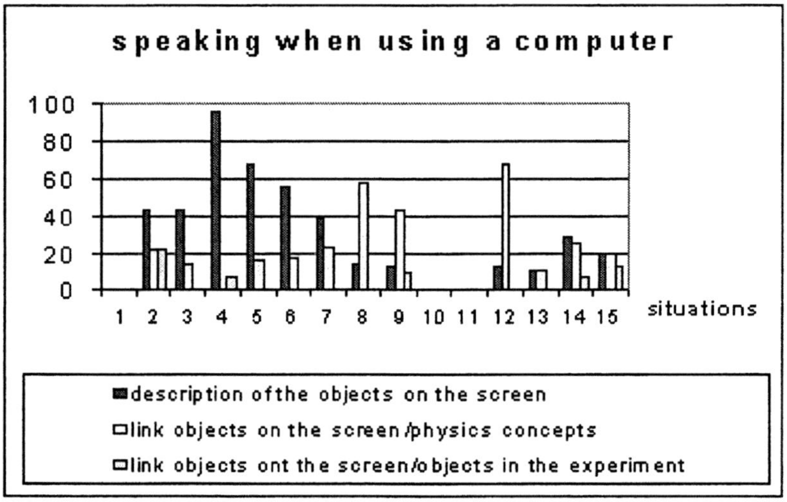

Figure 6: Speaking when using a computer

From this graph, we can see:

- Very often, at the beginning of the sequence, the main verbalisation is related to a plain description of the events on the screen. This category declines at the end of the sequence, which could indicate that the student gets accustomed to the handling of the computer and to the interface.
- In most of the situations, the three categories are effectively used. The student often establishes links between what he sees on the screen and physics theory (second category).
- At the contrary, the links between what the student sees on the screen and the experiments (third category) are rather rare. It can be explained by the fact that the aspect of the materialised model is (at least at the end of the sequence) nearer from the traditional schemas in geometrical optics than from objects.

Discussion

When comparing the two last graphs (Figures 5 and 6), we can derive a certain number of ideas.

It seems to be rather clear that when using a computer-based materialised model the student uses a wider range of knowledge categories than when he is handling the experimental device. It was one of the purposes of the introduction of the materialised model, and it is globally reached.

This enlargement of the verbal categories used by the student concerns mainly the use of physics theory when describing the objects on the screen of the computer (Figure 6). By contrast, we can observe that the use of physics theory is very poor when he handles the physical device (Figure 5).

On the contrary, the student makes rather rare references to objects of the real world when using the computer (Figure 6). That was not a wished effect. We can observe nevertheless that it is not the case for situation 2, where it was explicitly asked to students to say whether they saw similarities or differences between the aspect of the screen and the phenomena they had observed during the experiment they had performed before. We can see here the importance of the instructions given explicitly by the teacher, as it was observed by Becu-Robinault (1997, p. 189-190) in the domain of energy.

Some situations involve a greater variety of categor ies than others, when Emmanuel uses the computer and even during the experiments 2, 9, 12, 15. In the situations 9, 12, 15, as mentioned above (Table 1), the student uses first the computer-based model and after he performs the corresponding experiment.

Conclusions

If we look back to our research question aimed at in this study ("Did the use of a materialised model help students to make relations between the world of objects and events and the world of theories and models?"), we can say that these results argue with a reasonable plausibility that:

- This use of computer has favoured the use of physics theory by the student.
- It has also favoured the link between the world of objects/events and the world of model/theory during the experimental activities, when the computer was used before the experiment, in a somehow predicting way.
- Linking the dynamic representation on the screen of the computer to the real world when using the computer has not been very well realised, except when the instructions asked explicitly to.

Recommendations

The remarks and conclusions above authorise to put forward a certain number of recommendations, which should allow a more efficient use of computers in science education:

- The use of unknown software needs a certain effort from students, as can be seen in Figure 6. When simply looking at the videotapes, we can record some loose of time/energy/motivation and some mental confusion due to the necessity to learn the procedures and commands of the software; it should be avoided as far as possible;

- Modelling activities such as those shown in this study can facilitate the verbalisation of physics concepts by students during practical activities; they should play a greater part of computer use in science education, which should not be restricted to data acquisition and computation;
- Modelling activities have to be carefully embedded in appropriated instructions that prompt learners to externalise their knowledge; they can be a very powerful tool in situations implying predictions, performance of the experiment, discussion and formulation of findings in comparison with predictions;
- Finally, it was mentioned above that this teaching sequence took place in the last class of upper secondary school, and would be assessed in the final exam ("baccalauréat"). We could observe in a qualitative way that the perspective of examinations passed in a classical way (e.g. written tests) is not a help to the development of the use of new technologies in science education. It should lead to question the types of the final secondary exam.

References

Bécu-Robinault K. (1997). Rôle de l'expérience en classe de physique dans l'acquisition des connaissances sur les phénomènes énergétiques. Thèse, Université Claude Bernard-Lyon I.

Buty C. (2000). Etude d'un apprentissage dans une séquence d'enseignement en optique géométrique à l'aide d'une modélisation informatique. Thèse de Sciences de l'éducation, Université Lumière-Lyon II.

Durey A., Beaufils D. (1998). L'ordinateur dans l'enseignement des Sciences Physiques : questions de didactique. Actes des huitièmes Journées Informatique et Pédagogie des Sciences Physiques, Montpellier mars 98. Union des Physiciens, Institut National de Recherche Pédagogique, Inspection Générale de Sciences Physiques, 63-74.

Lunetta V. N. (1998). The school science laboratory: historical perspectives and contexts for contemporary teaching. In Tobin K., Fraser B. (eds.): *International Handbook of Science Education.* Dordrecht, Boston, London: Kluwer, 249-262.

Niedderer H., Schecker H. P., Bethge T. (1991). The role of computer aided modelling in learning physics. *Journal of Computer Assisted Learning,* n° 7, 84-95.

Pernot C. (1993). Une gestion de l'apprentissage de la chimie expérimentale en premier cycle universitaire. Didaskalia n° 2, décembre 1993, Bruxelles-Paris: De Boeck Université, 101-109.

Quintana Robles M. (1997). Etude didactique de films comme aide pour l'enseignement de la physique. Cas de l'expansion des gaz. Thèse, Université Claude Bernard-Lyon I.

Saltiel E. (1994). Un enseignement concret et attractif de la physique doit - il être avant tout expérimental, théorique ? Didaskalia supplément au n° 3, avril 1994, Bruxelles-Paris: De Boeck Université, 15-25.

Schecker H. P. (1990). The didactic potential of computer aided modelling for physics education. In: Ferguson D. L. (ed.): *Advanced technologies in the teaching of mathematics and science.* Berlin: Springer Verlag.

Tiberghien A. (1994). Modelling as a basis for analyzing teaching-learning situations. *Learning and Instruction,* Vol 4, 71-87.

Viennot L. (1996). *Raisonner en physique (la part du sens commun)* . Paris, Bruxelles, De Boeck Université.

Evolution of Students' Reasoning about Microscopic Processes in Electrostatics under the Influence of Interactive Simulations

Alexandros Barbas & Dimitris Psillos, Aristotle University of Thessaloniki, Greece

Abstract

A succession of cognitive states characterises the evolution of causal explanations of simple electrostatic interactions from a macroscopic point of view towards a deep and adequate microscopic one. Interactive simulations of microscopic processes facilitate, or even provoke, this evolution by supporting the transformation of learners' mental representations and patterns of reasoning.

Introduction

In 'circuit electricity', ongoing research is clarifying students' mental representations and patterns of reasoning and their differences with respect to the conceptual models and reasoning of physics. Recent research has also focussed on the ways learners' mental representations and patterns of reasoning are influenced by computer simulations and hands on experiments (Niedderer & Goldberg 1996; White, Frederiksen & Spoehr 1993). In 'static electricity', which is considered more close to students' reasoning (Frederiksen & White 1992), such research is scanty (Welzel 1998). At the teaching level, the study of electrical interactions, which constitute the core of static electricity, has not been extensively investigated, although it could support the study of dynamic electricity (Chabay & Sherwood 1995). Furthermore, conventional teaching of static electricity utilises mainly static representations (pictures, drawings, etc.), although physicists' and students' reasoning about electrical interactions is based, to a great extent, on particulate models of matter involving a dynamic behaviour of particles.

Following a unifying approach to static and circuit electricity (Psillos, Barbas, & Koumaras 1995) we have developed a teaching sequence about simple electrical phenomena based on short interactive computer simulations in close relation with simple hands-on experiments. In this study we investigate whether and how interactive computer simulations of the dynamic behaviour of particle interactions may facilitate learning. We focus on two teaching episodes of the sequence. They deal with electrical interactions between charged and neutral insulators, a familiar subject routinely studied in all secondary education electricity courses.

Theoretical framework and teaching approach

We consider that a learner's mental representation (interacting entities and rules of their interactions) of a material situation under examination and the associated mechanism, characterised by its reasoning pattern, for manipulating these interacting entities (de Kleer & Brown 1983) constitute a learner's cognitive state. In order (a) to describe and analyse the evolution of student-teachers' cognitive states from the point of view of reasoning patterns and (b) to compare the main features of this evolution with the design assumptions of the teaching sequence, we refer to

D. Psillos and H. Niedderer (eds.). Teaching and Learning in the Science Laboratory. 243-254.

epistemological work on scientific explanation and in particular to the work of Halbwachs as a common frame of analysis.

Halbwachs (1971) distinguishes three types of scientific explanation developed and adopted in the history of science: Heterogeneous or causal explanations, homogeneous or typical explanations and bathygeneous or multi-level explanations. In causal explanations, which are of interest here, the cause of changes observed in a system is attributed to changes outside the system and this integration of qualitatively different agents constitutes their basic explanatory strength. A sub-categorisation of causal explanations includes simple, linear and circular explanations: Simple explanations establish a causal relation between a certain cause and a certain effect and offer an elementary explanation of phenomena. A number of juxtaposed simple causal relations form a new ordered relation, a causal chain, where every change is the result of the immediately previous change and the cause of the immediately following change. This system of linear causality constitutes a partial and unilateral representation of reality, but also a higher degree of knowledge than simple causality. Finally, circular causality is in general an iterative process, that is a chain of reversible causal relations. By introducing a principle of reversibility, explanations based on circular causality constitute a necessary intermediate level en route from simple or linear causal explanations to homogeneous explanations. The emergence of circular causality generally enables a decisive progress in understanding physical processes.

The context and structure of teaching

The teaching sequence has been specifically designed for, developed and taught to prospective primary school teachers, at the School of Education of Aristotle University of Thessaloniki (Barbas & Psillos 1993). Its backbone is a set of qualitative and semi-quantitative microscopic conceptual models, which are based on Coulomb interaction and form a unifying approach to static and circuit electricity. They follow the developmental pattern towards more elaborate types of explanations provided by the Halbwachs classification. They are presented progressively through short interactive computer simulations in close relation with material situations involving simple hands-on experiments (Barbas & Psillos 1997). The simulations scaffold student-teachers (a) to infer the limited set of assumptions underlying these models through the qualitative and/or semi-quantitative processing of observations of the simulated behaviour of microscopic entities and (b) to develop mental representations appropriate for producing explanations of electrical phenomena at a qualitative and/or semi-quantitative level. Student teachers work in pairs following written instructions.

The teaching sequence consists of eight weekly 2-hour laboratory sessions in a three-part structure reflecting: (a) the developmental pattern of explanations based on Halbwachs hierarchy and (b) a three-level modelling of interactions; The first part refers to simple electrostatic phenomena at the level of interactions between individual particles and focuses on explanations based on simple and linear causality. The second part refers to electrical phenomena involving charge movement at the level of interactions between classes of particles exhibiting the same behaviour and focuses on explanations based on circular causality. The third part refers to simple dc circuits at the level of steady-state laws and focuses on explanations based on iterative processes (Table 1).

Table 1: Teaching sequence

The three-part structure	Content of the eight weekly laboratory sessions
electrostatic interactions individual particles simple & linear causality	1.Attraction/repulsion between charged/uncharged bodies
	2.The microscopic view: the single atom behaviour
	3.Polarisation of insulators & properties of metals
charge movement classes of particles circular causality	4.Polarisation of metals
	5.Transient flow of electrons and the role of the battery
simple dc circuits steady-state laws iterative processes	6.Resistance and the role of potential difference
	7.Circuit with different conductors in series
	8.Semi-quantitative study of changes in a circuit

A laboratory session consists of 2 - 3 teaching episodes. Each teaching episode comprises a number of tasks, is structured on an iterative constructivist cycle of "prediction-observation-explanation" and concludes with a class discussion on student teachers' findings. The two consecutive teaching episodes under study here are the last of the 2nd laboratory session and the first of the 3rd. Due to Easter vacations, in the application reported here, the 2nd episode followed after 3 weeks (Table 2).

Table 2: Teaching episodes in sessions 2 and 3

Laboratory session	Teaching episode
2. The microscopic view: The single atom behaviour	1. The charged / uncharged insulating bodies episode
	2. The single atom episode
3. Polarisation of insulators & properties of metals	1. The lattice episode
	2. The charged / uncharged metallic bodies episode

The interactive simulations

In the "single atom episode" students work on the "single atom simulation", an external charge interacting with an atom (Figure 1a), while in the "lattice episode" they work on the "lattice simulation", an external charge interacting with a 3-D arrangement of atoms in the lattice of an insulator (Figure 1b & 1c).

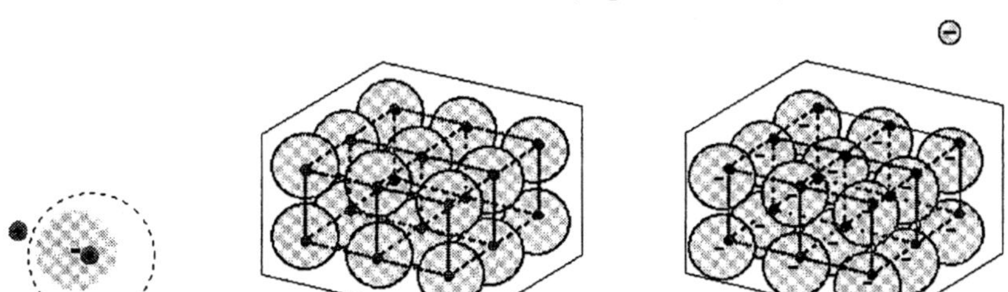

(a) (b) (c)

Figure 1: Screen instances of the "single atom simulation" (a), and of the "lattice simulation" (b), under the influence of an external negative charge (c)

A simulated atom consists of two entities interacting with the external charge: A red sphere, which depicts the positive nucleus, and a green one, which depicts the negative "electron cloud", the entity comprising all the electrons of the atom. The dashed circle (Figure 1a) defines the nearest possible approach of the external charge to the atom.

In the "single atom episode", students charge the mouse pointer positively, place it at certain positions and move it around the atom; then they charge the pointer negatively and repeat the same actions. They are asked to observe, describe and explain the simulated behaviour of the interacting entities and state the basic rules which, according to their view, determine this behaviour. The episode is completed with a teacher-led class discussion on students' answers.

The "lattice episode" consists of two tasks. In the first, students are given the information that "a solid electrically neutral body consists of many electrically neutral atoms regularly arranged in space. The nuclei of the atoms can not leave their positions nor can the electrons get out of their electron clouds". Then, they are asked to sketch annotated drawings of the atoms of a piece of a solid body (a) when there is no external charge, (b) when there is an external charge, first a positive and then an equal but negative one, at least at two different positions. In the second task, they work on the "lattice simulation". They are asked to do the same actions as in the "single atom simulation", then write down and justify any differences they have noticed between the simulated process and their conceptions as depicted in their drawings. The episode is completed with a class discussion on their answers.

Research design

The main research question is: What changes do students-teachers' mental representations and reasoning patterns about particle and body interactions undergo while interacting with the simulated interactions of microscopic entities?

Data were collected from: (a) written answers to the questions on the worksheets, as students worked on the tasks, at specific predefined moments of the learning process; (b) tape-recordings of the class discussions on students' answers; (c) semi-structured interviews, 2 - 3 weeks after the completion of the teaching episodes, focusing on the written answers students had given. We have analysed 32 students' written answers, 20 interviews, and the tape recordings of class-discussions. At the end of every teaching episode, we form a description of students' cognitive state. Changes to cognitive states are attributed to the teaching-learning process. Elements of the teaching process were taken from the tape recordings of class discussions. Data from interviews have been used only to illuminate the content and the context of written answers.

Results

Initial representations and reasoning patterns

In the "charged/uncharged insulating bodies episode", which preceded the "single atom episode", student teachers had constructed their explanations for the attraction between a charged (a rubbed ball-pen cap) and an uncharged body (small pieces of paper). It should be noted that an adequate causal explanation should be based on at least a two-step linear reasoning pattern and use three interacting entities in both steps: The first step should consider the polarisation of the neutral body, that is, the interactions between the external charge and both equal but opposite charges on the neutral body, their displacement, and the induction of minute, equal but opposite, charges on the body's surfaces, one facing the external charge and the other on the opposite side of the body. The second step should compare the forces of interaction between the external charge and both induced minute charges, on the basis of the distance between the interacting charges, the attraction always being greater since the distance between the opposite charges is always smaller. From our analysis of these explanations, according to the mental representations and reasoning patterns identified, three main groups of students emerge which correspond to three cognitive states: The 'macroscopic', the 'superficial microscopic' and the 'microscopic'.

Students of the first group lie at a 'macroscopic' cognitive state: They do not concern themselves with what is happening inside the neutral body. They either use the general rule "a charged body attracts an uncharged one" or they consider the charge as a macroscopic feature of the interacting bodies by inventing a 'neutral charge' on the uncharged body and by modifying the rule "opposite charges attract each other" to form a new rule of interaction, the "attraction of dissimilar charges".

For the second group, at a 'superficial microscopic' cognitive state, the interacting entities are two opposite charges, one on the charged body and the other on the uncharged body, and the rule of their interaction is "opposite charges attract each other". These students create a 'superficial microscopic' view of the material situation, since they concern themselves only with the interaction between the external charge and the opposite charge of the neutral body, neglecting the similar charge of the neutral body and the repulsion aspect of the interaction rule between similar charges. Both groups use two interacting entities and follow a simple one-step 'one cause - one effect' reasoning pattern.

For the third group, at a 'simple microscopic' cognitive state, the rule of interaction is "unlike charges attract while alike charges repel each other", with no reference to the dependence of the force on the distance between the interacting charges. The reasoning pattern is a two-step linear reasoning: First, the external charge causes a charge displacement inside the neutral body by interacting only with the opposite charge, in most explanations, or with both equal but opposite charges on the neutral body. In all explanations, this charge displacement results in the essential charging of the neutral body's surface, either that facing the external charge or the whole surface, with the opposite charge. In the second step, the opposite charge on the neutral body is attracted by the external charge while the alike charge is neglected. The polarisation process is used only to move the alike charge off stage, facilitating students to focus on the attraction of opposite charges. Only in a few explanations, and only in the first step of the reasoning pattern, do the interacting entities comprise three charges, one on the charged body and the other two equal but opposite charges on the neutral body. We suggest that this cognitive state may be seen as a 'superficial microscopic' cognitive state which has assimilated the polarisation process.

1st Episode: Students' engagement with the "single atom simulation"

While working with the simulation the following observations are possible:

1. When the external charge is positive, the electron cloud is displaced towards it.
2. When the external charge is negative, the electron cloud recedes away from it.
3. The displacement of the electron cloud becomes greater, when the external charge gets closer to the atom.

4. The displacement of the electron cloud becomes smaller, when the external charge recedes from the atom.
5. The nucleus of the atom is not displaced.

By correlating observations the following conclusions may be reached:

a. "Like charges repel while opposite charges attract each other" (qualitative aspect of Coulomb's law: Rule L.q) by correlating possible observations 1 and 2.
b. "The force of interaction between two charges increases, when the distance between them decreases, and the force decreases when the distance increases" (semi-quantitative aspect of Coulomb's law: Rule L.s) by correlating possible observations 3 and 4.
c. "The nucleus of the atom is not displaced or is least displaced in comparison to the electron cloud" by correlating observations 1, 2 and 5.

All students record observations 1 - 2 and reach conclusion a. Yet only half the students of the 1st and 2nd group go further to record observations 3 - 4 and reach conclusion b to some extent, since only a few of them mention it as a rule (1st student: rules). Regarding the 3rd group, 2/3 of the students record observations 3 - 4, reach conclusion b and mention it as a rule. It seems that when the causality of the interaction is known – here, L.q is a well-known rule – the simulated behaviour is readily observed. When it is not, as in the case of rule L.s, a significant percentage of students, who lack an elaborate mental representation of microscopic processes, may not observe the simulated events at all.

1st student, from the 1st group ('macroscopic' cognitive state):

Description: *When the pointer (external charge) is positive and close to the atom, then we have strong attraction. When the pointer is positive and far from the atom then the attraction is minimal. Consequently, the closer the pointer is to the atom then [the more] the atom is attracted, while the farther the pointer recedes from the atom, the less the attraction is visible. When the pointer is negative and close to the atom, then we have repulsion. When the pointer is negative and far from the atom, then we do not have any repulsion.*

Explanation: *The atom is neutral and the pointer is positive; then the two bodies are attracted and the nucleus is displaced. The atom is neutral and the pointer is negative; then the two bodies are repelled and the direction of the electron cloud is changed.*

Rules: *As we move far from the atom, the attraction decreases when the pointer is positive and the repulsion decreases when the pointer is negative.*

We note that almost all the students in the 1st and 2nd group do not distinguish observed events from conclusions when describing the behaviour of the interacting entities: they think in terms of attractions or repulsions, i.e., cause, while they are looking at displacements, i.e., effect (1st student: description). We suggest that this may favour a confusion between forces and displacements, cause and effect, which in turn allows the justification of the rule "no displacement, no force", a specific instance of "no effect no cause", a rule of "common sense reasoning" (Gutierrez & Ogborn 1992). Such a consideration is supported by data showing that, for some students, the non-movement of the nucleus does not attract their attention and consequently does not require any explanation (1st student: description). Furthermore, the movement of the electron cloud may be identified with the movement of the atom, a situation which may lead to the erroneous conclusion that the atom may also be repelled by an external charge (1st student: explanation). On the other hand, most students of the 'simple microscopic' group seem more concerned about the differentiation between observed events and conclusions.

The class discussion, which concluded the first teaching episode, was mainly engaged with the dependence of the forces of interaction, and also of the polarisation, on the distance of the external charge from the atom. Some students, from the 'simple microscopic' group, suggested that the shape of the electron cloud should also be affected: it should be ellipsoid and not circular, as in the simulation, since the electrons of the atom facing the external charge should be attracted or repelled more strongly than the other electrons of the same atom.

2nd Episode: Students' annotated drawings and the lattice simulation

Analysis of students' annotated drawings leads us to distinguish two main groups, the E-1 and E-2 groups. In the E-1 group, we find almost all students of the 1st group and a few from the 2nd group. They use the electron cloud representation, proposed in the "single atom simulation", with minus signs representing individual electrons (Figure 1a & Figure 2) and they arrange the atoms randomly, in a rather indicative manner, inside a parallelogram which defines the limits of the body.

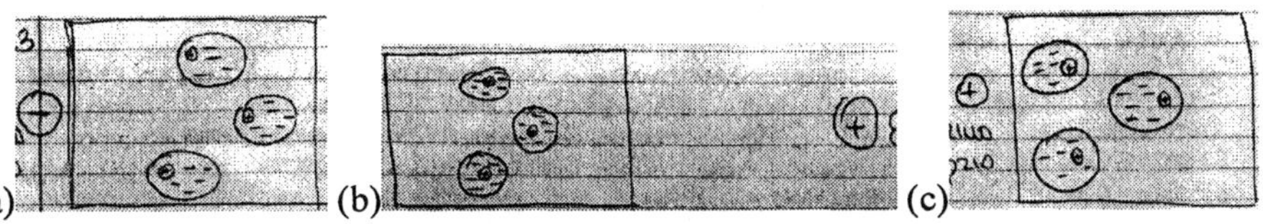

(a) (b) (c)

Figure 2: Drawings by the 2nd student

From the point of view of interactions, in the deformations and their dependence on distance, they consider only the distance between the external charge and the body: All atoms are equally deformed for a given position of the external charge, due to the relative displacements of nuclei and circular electron clouds. The dependence on distance is visualised by moving the external charge, that is through a succession of instances. After their interaction with the simulation, these students do not find - correctly regarding deformations - any differences between their annotated drawings and the simulation (Figure 2 / 2nd student: annotations & comments).

2nd student, from the 1st group ('macroscopic' cognitive state):

Annotations on the drawings: *When we bring near a positive charge (Figure 2c), then the electron clouds move towards, while the nuclei move away from the charge. When we bring near a charge but at a greater distance from the atoms of the body (Figure 2b), the attraction or repulsion is smaller. The rules are the same as in the case of one atom, i.e., the atoms of the neutral body are deformed into dipoles which have a positive and a negative charge.*

Comments on differences, after the interaction with the simulation: *I do not find any differences.*

In the E-2 group, almost all the students of the 3rd group, i.e., at the 'simple microscopic' cognitive state, and most from the 2nd group use the electron cloud representation without any signs for individual electrons and they arrange the atoms more or less in rows and columns. Students of the 3rd group are more precise in their drawings, which are more ordered and reminiscent of the structure of a lattice (Figure 3). From the point of view of the deformations, the students of this group consider (a) the distance between the external charge and each individual atom, but also (b) the distance between the external charge and the individual electrons in each electron cloud. Here, the dependence on distance is visualised through only one instance: For a given position of the external charge (a) every atom is differently deformed with respect to the relative displacement of its nucleus and electron cloud and (b) every electron cloud is differently deformed with respect to its ellipsoid deviation from its initial circular shape. While this double deformation is apparent in their drawings (Figure 3), their annotations are focused on the displacement deformation (3rd student: annotations). Comments on the ellipsoid deformation of the electron clouds are evoked only when they compare their annotated drawings with the simulation, after their interaction with the simulation (3rd student: comments).

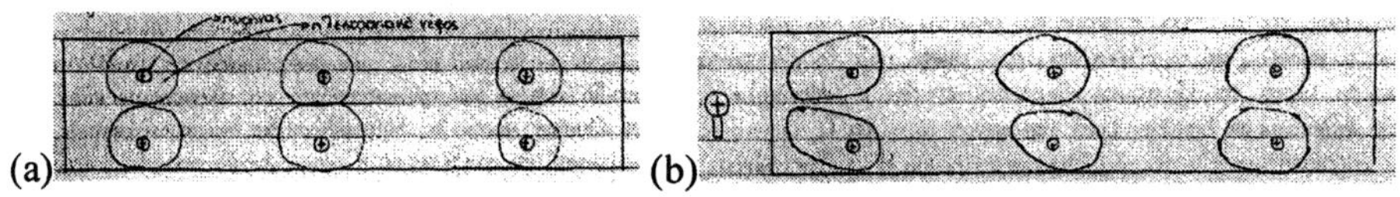

Figure 3: Drawings by the 3rd student

3rd student, from the 3rd group ('simple microscopic' cognitive state):

Annotations on the drawings: *[...] The positive external electrical charge attracts the electron cloud of the atoms. The greater the deformation of the electron cloud the greater the attraction of the electron cloud by the external electrical charge (Figure 4b).*

Comments on differences, after the interaction with the simulation: *In the simulation, the electron clouds keep their circular shape even though they are under the influence of the external charge. So, the simulation does not show that the force acting on the electron clouds varies with the distance. To show these, I gave an ellipsoid shape to the electron clouds near the external charge whereas the shape of the electron clouds far from the external charge have a more circular shape.*

Finally, we find again students reaching the erroneous conclusion that the atoms in a solid may move (Figure 4b) and even be repelled by an external charge (Figure 4c). They neglect interactions of the nuclei and they identify the movement of the electron cloud with the movement of the atom (4th student: annotations). As we have already suggested, this conclusion may be the result of the confusion between cause and effect. The comparison between their drawings and the simulation seems to initiate the deconstruction of this identification (4th student: comments).

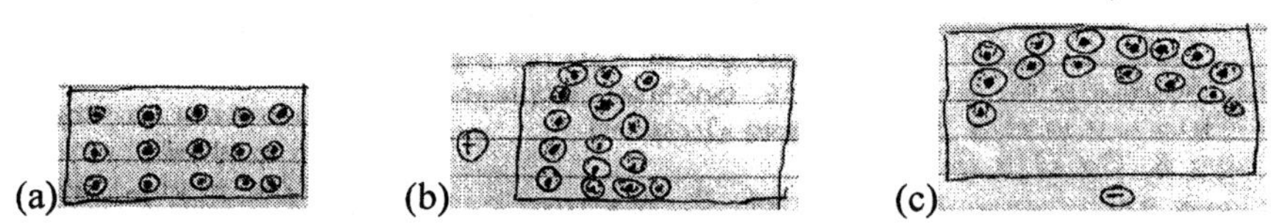

Figure 4: Drawings by the 4th student

4th student, from the 2nd group ('superficial microscopic' cognitive state):

Annotations on the drawings: *When there is no external charge the atoms are arranged in rows (Figure 4a). [...] When we bring a positive charge close to the body, then the atoms move towards this charge, that is, they are attracted (Figure 4b). When we bring a negative charge close to lower side of the body, the atoms are repelled, that is, they move to the upper part (Figure 4c). [...]*

> Comments on differences, after the interaction with the simulation: *In my drawings, the atoms are transferred towards the charge or away from the charge without showing that only the electron clouds of the atoms are moving and not their nuclei. This behaviour is clearly shown in the simulation.*

It seems that the electron cloud is a convenient entity for the description of the interactions of the atoms, since it is correctly used by all students. The minus signs added by the students of the E-1 group do not indicate any active role of individual electrons; rather, they are used as a reminder of the content of the electron clouds. On the contrary, although they do not depict individual electrons, the students of the E-2 group envision additional interactions, between individual electrons and the external charge, resulting in the ellipsoid deformation of the electron clouds. These interactions, although correct in their conception, may be seen as second level interactions, which do not affect the overall result. However, many students seem to be preoccupied with these interactions to such an extent that they consider them to be the only indication that the force of interaction depends on the distance (3rd student: comments).

Finally, the fact that all students recognise the dependence of the interaction force on distance in their drawings is more the result of the class discussion at the end of the 1st teaching episode, since most students of the 1st and 2nd groups had not reached such a conclusion while working with the "single atom simulation".

Mental representations and reasoning patterns after the teaching episodes

In the first task following the two teaching episodes, students are asked (a) to explain the attraction between a charged and a neutral body A and to state the entities they use and the rules of their interactions and (b) to predict the interactions and their effects when a second neutral body B is brought close to A (Figure 5). While part (a) is identical with, though more formal than, the task given before the two teaching episodes, part (b) is more demanding: Since A and B are neutral, to predict / explain an interaction between them requires to visualise the polarisation of both A and B, to consider interactions between minute charges on A and B and to structure a linear reasoning, with more than two steps, which should compare the forces of attraction, between δ_A^-/δ_B^+ and δ_A^+/δ_B^-, with the forces of repulsion, between δ_A^-/δ_B^- and δ_A^+/δ_B^+. Since distance x is many times smaller than d, the attraction between δ_A^-/δ_B^+ alone is many times greater than both forces of repulsion.

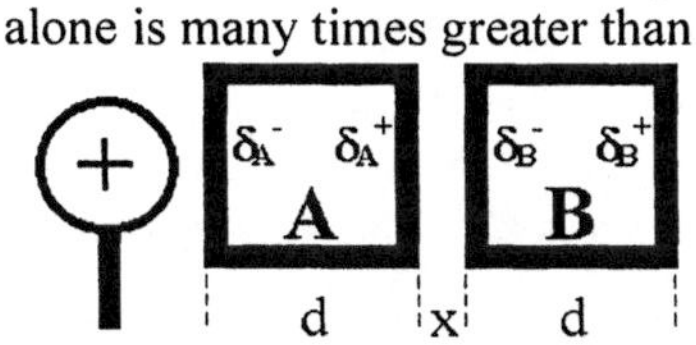

Figure 5: Interaction between charged and uncharged (A & B) bodies with minute charges δ- and δ+ induced through polarisation

According to students' responses, we distinguish three main groups of students:

- More than 1/3 of all the students, some from the 1st and 2nd groups and almost the entire 3rd group, form the 'advanced microscopic' group. They consider the polarisation of A and B and the interactions between the minute charges on their surfaces: Most of them consider the interactions of δA+ with δB- and δB+ and a few the interactions of δB- with δA- and δA+. They all organise a linear type of causal reasoning with more than two steps and reach the proper conclusion by comparing the forces of interaction on the basis of the distance between the interacting entities. In the second teaching episode, they were among the students of the E-2 group with the more organised annotated drawings. They have deepened their microscopic insight by assimilating the polarisation process in a way, which focuses on a comparison between the forces, and they have arrived at a cognitive state very close to that required for an adequate explanation of the material situation.

- Some students from the E-2 group, less than 1/3 of all, almost all deriving from the 1st or 2nd groups, form the 'intermediate microscopic' group. They refer to both attractions and repulsions between the charges of the charged and the neutral bodies and the deformations they induce, in many cases in a more detailed manner than the advanced group. They organise a linear causal reasoning with more than two steps but they reach their conclusions without comparing the opposite forces of interaction. We suggest that these students have not grasped, during the preceding episode, the interactions between individual electrons and external charge in their proper perspective. They were preoccupied with the ellipsoid deformation of electron clouds and deepened their microscopic insight by assimilating the polarisation process mainly as a distortion of atoms, arriving at a cognitive state behind the required one, a transitional state between the 'simple microscopic' and the desired one.
- About 1/3 of all students, almost all from the E-1 group of the preceding episode and from the 1st and 2nd groups, arrived at the 'simple microscopic' cognitive state: They consider the polarisation of A and B and the attractions between minute charges δA+ and δB-, on the facing sides of A and B, only when they are challenged by the possible interaction between two neutral bodies. They organise a linear type of causal reasoning with two steps and reach their conclusions by taking into account only the forces of attraction. In the preceding teaching episodes, they deepened their microscopic insight by assimilating the polarisation process as a means to bring closer the opposite charges and implicitly justify the non-consideration of alike charges.

Conclusions

Five cognitive states (the 'macroscopic', the 'superficial microscopic', the 'simple microscopic', the 'intermediate microscopic' and the 'advanced microscopic') seem to characterise the evolution of students' causal explanations of simple electrostatic interactions from a macroscopic point of view towards a deep and adequate microscopic one. Transitions from the 'macroscopic' to the 'superficial microscopic' and vice versa seem to be rather easy. Students who have adopted the 'superficial microscopic' viewpoint do not hesitate to return to the 'macroscopic' one later on. The transition to the 'simple microscopic' cognitive state is the main change which students, with initial 'macroscopic' or 'superficial microscopic' views, undergo: (a) by deepening their insight into the microscopic structure of matter, by considering more than two interacting entities and thus by modifying their mental representations; (b) by adopting a linear causal pattern of reasoning, in place of simple causality, in order to take into account more than one simultaneous interactions; and, (c) by associating the dependence of forces on distance with the deformation of atoms. For many students the use of linear causality is triggered off by a material situation which does not fit at all to their mental representation, the interaction between two uncharged bodies under the influence of a charged one (Figure 5).

The formation of deep and adequate microscopic explanations requires a second critical transition from the 'simple microscopic' cognitive state, where there are mainly attractions, through intermediate states, where the dependence of forces on distance is associated mainly with the various deformations of atoms, to the 'desired' or 'advanced microscopic' state, where this dependence is associated mainly with the comparison of forces and the variety of deformations is recognised as irrelevant to the end result. It seems that, for many students, this insight needs more time to mature and thus intermediate states may emerge some steps behind the 'desired microscopic' cognitive state.

These evolutionary transitions are facilitated, or even provoked, by students' interaction with the designed simulations of microscopic processes. These simulations offer: (a) the iconic bases for enriching the mental representation of microscopic processes: in most annotated drawings we find the iconic representation of atoms proposed by the simulations; (b) events with more than two interacting entities which may not otherwise be 'observed'; (c) the underlying causality through the correlation of simulated events.

The analysis of students' interactions with these simulations suggest that there is a deep "causality-observation" interrelation: when the causality of the interaction is known the simulated events are readily observed, while the reverse path, from events to the underlying causality, is difficult, especially for students with no or a poor mental representation of microscopic processes. In this "causality-observation" relationship we find elements of "common sense reasoning" which obstruct the establishment of the intended relations between simulated events. Also, the differentiation between cause and effect plays a significant role by diminishing the impact of "common sense reasoning". This differentiation is more evident when mental representation of microscopic processes is more elaborate and facilitates the upgrade of conclusions to the status of rules, which, in turn, may further diminish the confusion between cause and effect.

The evolution of students' causal explanations is, for every student, a personal pathway influenced by the status and development of such factors as their mental representations of the material situation, their reasoning patterns, their conceptual and reasoning obstacles. The combined influence of these factors may result in divergent individual evolutionary pathways, which in the laboratory environment seem to converge and intersect at some more or less stable cognitive states, which may be crucial for instructional design. The analysis of the annotated drawings suggests that this convergence may be facilitated by class discussion between teaching episodes.

Recommendations

Since learners' mental models and patterns of reasoning influence and constrain learners' interactions with material situations, the instructional design should take into account the transformation process of learners' mental representations of material situations and use tools for influencing this process on the levels of entities, rules of interaction, and reasoning patterns:

- A modular laboratory-based design comprising a number of inter-related tasks of the type "predict – observe – explain" in the form of targeted teaching episodes may focus on students' cognitive states and on explicitly supporting students' evolution regarding microscopic electrical interactions.
- Small-scale interactive simulations supporting the intended mental representations and reasoning patterns are appropriate tools for influencing learners' transformation processes. Interactions with such simulations should be embedded in teaching episodes and should encourage learners to materialise specific features of their mental representations, for example through annotated drawings, to record their differences with the simulated ones and explain them in writing or orally.
- Class discussion should come in between teaching episodes to increase homogeneity and create a better and more stable base for the next teaching episode. We suggest that the increased homogeneity, or the discrete heterogeneity, brought up by such discussions may constitute a better and more stable base for adapting laboratory materials, including simulations, to students' reasoning. To consolidate elements gained through discussion, students should return to selected simulation aspects in order to 'observe' or to correlate elements they had previously missed.

References

Barbas, A., & Psillos, D. (1993). Designing a computer-based course on basic electricity for prospective primary school teachers. In P. L. Lijnse (Ed.), *European Research in Science Education. Proceedings of the first Ph.D. Summerschool* (pp. 215-223). Utrecht: Cd-ß Press.

Barbas, A., & Psillos, D. (1997). Causal reasoning as a base for advancing a systemic approach to simple electrical circuits. *Research in Science Education*, 27(3), 445-459.

Chabay, R. W., & Sherwood, B. A. (1995). *Electric & Magnetic Interactions*. New York: John Wiley.

de Kleer, J., & Brown, J.S. (1983). Assumptions and ambiguities in mechanistic mental models. In D. Gentner, & A. L. Stevens (Eds.), *Mental Models* (pp. 155-190). Hillsdale NJ: Lawrence Erlbaum.

Eylon, B. S., & Ganiel, U. (1990). Macro-micro relationships: The missing link between electrostatics and electrodynamics in students' reasoning. *International Journal of Science Education*, 12(1), 79-94.

Frederiksen, J. R., & White, B. Y. (1992). Mental models and understanding: A problem for science education. In E. Scanlon & T. O'Shea (Eds.), *New Directions in Educational Technology* (pp. 211-226). NATO ASI Series F: Computer and Systems Sciences, Vol. 96. Berlin, Heidelberg: Springer-Verlag.

Gutierrez, R., & Ogborn, J. (1992). A causal framework for analysing alternative conceptions. *International Journal of Science Education*, 14(2), 201-220.

Halbwachs, F. (1971). Causalité linéaire et causalité circulaire en physique. In M. Bunge, F. Halbwachs, Th. S. Kuhn, J. Piaget, & L. Rosenfeld (Eds.), *Les Théories de la Causalité* (pp. 39-111). Paris: Presses Universitaires de France.

Niedderer, H., & Goldberg, F. (1996, April). Learning processes in electric circuits. Paper presented at the 1996 NARST Annual Meeting in St. Louis, Missouri.

Psillos, D., Barbas, A., & Koumaras, P. (1995, April): A causal approach for relating equilibrium to change in electrical circuits. Paper presented at the symposium "Learning Electricity", European Conference on Research in Science Education, Leeds, UK.

Welzel, M. (1998). The emergence of complex cognition during a unit on static electricity. *International Journal of Science Education*, 20(9), 1107-1118.

White, B., Frederiksen, J., & Spoehr, K. (1993). Conceptual models for understanding electrical circuits. In M. Caillot (Ed.), *Learning Electricity or Electronics with Advanced Educational Technology* (pp. 77-95). NATO ASI Series F: Computer and Systems Sciences, Vol. 115. Berlin, Heidelberg: Springer-Verlag.

Epilogue

TOWARDS TARGETED LABWORK

Marie-Geneviève Séré, University Paris XI, France

Study by study, this book constitutes a sort of tour, taking the reader all around Europe. More significantly, it reveals that it is worth exploring a variety of ways of associating lab work with different teaching contexts. Never again will the reader be tempted to consider 'lab work' as a whole, as if it were a single teaching method, globally carrying standard effects and benefits (motivation, familiarity with scientific devices, for example), and recognised drawbacks (time-consuming, poor learning). Chapter by chapter, the reader is offered studies of the effectiveness of lab work in two senses of the word.

In the usual sense of what is learned and acquired after a session, this means the discovery of numerous learning possibilities that go beyond other teaching methods. For example, joining experiments and simulation, modelling data obtained by experiment, addressing erroneous conceptions by an experiment: these are all powerful and effective ways for students to learn about concepts and procedures.

Other studies reveal benefits specific to lab work and open further possibilities: during lab work students may use conceptual knowledge not focusing directly on the task in hand but in order to design experiments, to be exposed to scientific method and become conscious of its characteristics, to understand the role of autonomy and decisions as parts of a scientific experiment - in a nutshell, to understand science. This is a second type of effectiveness, relating to the intent of actions, to what is experienced and achieved in the students' minds. This is treated as a separate aspect of the matter in this book, and consequently it constitutes a demonstration of the wealth of lab work based on research studies, being different from the numerous classifications that have been produced during the last decade (see Chapter 1 *of this volume*).

Obviously this is of interest to lab work developers. They will find in this book a large range of objectives and a large range of strategies, as well as arguments to match them. In fact, lab work developers have choices to make when confronted with the wealth of learning potentialities of each experimental situation, here fully illustrated. The challenge here is to prevent a wealth of objectives from turning to confusion, a choice among objectives from turning to impoverishment, leading to

D. Psillos and H. Niedderer (eds.). Teaching and Learning in the Science Laboratory. 255-260.

diminished effectiveness. By opening out so many learning possibilities, this book also intends to answer the following legitimate questions:

- how to help students cope with so much supporting knowledge (conceptual, procedural, epistemological)?
- how to encourage them to achieve so many different objectives?
- how many aspects of science, emerging from experimental situations, is it possible to understand?

The answer lies in terms of choices among objectives. We propose replacing the notion of objective by the idea of 'target'. Targeted lab work means that the number of objectives has been restricted, the selection having been made according to research outcomes: the type of expected effectiveness is clearly identified, the frequent intertwining between conceptual/epistemological/procedural objectives is taken into account, the teaching strategies are appropriate to the objectives, the target is made crystal clear to the students.

The next sections will develop the main aspects of targeted lab work.

A twofold frame for the organisation of lab work sessions: aiming at intended actions during lab work, versus aiming at acquisitions after lab work.

Obviously, some of the research in this book did shed light on certain types of knowledge which, in current practice, are not identified as possible objectives. A relevant way to organise and structure lab work is to consider within which frame of effectiveness it is meaningful.

Data processing is a telling example of a single activity that can be organised towards one or the other type of effectiveness. It may be presented as a routine, to a certain extent independent from the concepts in play. In this case, students are required to learn the routine in order to be able to apply it later. But it may be presented in a very different way, justifying the method itself, as well as the link between the nature of the data (reproducible or not) and the expected type of conclusion, a link which will probably influence the method. This is an example of a target, which deeply modifies the type of practice in the course of lab work and the type of final acquisition, even when the theme, the apparatus and the concepts in play are not modified.

It has been demonstrated that a similar duality exists for conceptual objectives. A classical target may be to obtain learning, with acquisition of conceptual knowledge at the end of the session. Nevertheless, the target may in fact be rather different. It may be to lead students to use a certain level of theory properly during the session. Knowing that students avoid that as much as possible, this requires a specific organisation. In particular, it requires reducing the time devoted to measurement, abandoning the constraint of rigorous data processing, in favour of making predictions, making rapid calculations to give sense to formulae, calculating orders of magnitude, etc.

With regard to the role of conceptual knowledge during lab work, the difficulty of applying models to the real world (linking theory and practice) has been pointed out. A target here may be to make sure that students not only 'do' what they are intended to do (for example with a modelling software), but also 'think' that the real world is 'understandable' through models. This requires a specific underlying epistemological dimension, through guidance for instance.

Some objectives may exist with respect to only one sort of effectiveness. For instance, it may be decided to confront students with the difficulty of a measurement device interfering with a given phenomenon, when a quantitative study necessarily distorts the corresponding qualitative experiment. This concerns comprehension and requires the following intellectual activity during the session: addressing the complexity of a possible modification of an experiment in order to collect data and to insure the feasibility of measurement. This is a sort of comprehension, which is helpful in experiment designing. It is difficult to elicit the learning outcome of designing an experiment addressing a particular question. But it is obviously a creative, fruitful experience, often demanded by students, and it constitutes a possible target.

These examples show why the results of our studies, in terms of two types of effectiveness, lead to a different way of organising lab work sessions around targets.

Targets and objectives: the consequence of adopting an alternative framework to the development of lab work sessions is to promote some newly defined objectives and to put aside some others

The notion of objective is useful in the development of lab work, in checking consistency and in ensuring connections with other teaching strategies. Objectives are classically categorised in three broad sets, generally designated conceptual/procedural/epistemological. The challenge here is that these categories are intertwined. The higher the academic level, the more interdependent the objectives are. At the main level addressed in this book (end of upper secondary school and undergraduate level), there are obviously mutual relations inside the general classification within these three categories. The notion of targeted lab work helps to take this complexity into account.

Taking the same example as above, it can be said that measurement and data processing can be taught as a procedure and, as such, constitutes a routine. But it may also be an opportunity for epistemological learning, because it allows understanding of the mutual dependency of theory and data, and understanding of how to evaluate confidence in values obtained through measurement. This is a part of epistemology, as is understanding of the relevance and specificity of statistical reasoning, even if not totally and precisely carried out. This appears to be an original objective, and is explored by some of these studies.

Another example is the various objectives that can be targeted by the activity of modelling. This has been studied in different forms in a number of studies. What has been highlighted are the possibilities of learning content by handling a given model, by the building and adjustment of models, by simulation by models, by comparison of the pros and cons of various models. This put forth the idea of competing and alternative models, which is an important basis for posing epistemological questions. With regard to any activity concerning models, recognition of the various roles they play in science appears to be facilitating understanding of scientific experimental approaches. In addition, models can be used during lab work both quantitatively, with measurement, and also qualitatively, to obtain interpretation.

The same can be said about procedural knowledge. In fact, procedures are embedded in a given content, not existing on their own. They can, however, be taught for themselves, strictly with the aim of imparting skills, but always as an intrinsic part of an experimental process using theory.

Experiment designing, a central skill which may be obtained through open-ended lab work also implies a large span of knowledge, of all three generally recognised types, as mentioned above.

Consequently, when reviewing how the idea of target is compatible with the three usual categories of objectives, it appears not only that these categories are intertwined, but also that new objectives may be promoted and others abandoned because they are not specific to lab work.

It appears from research-based results that items of knowledge may eventually be acquired, even when these are not directly focused on as an objective. This is the case, as already underlined, with conceptual knowledge. Studying aspects of modelling, pointing out the conditions of use of a given procedure, pondering the relevance of taking a theory for granted in order to design an experiment, all these are powerful tools for learning theoretical knowledge, more than simply 'verifying a theory', a reportedly rather ineffective method.

Several studies in this book put forth the idea that aiming at eliciting, in particular situations, the threefold relationship between theory, experiment and data processing provides a structure for lab work sessions, in an especially motivating way. This is more effective than lectures about epistemology and not an obstacle to content learning. This raises the issue of the extent to which a unique epistemology can and should be presented to students through lab work, and indeed through the science curriculum more generally. It is necessary to address at a policy level the question of the relative placing of examples from the history of science in the curriculum, and the treatment of epistemology in student lab work. Various studies in this book support the idea of conveying epistemology in the laboratory.

Finally, the art of developing lab work requires to a prior sound knowledge of different teaching strategies in order to match targeted lab work.

The obstacles encountered by developers when trying to modify lab work sessions are a sad reality and are all too well known. The in-depth studies in this book should help to avoid engaging in a trial and error approach in this domain. Examples here are the results obtained from studies of open-ended sessions. In one such study, Projects are demonstrated to require students to draw upon conceptual knowledge in order to solve a given problem, even if the Project is introduced before formal teaching of 'theory'. In another, students realise and become conscious that they have to judge the quality of their data. In another one, autonomy is difficult to manage by students. In any case, Projects are particularly useful in ensuring that students work under their own direction. If this is to happen, a generous time allocation has to be given to project work, possibly several weeks. This presupposes accepting the paring down of a curriculum already overcrowded with content.

Another strategy that has been studied is to ask students to make predictions about the behaviour of events, or alternatively about orders of magnitude, before actually making measurements. This suggests revised types of organisation.

Research-based studies are also of help in attributing a relevant place to computers during lab work. It is already accepted that computers and sensors can play an important role in saving time during these tasks and, moreover, that in some cases it is only possible to make measurements with the aid of computers. But the role of computers is hereby recognised as going beyond time saving. For example, the routine part of data processing may be handled by computer, making it possible to understand the preliminary unavoidable options.

Last but not least, it has been shown that the role of tutors and/or written guidance during lab work has a great impact on effectiveness. A first requirement is the students' consciousness of the targets. They must be well defined, both for students and for teachers. This is not an easy task. In order to help teachers, and for compatibility with the actual practice of lab work throughout Europe, standard lab work has been studied (small groups of students, written guidance by lab sheets, apparatus available, uniqueness of the tasks to be achieved, etc.). It has been shown that letting students manage by themselves provides an illusory autonomy, because students follow written instructions pretty passively. This of course does nothing to promote initiative. If a classical practical task has to be achieved, its educational value is enhanced by a deep comprehension of the device, by pointing out at a procedure, by posing questions on the link between theory and data. It is what we mean by defining targets and intervening accordingly.

Written or verbal, the teachers' interventions and instructions during targeted lab work, using a variety of strategies, are multiple. This suggests that innovation and research are presently necessary in two directions: firstly in a revision of the lab sheets offered to students during standard lab work, and secondly in specific input during initial and in-service training.

This is a short overview of the suggestions resulting from the studies presented in this tour of Europe, of teaching strategies, of lab work sessions. In order to avoid the frequent mismatch between teachers' objectives and what is achieved, done and thought by students, each lab work session should be reasonably ambitious and targeted, the strategy being a clear orientation towards certain selected objectives. The notion of target could replace the notion of objectives, acknowledging the frequent overlapping of conceptual / procedural / epistemological objectives at the academic level considered. Lab work should have a structure, in terms of target, which is made clear to students, supported by a given strategy, and organised within a coherent long-term programme with varied types of lab work. In conclusion, what is promoted throughout this book is the idea of lab work targeted towards all every dimension of an understanding of science.

Authors

Bandiera, Milena

Milena Bandiera is Professor at the Biology Department of "Roma Tre" University. Over the years she has added research and teaching of "Natural Science Education" to her original base of "Microbial Genetics", both in the context of the Biology Degree programme as well as that aimed at the initial teacher training for primary and secondary school teachers.
e-mail: bandiera@uniroma3.it

Barbas, Alexandros

Alexandros Barbas is Science/physics Teacher in Greek upper secondary schools (Lyceum) since 1975. He is carrying out his PhD at the Department of Primary education, School of Education, University of Thessaloniki. His thesis focuses on the teaching and learning of simple electricity, supported by computer simulations. His research interests are in modelling students' reasoning and on ICT supported science teaching.
e-mail: albarbas@otenet.gr

Bécu-Robinault, Karine

Karine Bécu-Robinault studied physics and chemistry at university. Then she choose to study physics education, and did a thesis within COAST research team, directed by A. Tiberghien. She studied the role of labwork activities in learning energy. This thesis was finished in 1997. Her research interests relate to teaching and learning physics modelling with a particular focus on the role of new technologies in physics education.
e-mail: karine.robinault@upmf-grenoble.fr

Beney, Michel

Michel Beney is a member of the DIDASCO group (Didactique des Sciences d'Orsay) at the University Paris Sud XI. He teaches physics at the University of Bretagne Occidentale (UBO-Brest- France). After having taught physic and chemistry in French secondary schools, in France and abroad, he completed a PhD in science education. His research concerns teaching and learning in science labwork during the first two years at university with a particular focus on effects of guidance on students' learning performances.
e-mail: Michel.Beney@univ-brest.fr
http://formation.etud.u-psud.fr/didasco/index.htm

Bisdikian, Garabet

Garabet (Garo) Bisdikian is a Science/physics Teacher in Greek secondary schools since 1987 and head of a regional science teacher centre. He has a PhD on the role of multimedia applications in enhancing primary education student teachers' ability to construct links between heat theory and phenomena with their graphical representation. He is currently involved in the development of ICT applications in science education and in teacher training.
e-mail: garo@vergina.eng.auth.gr

Buty, Christian

After his first formation in Physics in the Ecole Normale Supérieure (Saint Cloud), Christian Buty has taught physics and chemistry during some twenty years at upper secondary school and university levels. He has completed a PhD in Science Education about the efficiency of computer-based modelling, in the field of image formation in Optics. His research interests are in the relations between teaching situations, interactions and learning; he is currently working in the GRIC laboratory (UMR 5612 Université Lyon II - CNRS), and in the COAST team (directed by Andrée Tiberghien).

e-mail: Christian.Buty@univ-lyon2.fr
http://sir.univ-lyon2.fr/GRIC-COAST/Home/buty/default.html

Evangelinos, Dimitris

Dimitris Evangelinos has studied Physics at the Physics Department University of Thessaloniki. He has taught for several years introductory experimental Physics in the same Department. Evangelinos is carrying out his Doctoral Degree in students understanding of introductory treatment of measurement and data handling. His research interests include modelling of students understanding and conceptual evolution regarding introductory data handling, the development of research based laboratory activities and application of ICT in laboratory teaching.

e-mail: evangelinos@physics.auth.gr

Fischer, Hans E.

After teaching mathematics, physics and pedagogy in a German school Hans E. Fischer became a researcher in the Institute of Physics Education at the University of Bremen and completed his PhD in physics education. After his habilitaion he became Professor of Physics Education and head of the centre of teacher education at the University of Dortmund, Department of Physics, D-44221 Dortmund, Germany. His research addresses learning and teaching of physics, focussing on teaching concepts and its influence on both the structure of physics lessons and the students' performance.

e-mail: hans.fischer@uni-dortmund.de
http://www.physik.uni-dortmund.de/didaktik

Guillon, Alain

Alain Guillon is a member of the DIDASCO group (Didactique des Sciences d'Orsay) at the University Paris Sud XI. After having taught chemistry and physics in French secondary schools, he teaches physics at the University of Cergy - Pontoise (France). He has completed a PhD in science education (didactics of experimental sciences) within the european project "Labwork in Science Education". His research addresses teaching and learning in physics, with a particular focus on the role of epistemological knowledge in physics laboratory work.

e-mail: alain.guillon@phy.u-cergy.fr
http://formation.etud.u-psud.fr/didasco/index.htm

Haller, Kerstin

After studying physics at university, Kerstin Haller did her PhD in physics didactics within the european project "Labwork of Science Education". She studied the gools and objectives of labwork in a physiscs major cours.
She is now working in the research and exhibition development department of the universum science center in Bremen, Germany.
e-mail: khaller@universum-sc.de
http://www.universum.bremen.de

Hucke, Lorenz

Lorenz Hucke studied physics in Germany and France and completed a PhD in science education within the european project "Labwork of Science Education". He then designed web-based multimedia learning environments for the German "Virtual University of Applied Sciences" and now works as a consultant for e-learning and knowledge management for Arideon in Düsseldorf, Germany.
e-mail: lhucke@arideon.com
http://www.oncampus.de and www.arideon.com

Kariotoglou, Petros

Petros Kariotoglou is an Assistant Professor of Science Education, in the School of Education, Department of Primary Education, Aristotle University of Thessaloniki, Greece. He studied Physics, then obtained his Msc in Electronics and his PhD in Science Education. He has taught Science in Secondary schools for ten years. His main research interests currently focuses on the design, development, application and evaluation of teaching – learning sequences, informal science education and pedagogical content knowledge in science.
e-mail: kariotog@eled.auth.gr

Le Maréchal, Jean-François

Jean-François Le Maréchal is a researcher of the CNRS U.M.R. GRIC of Lyon 2 University. He belongs to Andrée Tiberghien's team known as COAST. He works in chemical education with a group of students; most of his work is labwork oriented and centred on learning. He got a PhD in organometallics chemistry in 1986. He also teaches inorganic chemistry at the Ecole Normale Supérieure de Lyon to senior students, and is at the head of the initial formation of Physic and Chemistry teachers.
e-mail: jflm@ens-lyon.fr
http://www.ens-lyon/chorale/direction.html

Leach, John

John Leach is Professor of Science Education and Director of the Centre for Studies in Science and Mathematics Education at the University of Leeds, Leeds LS2 9JT, UK. After having taught chemistry and science in British secondary schools, he worked as a researcher in the Learning In Science Research Group, and completed a PhD in science education. His research addresses teaching and learning in science, with a particular focus on the role of epistemological knowledge in science learning.
e-mail: j.t.leach@education.leeds.ac.uk
http://edu.leeds.ac.uk/projects/lis/john_leach.htm

Lewis, Jenny

Jenny Lewis is Lecturer in science education at the Center for Studies in Science and Mathematics Education, University of Leeds. After completing a PhD in human genetics she taught in Further Education (age 16+) for several years before returning to research - this time within science education. Her research interests relate to teaching and learning in science and include the consideration of social issues within the science classroom and the use of research evidence to inform practice.
e-mail: j.m.lewis@education.leeds.ac.uk

Millar, Robin

Robin Millar is Professor of Science Education in the Department of Educational Studies at the University of York. He completed a PhD in medical physics before training as a teacher. He then taught for eight years in secondary schools before moving to a university post. His research interests focus on science concept learning, and learning *about* science as a form of enquiry. He is also interested in the role of school science in enhancing scientific literacy and has been involved in several curriculum development projects exploring this area.
e-mail: rhm1@york.ac.uk
http://www.york.ac.uk/depts/educ/staff.htm

Niedderer, Hans

Hans Niedderer is Professor of Physics Education at the University of Bremen since 1974. He did his PhD at the University of Kiel (IPN) in 1972 with empirical work on students conceptions and learning processes about simple electric circuits. His main interests are content based analysis of learning processes in physics, curriculum development and evaluation in quantum atomic physics, history and philosophy of science and teaching strategies of student oriented science teaching, e.g. in open-ended projects.
e-mail: niedderer@physik.uni-bremen.de
http://didaktik.physik.uni-bremen.de/niedderer

Psillos, Dimitris

Dimitris Psillos is Professor of Didactics of Science at the Department of Elementary Education University of Thessaloniki. He did his degree in Physics at Thessaloniki and his PhD in Science Education at the University of Exeter. He was staff member for several years in the Physics Department of the University of Thessaloniki before moving in his present position. He has taught courses in physics, didactics of science, educational technology and was for several years in charge of a Postgraduate course in "Didactics of Science and New Technologies". His research interests involve modelling of students' conceptual evolution, the development of research based teaching sequences and the applications of ICT in improving the linking of theory with practice in science teaching . He has run several research and development projects.
e-mail: psillos@skiathos.physics.auth.gr

Ryder, Jim

Jim Ryder is Lecturer at the Centre for Studies in Science and Mathematics Education (CSSME) at the University of Leeds, Leeds, LS2 9JT, UK. He has a PhD in condensed matter physics and has taught science/physics in secondary schools. His research interests are in teaching/learning science at school and university level, with a particular focus on the development of understandings about the nature of science in science education.
e-mail: j.ryder@education.leeds.ac.uk
http://education.leeds.ac.uk/~edu-jr/

Sander, Florian

Florian Sander studied physics at university. He completed a doctoral dissertation in physics education at Bremen in the group of Hans Niedderer. He studied students' actions in the university physics laboratory and in addition the effects of computer-based model building systems integrated into labwork. After finishing his thesis in 1999 he was working at the Institute for Scientific Media (IWF) in Goettingen. He is now a writer for a scientific magazine.
e-mail: sanderflorian@gmx.de

Schecker, Horst

Horst Schecker is Professor of Physics Education at the University of Bremen since 2002. He did his PhD at the University of Bremen in 1986 with a dissertation on students alternative frameworks in Newtonian mechanics. In 1995 he qualified for university professorship with work on systems dynamics modelling in physics education. His main interests are multimedia in physics education on school and university level (development and evaluation) and science curricula development.
e-mail: schecker@physik.uni-bremen.de
http://didaktik.physik.uni-bremen.de/niedderer

Schumacher, Dieter

Dieter Schumacher is head of the group "Physikalische Grundpraktika" (Introductory Physics Labs) at the University of Düsseldorf, Germany. He is an Associate Professor in experimental physics and works on the development and evaluation of labwork courses in physics. During the last three years he has been the speaker of the group "AG Physikalische Praktika" in the German Physical Society (DPG). At the moment he is involved in the development and evaluation of a hypermedial learning environment in physics for medical students.
e-mail: Schumac@uni-duesseldorf.de
http://www.gpphy.uni-duesseldorf.de

Séré, Marie-Geneviève

Marie-Geneviève Séré is Professor of Science Education, responsible of the DIDASCO group (Didactique des Sciences d'Orsay) at the University Paris Sud XI. She graduated first in Physics and completed a PhD in Quantum Optics. Then she studied cognitive psychology and completed a PhD thesis in Science Education. She has taught physics at Undergraduate level and pedagogy at doctoral level and for teachers'training. She has been invited professor in Spain and various countries in Africa and South America. She was the co-ordinator of the LSE project. Her research deals with labwork, epistemology teaching and students'causes of failure.
e-mail: marie-genevieve.sere@didasco.u-psud.fr
http://formation.etud.u-psud.fr/didasco/index.htm

Theßyen, Heike

Heike Theyßen is a member of the group "Physikalische Grundpraktika" (Introductory Physics Labs) at the University of Duesseldorf, Germany. She has a PhD in physics didactics and worked on the development and evaluation of a labwork course in physics for medical students with special regard to the students learning conditions. In a subsequent project she currently develops and evaluates a hypermedial learning environment in physics for medical students.
e-mail: theyssen@uni-duesseldorf.de
http://www.mm-projekt.uni-duesseldorf.de

Tiberghien, Andrée

Andrée Tiberghien is Director of Research in the National Center of Scientific Research (CNRS). She is currently Head of the COAST Research group (COmmunication et Apprentissage des Savoirs Scientifiques et Techniques) of the GRIC laboratory. She is in charge of the "Formation Doctorale - Didactiques et Interactions" (Université Lyon 2). Her research is focused on the relations between students' learning in the physics domain and the teaching situations.
e-mail: Andree.Tiberghien@univ-lyon2.fr
http://gric.univ-lyon2.fr/gric3/

Valassiades, O.

Odysseas Valassiades is Associate Professor in the Physics Department University of Thessaloniki. He took his PhD in solid state physics and has taught for several years undergraduate and postgraduate lessons in the Physics Department. His research interests focus on electrical measurements. He is also involved in teaching and developmental activies concerning the introductory physics laboratory.

Von Aufschnaiter, Stefan

Stefan von Aufschnaiter is Professor of Physics Education at the University of Bremen (Germany) since 1971. After having completed his PhD in solid state physics he started working as a science educator at the Institute for Science Education (IPN) at the University of Kiel (Germany). His research addresses video based analysis of learning and teaching processes in physics in school as well as in university. In addition, he is doing curriculum development with a particular focus on the development of activity and play oriented tasks on which learners can work self-organised.
e-mail: aufschna@physik.uni-bremen.de
http://didaktik.physik.uni-bremen.de/aufschnaiter

Index

Action 3, 8, 11, 14, 22-28, 41-47, 49, 64, 65-78, 80, 97, 126-138, 152, 195, 205-216, 223, 229, 232-238, 246, 255, 256
- action networks 78
- action regulation 205-215
- action theory 205

Bayesian approach 181
Categories 3, 8, 10-17, 23, 27, 31-40, 56, 57, 74, 81, 96, 98, 107-115, 142, 155, 170-175, 191, 194-203, 206-214, 223, 227, 237-241, 257, 258
- Category-based analysis (CBAV) 8, 28, 31-38, 96-102, 191, 205, 222-230

Classification 9-16, 41, 69, 72, 108-113, 155, 184, 244, 255, 257
- of labwork tasks 9-11

Cognitive 3, 13-15, 23, 28, 32, 34, 50, 51-64, 65-78, 82-89, 91-103, 134, 136, 182, 193-204, 205-217, 225, 229, 233, 236
- Cognitive state 206, 243-253
- Mental representation 192, 243-254

Complexity 4, 28, 40, 49, 50, 93-103, 134, 206-210, 257
Computer 5, 8, 18, 34, 36, 44, 123, 125, 171, 191, 192, 193-204, 205-217, 219-229, 231-242, 254, 259
- Computer model building (MBS) 36, 38, 191, 208-216, 219-230, 237
- Computer simulation 243, 244
- Computer-based laboratory (MBL) 32, 36, 191, 208-216, 219-230

Concept map 24, 28, 90, 191, 205-217, 221
Content analysis 32, 39, 79, 163, 165-169
Doing and learning 65, 67
Doing and thinking 65
Effectiveness of Labwork 6, 7, 9-19, 21-30, 33, 38, 50, 219
Epistemology 48, 126, 131, 151-161, 162, 257, 258
- Epistemological understanding 4, 24, 27, 41, 48, 178
- Image of science 105-115, 126
- Metaknowledge 128, 135
- Nature of theory 8, 42-46, 163, 181

Error 29, 48, 73, 75, 87-90, 123-138, 163, 168-178, 179-190, 212, 216, 258
- Instrument readings 179, 183, 184
- Metrological approach 190

Experiment 4, 28, 32-39, 45, 49, 51-63, 65-77, 79-89, 92-96, 105-116, 121-138, 144, 145, 163, 173, 182-185, 191, 192, 194-204, 205-216, 219-229, 231-242, 243, 255-258
- Experimental activities 21, 63, 83, 105-118, 194, 203, 241
- Experimental process 121-131, 258

Explanation 13-16, 27, 28, 41-47, 63, 67-75, 84, 101, 109-116, 124, 142, 189, 198, 199, 236, 244-252
- Causal explanation 243-253
- Nature of explanation 8, 42, 46
- Scientific explanation 16, 244

Graphs 22, 24, 36, 43, 44, 108, 123, 138, 166-171, 183, 193-204, 212, 219-228, 239-241
- Computer-based graghing 204
- Graphing skills 193-204

Intended learning outcomes 13, 17, 157
Interactive 191, 192, 193-202, 219-224
- Interactive simulation 243-253

Interviews 24-28, 41, 42, 65-75, 83, 84, 92, 122, 140-147, 151-161, 165-173, 183, 192, 197-203, 246
Investigation strategy 121, 124
Knowledge
- Conceptual knowledge 2, 4, 24, 66-75, 79, 88, 121, 131, 192, 210, 255-258
- Epistemological knowledge 3, 24, 119, 121-135
- Mathematical knowledge 36, 91, 217, 223
- Metaknowledge *See* Epistemology
- Physics knowledge 8, 33-38, 91, 92, 223, 226, 238, 239
- Procedural knowledge 3, 21, 28, 65-75, 79-89, 162, 193, 258
- Technical knowledge 33-36, 223
- Verbalised knowledge 31-36, 223, 237-239

Labguide 4, 32-38, 94-102, 207-216, 219-229
Laboratory Dimensions Inventory (LDI) 11
Laboratory Structure and Task Analysis Inventory (LAI) 10-19, 32

Labwork
aims, goals 41, 51, 56
Context 28, 31-38, 94, 223, 225
Role of 1, 51
Traditional labwork 51-63, 92
Levels
of modelling 54-62
Linking of theory and experiment 193, 201, 219-229
Measurement
Repeated 44, 182
Single 163, 181-190
Uncertainty 43, 144, 163, 172, 179-190
Model
Behaviour model 123-137
Black box model 123, 132
Materialised model 192, 232-241
Mathematical model 16, 123
Model of object 123-138
Modelling 4, 5, 23, 36, 40, 50, 51-64, 80, 129, 162, 163, 179-184, 191, 192, 207-217, 219-229, 231-242, 244, 255-258
Simulated model 22, 123
Objects and events 28, 34, 54-63, 105-116, 192, 219, 231-241
Open-ended
Investigation 42, 120, 139-149
Labwork 4, 45, 47, 119, 120, 124, 134, 144
Process
Experimental *See* Experiment
Scientific 121-128
Simulated 246
Theoretical 123-131
Profile of a labwork task 9-19
Project 2-5, 7, 11, 19, 24, 27, 33, 42, 47, 51, 56, 84, 92, 96, 119, 121-137, 139-150, 167, 237, 258
Mini-Projects 24, 120, 139-150, 220, 221
Textbooks 4, 11-20, 39, 50, 76, 88, 105-118, 127, 131, 132, 165-167, 181
Theories and models 54, 63, 192, 231-236, 241
Theory and Experiment 130, 137, 182, 186, 219-229
Transcript-based analysis 97, 102
Tutor 28, 35-38, 65, 66, 92-101, 120, 151-161, 173, 207-215, 219-229, 259
Verbalisation 28, 33-38, 96-102, 212, 223-226, 237-242
Video data of labwork 31, 209-211
Categorie-based analysis (CBAV) *See* Categories

Science & Technology Education Library

Series editor: William W. Cobern, *Western Michigan University, Kalamazoo, U.S.A.*

Publications

1. W.-M. Roth: *Authentic School Science.* Knowing and Learning in Open-Inquiry Science Laboratories. 1995 ISBN 0-7923-3088-9; Pb: 0-7923-3307-1
2. L.H. Parker, L.J. Rennie and B.J. Fraser (eds.): *Gender, Science and Mathematics.* Shortening the Shadow. 1996 ISBN 0-7923-3535-X; Pb: 0-7923-3582-1
3. W.-M. Roth: *Designing Communities.* 1997 ISBN 0-7923-4703-X; Pb: 0-7923-4704-8
4. W.W. Cobern (ed.): *Socio-Cultural Perspectives on Science Education.* An International Dialogue. 1998 ISBN 0-7923-4987-3; Pb: 0-7923-4988-1
5. W.F. McComas (ed.): *The Nature of Science in Science Education.* Rationales and Strategies. 1998 ISBN 0-7923-5080-4
6. J. Gess-Newsome and N.C. Lederman (eds.): *Examining Pedagogical Content Knowledge.* The Construct and its Implications for Science Education. 1999 ISBN 0-7923-5903-8
7. J. Wallace and W. Louden: *Teacher's Learning.* Stories of Science Education. 2000 ISBN 0-7923-6259-4; Pb: 0-7923-6260-8
8. D. Shorrocks-Taylor and E.W. Jenkins (eds.): *Learning from Others.* International Comparisons in Education. 2000 ISBN 0-7923-6343-4
9. W.W. Cobern: *Everyday Thoughts about Nature.* A Worldview Investigation of Important Concepts Students Use to Make Sense of Nature with Specific Attention to Science. 2000 ISBN 0-7923-6344-2; Pb: 0-7923-6345-0
10. S.K. Abell (ed.): *Science Teacher Education.* An International Perspective. 2000 ISBN 0-7923-6455-4
11. K.M. Fisher, J.H. Wandersee and D.E. Moody: *Mapping Biology Knowledge.* 2000 ISBN 0-7923-6575-5
12. B. Bell and B. Cowie: *Formative Assessment and Science Education.* 2001 ISBN 0-7923-6768-5; Pb: 0-7923-6769-3
13. D.R. Lavoie and W.-M. Roth (eds.): *Models of Science Teacher Preparation.* Theory into Practice. 2001 ISBN 0-7923-7129-1
14. S.M. Stocklmayer, M.M.. Gore and C. Bryant (eds.): *Science Communication in Theory and Practice.* 2001 ISBN 1-4020-0130-4; Pb: 1-4020-0131-2
15. V.J. Mayer (ed.): *Global Science Literacy.* 2002 ISBN 1-4020-0514-8
16. D. Psillos and H. Niedderer (eds.): *Teaching and Learning in the Science Laboratory.* 2002 ISBN 1-4020–1018-4

KLUWER ACADEMIC PUBLISHERS – DORDRECHT / BOSTON / LONDON

Printed in the United States
68438LVS00002B/128

9 781402 010187